Advances in

ORGANOMETALLIC CHEMISTRY

VOLUME 20

CONTRIBUTORS TO THIS VOLUME

Gordon K. Anderson

F. E. Brinckman

R. J. P. Corriu

J. A. Gladysz

C. Guerin

Wolfgang A. Herrmann

John S. Thayer

Shen Yanchang (Y. C. Shen)

Huang Yaozeng (Y. Z. Huang)

Advances in Organometallic Chemistry

EDITED BY

F. G. A. STONE

DEPARTMENT OF INORGANIC CHEMISTRY
UNIVERSITY OF BRISTOL
BRISTOL, ENGLAND

ROBERT WEST

DEPARTMENT OF CHEMISTRY
UNIVERSITY OF WISCONSIN
MADISON, WISCONSIN

VOLUME 20

1982

ACADEMIC PRESS

A Subsidiary of Harcourt Brace Jovanovich, Publishers

New York   London
Paris San Diego San Francisco São Paulo Sydney Tokyo Toronto

ACADEMIC PRESS, INC.
111 Fifth Avenue, New York, New York 10003

United Kingdom Edition published by
ACADEMIC PRESS, INC. (LONDON) LTD.
24/28 Oval Road, London NW1 7DX

LIBRARY OF CONGRESS CATALOG CARD NUMBER: 64–16030

ISBN 0–12–031120–8

PRINTED IN THE UNITED STATES OF AMERICA

82 83 84 85   9 8 7 6 5 4 3 2 1

Contents

Transition Metal Formyl Complexes

J. A. GLADYSZ

The Organic Chemistry of Gold

GORDON K. ANDERSON

Arsonium Ylides

HUANG YAOZENG (Y. Z. HUANG) and SHEN YANCHANG (Y. C. SHEN)

The Methylene Bridge

WOLFGANG A. HERRMANN

Nucleophilic Displacement at Silicon:
Recent Developments and Mechanistic Implications

R. J. P. CORRIU and C. GUERIN

The Biological Methylation of Metals and Metalloids

JOHN S. THAYER and F. E. BRINCKMAN

Contents

List of Contributors

Numbers in parentheses indicate the pages on which the authors' contributions begin.

GORDON K. ANDERSON[1] (39), *Department of Chemistry, University of Guelph, Guelph, Ontario, Canada*

F. E. BRINCKMAN (313), *Chemical and Biodegradation Processes Group, Center for Materials Science, National Bureau of Standards, Washington, D.C. 20234*

R. J. P. CORRIU (265), *Laboratoire des Organométalliques, Equipe de Recherche Associée au C.N.R.S. N° 554, Université des Sciences et Techniques du Languedoc, Montpellier, France*

J. A. GLADYSZ[2] (1), *Department of Chemistry, University of California, Los Angeles, California 90024*

C. GUERIN (265), *Laboratorie des Organométalliques, Equipe de Recherche Associée au C.N.R.S. N° 554, Université des Sciences et Techniques du Languedoc, Montpellier, France*

WOLFGANG A. HERRMANN[3] (159), *Institut für Anorganische Chemie, Universität Regensburg, D-8400 Regensburg 1, Bundesrepublik Deutschland*

JOHN S. THAYER (313), *Department of Chemistry, University of Cincinnati, Cincinnati, Ohio 45221*

SHEN YANCHANG (Y. C. SHEN) (115), *Shanghai Institute of Organic Chemistry, Academia Sinica, Shanghai, People's Republic of China*

HUANG YAOZENG (Y. Z. HUANG) (115), *Shanghai Institute of Organic Chemistry, Academia Sinica, Shanghai, People's Republic of China*

[1] Present address: Department of Chemistry, University of Missouri-St. Louis, St. Louis, Missouri 63121.

[2] Present address: Department of Chemistry, University of Utah, Salt Lake City, Utah 84112.

[3] Present address: Institut für Anorganische Chemie, Johann Wolfgang Goethe-Universität, D-6000 Frankfurt am Main 50, Bundesrepublik Deutschland.

ADVANCES IN ORGANOMETALLIC CHEMISTRY, VOL. 20

Transition Metal Formyl Complexes

J. A. GLADYSZ*

Department of Chemistry
University of California
Los Angeles, California

I

INTRODUCTION

A. Background

The synthesis, chemical characterization, and physical characterization of transition metal complexes of new carbon-containing ligands constitute important ongoing objectives of organometallic chemists. In recent

---

* Present address: Department of Chemistry, University of Utah, Salt Lake City, Utah 84112.

years, particular attention has been given to the study of "new" ligands that may play important roles in industrial catalytic processes. It is reasonable to expect that an understanding of the fundamental chemistry of such ligands can ultimately aid in increasing process efficiency or selectivity.

Single-carbon ligands of the formula CH_xO_y (x, $y \geq 0$) occupy a unique position in organometallic chemistry. There are only two carbon-bound ligands of this type, —CO and —CH_3, for which homogeneous transition metal complexes were known prior to the 1970s. Only a finite number of CH_xO_y ligands can be formulated, and in most cases they represent the parent member of an important family of homologous ligands. Examples include —CHO (formyl), =CHOH (hydroxymethylidene), —CH_2OH (hydroxymethyl), ≡C (carbide), ≡CH (methylidyne), =CH_2 (methylidene), ≡COH (hydroxymethylidyne), H_2C=O, and CO_2. The synthesis of transition metal complexes of these ligands was vigorously pursued by many research groups in the 1970s. This chapter will attempt to review comprehensively the synthesis and properties of transition metal formyl complexes.

The recent interest in CH_xO_y ligands can be traced in part to the unstable world crude oil and natural gas supply situation. CO/H_2 gas mixtures (synthesis gas) can be readily obtained from coal and other abundant sources. In the presence of metallic heterogeneous or homogeneous catalysts, synthesis gas can be converted to a variety of organic molecules (methane, methanol, higher alkanes and alcohols, glycols, etc.) normally derived from petroleum or natural gas (*1–2a*). Since CO is being reduced in these transformations, C—O bonds must be broken and C—H bonds must be formed. Although there exist many possible sequences of catalyst-bound intermediates, it is clear that single-carbon species of types discussed in the previous paragraph must be involved. In the following section, data bearing on the existence of catalyst-bound formyls will be examined.

B. *Formyl Intermediates in Catalytic Reactions*

Pichler and Schulz were the first to suggest that the initial step in metal-catalyzed CO/H_2 reactions might be the migration of catalyst-bound hydride to CO to yield a formyl intermediate [Eq. (1)] (*3*). The alkyl ligand in many isolable metal carbonyl alkyls can similarly be induced to migrate to coordinated CO (*4, 5*). Thus, Eq. (1) has some precedent in homogeneous chemistry. Other suggestions for the involvement of formyl intermediates have come from Wender (*6*) and Henrici-Olivé and Olivé (*7*).

$$\boxed{\begin{array}{cc} H & CO \\ | & | \\ \text{Catalyst} \end{array}} \longrightarrow \boxed{\begin{array}{c} H\diagdown_{C}\diagup^{O} \\ | \\ \text{Catalyst} \end{array}} \xrightarrow{H_2} \xrightarrow{\quad} \begin{array}{c} \text{organic} \\ \text{molecules} \end{array} \qquad (1)$$

Good evidence has been obtained that heterogeneous iron, ruthenium, cobalt, and nickel catalysts which convert synthesis gas to methane or higher alkanes (Fischer–Tropsch process) effect the initial dissociation of CO to a catalyst-bound carbide $(8-13)$. The carbide is subsequently reduced by H_2 to a catalyst-bound methylidene, which under reaction conditions is either polymerized or further hydrogenated (13). This is essentially identical to the hydrocarbon synthesis mechanism advanced by Fischer and Tropsch in 1926 (14). For these reactions, formyl intermediates seem all but excluded.

The important question remains as to the mechanism by which *oxygenated* organic molecules are produced from CO/H_2 gas mixtures. A double-labeling experiment utilizing $^{13}C^{16}O$ and $^{12}C^{18}O/H_2$ indicated that methanol formation over Rh/TiO_2 catalysts occurs by a nondissociative process (exclusive formation of $^{13}CH_3^{16}OH$ and $^{12}CH_3^{18}OH$) (15). Similarly, glycol formation (which can be effected using homogeneous rhodium catalyst precursors) (16) seems unlikely to involve carbide intermediates.

Over the past few years, indirect evidence has accumulated suggesting that the formation of oxygen-containing CO reduction products such as methanol, ethanol, and formate esters from *homogeneous* metal carbonyl catalyst precursors is diagnostic of a homogeneous active catalyst. Species such as $HCo(CO)_4$, $HMn(CO)_5$ $(17, 18)$, and $Ru(CO)_5$ $(19-21)$ serve as effective catalyst precursors and are believed to remain mononuclear under the reaction conditions. Clearly, these are transformations in which catalyst-bound formyls can be reasonably postulated. Furthermore, the chemistry of such catalyst-bound formyls should have some analogy with the chemistry of the formyl complexes described in this chapter.

II

SYNTHESIS OF TRANSITION METAL FORMYL COMPLEXES

A. General Considerations

In advance of any experiment, formyl complexes appear deceptively easy to synthesize. Scheme 1 dissects some obvious approaches. Surprisingly, only one route (A-ii) has any semblance of generality.

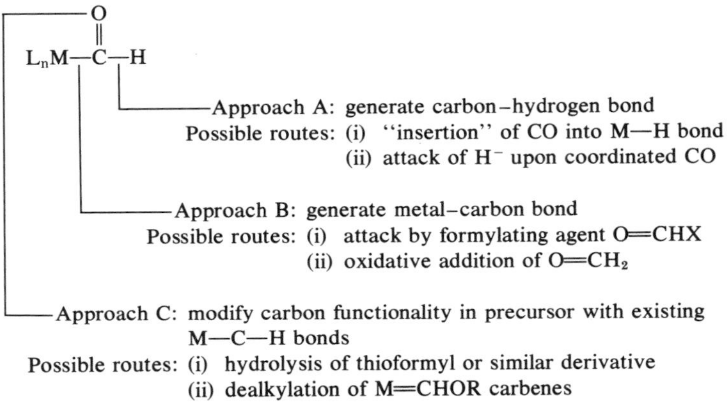

$$
\begin{array}{c}
\text{O} \\
\parallel \\
L_nM\!-\!C\!-\!H
\end{array}
$$

Approach A: generate carbon–hydrogen bond
Possible routes: (i) "insertion" of CO into M—H bond
(ii) attack of H⁻ upon coordinated CO

Approach B: generate metal–carbon bond
Possible routes: (i) attack by formylating agent O=CHX
(ii) oxidative addition of O=CH₂

Approach C: modify carbon functionality in precursor with existing
M—C—H bonds
Possible routes: (i) hydrolysis of thioformyl or similar derivative
(ii) dealkylation of M=CHOR carbenes

SCHEME 1.   Some synthetic approaches to transition metal formyl complexes.

Comparisons to methods for transition metal acyl synthesis are instructive. Although many transition metal alkyls can be readily carbonylated to transition metal acyls, the carbonylation of transition metal hydrides to transition metal formyls has not been observed [Eq. (2)] (*4, 5*). As will be seen, many transition metal formyls are thermodynamically unstable with respect to transition metal hydrides and CO. Thus approach A-i (Scheme 1) is not preparatively useful. (See, however, Addendum, p. 34.)

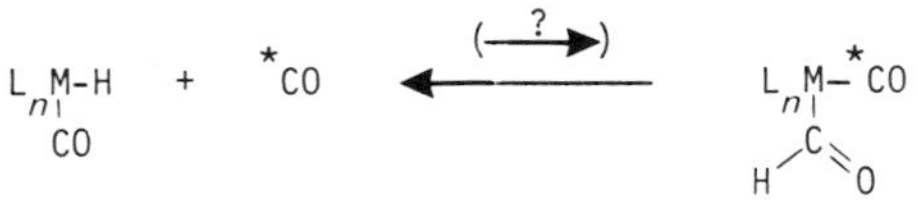

$$
\underset{\underset{CO}{|}}{L_nM\!-\!H} \;+\; \overset{*}{C}O \;\underset{\xleftarrow{\hspace{1.2cm}}}{\overset{(\xrightarrow{\;?\;})}{}}\; \underset{\underset{H}{\diagup}\overset{C}{\underset{\diagdown O}{}}}{L_nM\!-\!\overset{*}{C}O} \tag{2}
$$

Coordinated CO is readily attacked by a variety of carbon nucleophiles (*22, 23*). Therefore, the reaction of suitable hydride sources with metal carbonyls might be expected to yield formyl complexes (approach A-ii, Scheme 1). This route has proved successful with several types of borohydrides [Eq. (3)]. Since reaction commonly occurs below 0°C, formyl complexes which would be unstable at room temperature can be prepared.

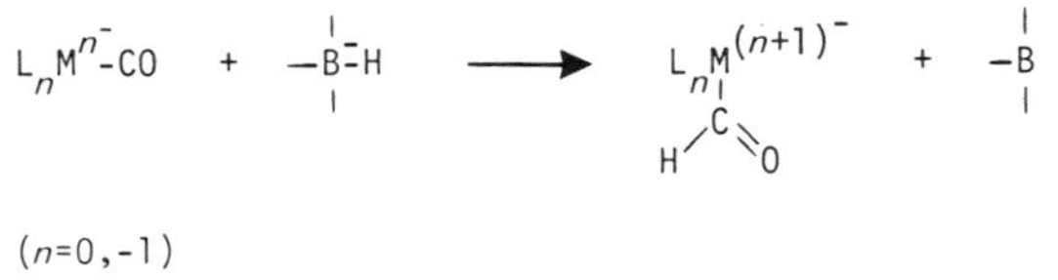

$$
L_nM^{n-}\!-\!CO \;+\; {-}\overset{|}{\underset{|}{B}}{}^-\!H \;\longrightarrow\; \underset{\underset{H}{\diagup}\overset{C}{\underset{\diagdown O}{}}}{L_nM^{(n+1)-}} \;+\; {-}\overset{|}{\underset{|}{B}} \tag{3}
$$

$(n=0,-1)$

Transition metal acyls are often synthesized by acylation of a transition metal anion (*24*). Also, transition metal acyls (sometimes transient) can be generated by the oxidative addition of aldehydes to coordinatively unsaturated metal species (25). As will be described, approaches of these types (B in Scheme 1) have been used in certain special cases to prepare transition metal formyl complexes. Transition metal acyls are seldom prepared by routes analogous to approach C in Scheme 1 (26). However, such strategies have been utilized to generate several neutral formyl complexes.

B. Anionic Formyl Complexes

All anionic transition metal formyl complexes described in the literature through the end of 1980 (21–47) are compiled in Table I. Since several of these have been prepared with more than one counterion, cations are not specified in the table. Geometric isomers are not assigned unless warranted by direct spectroscopic evidence. Also, the stability data in Table I should be regarded as qualitative, since decomposition rates have been shown to be dependent on both purity and counterion. When half-lives are specified, they are usually based upon measured rate constants.

The first synthesis of a formyl complex was described in 1973 (27, 28). In a landmark paper, Collman and Winter reported that the reaction of $Na_2Fe(CO)_4$ with formic acetic anhydride [Eq. (4)] afforded the anionic formyl $Na^+(CO)_4Fe(CHO)^-$ (subsequently isolated as its $[(C_6H_5)_3P]_2N^+$ or PPN^+ salt) in good yield. Formic acetic anhydride is an excellent formylating agent. This seemingly unusual reagent was selected because many HCOX species (X = Cl, O_2CH) are unstable at room temperature (48).

$$2Na^+ \ (CO)_4Fe^= \ + \ \underset{\substack{O \quad O \\ \| \quad \|}}{HC-O-CCH_3} \ \longrightarrow \ Na^+ \ (CO)_4Fe^- \underset{H}{\overset{}{\underset{}{\overset{C}{\|}}}} {}_{O}$$

$$Na^+ \ \mathbf{22}$$

(4)

Unfortunately, formic acetic anhydride is not a general reagent for formyl complex synthesis (29). One reason is that formylation of a transition metal monoanion would afford a neutral formyl complex. Insofar as comparisons are valid, neutral formyl complexes tend to be kinetically less stable than anionic formyl complexes. In cases where neutral formyl complexes are stable (vide infra), the corresponding transition metal monoanions are unknown. Whereas formic acetic anyhydride might be of greater use for the preparation of anionic formyl complexes from transition metal dianions, only a limited number of transition metal dianions [i.e., $(CO)_5Cr^{2-}$, $(\eta-C_5H_5)(CO)_3V^{2-}$] are known (49). These appear to

TABLE I

ANIONIC TRANSITION METAL FORMYL COMPLEXES

Complex	Stability	Characterization	Reference
(1) $(CO)_5Cr(CHO)^-$	half-life 40 min, 25°C	^{1}H NMR	29
(2) $(CO)_4(Ph_3P)Cr(CHO)^-$	No data	^{1}H NMR	29
(3) $(\eta\text{-}C_5H_5)(CO)_2Mo(COCO_2CH_3)(CHO)^-$	Rapid decomp, <0°C	^{1}H NMR	31, 32
(4) $(CO)_5W(CHO)^-$	No data	^{1}H NMR	29
(5) $(CO)_4(Ph_3P)W(CHO)^-$	No data	^{1}H NMR	29
(6) $cis\text{-}(CO)_4Mn(COC_6H_5)(CHO)^-$	half-life 7 min, 7°C	^{1}H NMR, IR	31, 32
(7) $cis\text{-}(CO)_4Mn(COCH_2OCH_3)(CHO)^-$	half-life 92 min, 7°C	^{1}H, ^{13}C NMR, IR	31, 32, 39
(8) $(CO)_4Mn(COCO_2CH_3)(CHO)^-$	Rapid decomp, 0°C	^{1}H NMR	31, 32
(9) $(CO)_4Mn(COCF_3)(CHO)^-$	half-life 7 min, −23°C	^{1}H NMR	31, 32
(10) $(CO)_4Mn(CF_3)(CHO)^-$	Rapid decomp, 0°C	^{1}H NMR	32
(11) $(CO)_4Mn(SnPh_3)(CHO)^-$	Rapid decomp, 0°C	^{1}H NMR	32
(12) $(CO)_5MnMn(CO)_4(CHO)^-$	half-life 1 h, 14°C	^{1}H NMR, IR	35, 39, 47
(13) $cis\text{-}(CO)_4Re(COCH_3)(CHO)^-$	half-life 8 min, 32°C	^{1}H NMR, IR	40
(14) $(CO)_4Re(CH_2C_6H_5)(CHO)^-$	Rapid decomp, 0°C	^{1}H NMR	32
(15) $cis\text{-}(CO)_4ReCl(CHO)^-$	Moderate stability 25°C	^{1}H NMR, IR	40
(16) $cis\text{-}(CO)_4ReBr(CHO)^-$	Moderate stability 25°C	^{1}H NMR, IR	32, 40
(17) $cis\text{-}(CO)_4ReI(CHO)^-$	Moderate stability 25°C	^{1}H NMR, IR	40

Compound	Stability	Characterization	References
(18) ReMn(CO)$_9$(CHO)$^-$	half-life 13 min, 32°C	^{1}H, ^{13}C NMR, IR	38, 47, 66
(19) cis-(CO)$_5$ReRe(CO)$_4$(CHO)$^-$	half-life several days, 50°C	^{1}H, ^{13}C NMR, IR, anal.	37, 38, 42, 47
(20) (CO)$_3$(Ph$_3$P)(Re)(CHO)$_2^-$	Rapid decomp, 0°C	^{1}H, ^{13}C NMR	57, 66
(21) (η-C$_5$H$_5$)Re(NO)(CHO)$_2$	half-life 77 min, 25°C	^{1}H, ^{13}C NMR	56, 57, 62, 66
(22) (CO)$_4$Fe(CHO)$^-$	>Several days, 25°C	^{1}H, ^{13}C NMR, IR, anal.	27–29, 33, 34, 36
(23) (CO)$_3$(Ph$_3$P)Fe(CHO)$^-$	No data	^{1}H NMR	29
(24) trans-(CO)$_3$[(PhO)$_3$P]Fe(CHO)	half-life ~1 h, 67°C	^{1}H NMR, IR	29, 37, 42, 63
(25) trans-(CO)$_3$[(ArO)$_3$P]Fe(CHO)$^-$	half-life ~1 h, 67°C	X ray	37, 42, 63

ArO =

Compound	Stability	Characterization	References
(26) (η-C$_5$H$_5$)(CO)Fe(COC$_6$H$_5$)(CHO)$^-$	half-life 18 min, 7°C	^{1}H, ^{13}C NMR, IR	30–32, 39
(27) (η-C$_5$H$_5$)(CO)Fe(CO-p-C$_6$H$_4$OCH$_3$)(CHO)$^-$	half-life 20 min, 7°C	^{1}H NMR, IR	30–32
(28) (η-C$_5$H$_5$)(CO)Fe(COCH$_3$)(CHO)$^-$	half-life 12 min, −18°C	^{1}H NMR	30–32
(29) Ru$_3$(CO)$_{11}$(CDO)$^-$	Rapid decomp, −40°C	^{2}H NMR	43, 46
(30) Os$_3$(CO)$_{11}$(CHO)$^-$	Rapid decomp, 0°C	^{1}H NMR, IR	43, 45, 46
(31) H$_2$Os$_3$(CO)$_8$S(CHO)$^-$	Rapid decomp, 0°C	^{1}H NMR	44
(32) HOs(CO)$_9$(O$_2$CCH$_3$)(CHO)$^-$	Rapid decomp, 0°C	^{1}H NMR	44
(33) Ir$_4$(CO)$_{11}$(CHO)$^-$	Rapid decomp, 20°C	^{1}H, ^{13}C NMR, IR	43, 46
(33a) Ir$_4$(CO)$_{10}$(CDO)$_2^=$	Rapid decomp, 20°C	^{2}H, ^{13}C NMR	46

be somewhat less nucleophilic than $(CO)_4Fe^{2-}$. However, even if formylation could be effected, the resulting anionic formyls (cf. 1 in Table I) would likely be of marginal stability.

As mentioned above, appropriate hydride nucleophiles are capable of attacking coordinated CO [Eq. (3)]. This route to anionic formyl complexes was reported in 1976 by Casey, Gladysz, and Winter (*29–34*). All of the anionic formyl complexes in Table 1, including **22** [Eq. (4)], can be prepared by hydride attack on neutral metal carbonyl precursors.

Several classes of hydride donors can be employed in Eq. (3). Trialkylborohydrides (R_3BH^-) have received wide use (*41*); $Li(C_2H_5)_3BH$ is commonly employed, but $K(C_2H_5)_3BH$ and $K(sec\text{-}C_4H_9)_3BH$ are also commercially available and are stronger hydride donors. The removal of R_3B by-products from the resulting formyls [Eq. (3)] can be troublesome or impossible. Unfortunately, $Li(CH_3)_3BH$ [which would produce $(CH_3)_3B$ gas as by-product] is a weaker hydrogen donor and does not efficiently transfer hydride to most metal carbonyls (*32*). Trialkoxyborohydrides have been extensively used by Casey and co-workers to generate formyl complexes (*29, 37*). Best results are obtained with $K(i\text{-}C_3H_7O)_3BH$.

Reagents such as $LiAlH_4$ and KH are not effective for the synthesis of formyl complexes. $LiAlH_4$ does react with many metal carbonyl compounds, but it can transfer more than one H^- and usually effects the formation of metal hydride products (*50*). Similar results are usually found with $NaBH_4$(*50*), although some neutral formyl complexes (*vide infra*) can be obtained under special conditions. KH will also react with some metal carbonyls. However, rates are not very rapid, and any formyl intermediates are likely to decompose faster than they form (*51*).

Other than $(CO)_4Fe(CHO)^-$ and derivatives thereof (**22–25**), only a few of the formyl complexes in Table I show significant stability at room temperature. Of these, the binuclear rhenium formyl *cis*-$(CO)_5ReRe(CO)_4(CHO)^-$ [**19**, Eq. (5)] has received the greatest attention (*37, 38, 42, 47*). The manganese and mixed manganese/rhenium homologs (**12, 18**) can be similarly prepared, but are less stable (*35, 38, 47*).

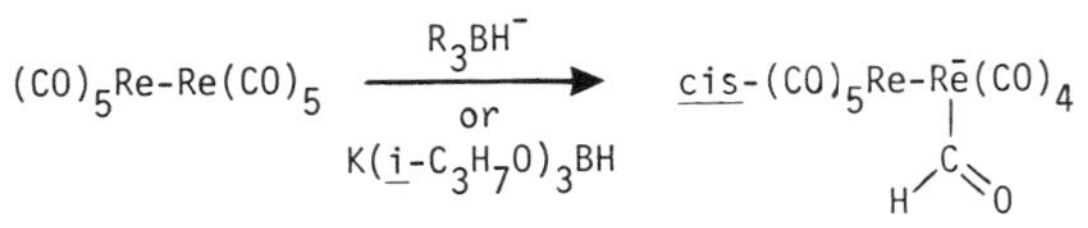

$$(CO)_5Re\text{-}Re(CO)_5 \xrightarrow[\substack{\text{or} \\ K(\underline{i}\text{-}C_3H_7O)_3BH}]{R_3BH^-} \underline{cis}\text{-}(CO)_5Re\text{-}\overset{|}{\underset{H\diagup C\diagdown_0}{Re}}(CO)_4 \qquad (5)$$

19

Other principal classes of formyl complexes include **6–9**, **26–28**, and **15–17**, which are shown below. Recently, several unstable cluster-bound formyls have been described (**29–33a** in Table I) (*43–46*).

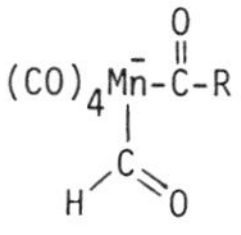

$$(CO)_4\overset{-}{Mn}\overset{\overset{O}{\|}}{-}C-R \qquad (\eta-C_5H_5)(CO)\overset{-}{Fe}\overset{\overset{O}{\|}}{-}C-R \qquad (CO)_4\overset{-}{Re}-X$$

$$\underset{H}{\overset{C}{\diagdown}}{}_O \qquad\qquad \underset{H}{\overset{C}{\diagdown}}{}_O \qquad\qquad \underset{H}{\overset{C}{\diagdown}}{}_O$$

6-9 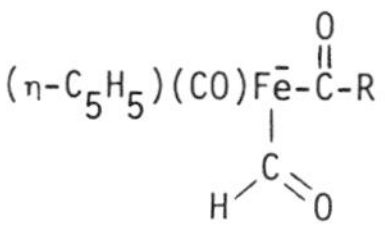   **26-28** 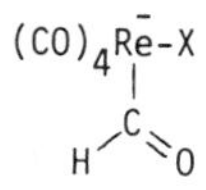   **15-17**

(X=halogen)

A special class of anionic formyl complexes are those with two formyl ligands. Examples are given in Eqs. (6) and (7). These complexes have analogy in anionic bisacyls previously synthesized by Lukehart (23). Very recently, evidence for the bisformyl cluster $Ir_4(CO)_{10}(CDO)_2^{2-}$ has been obtained (46).

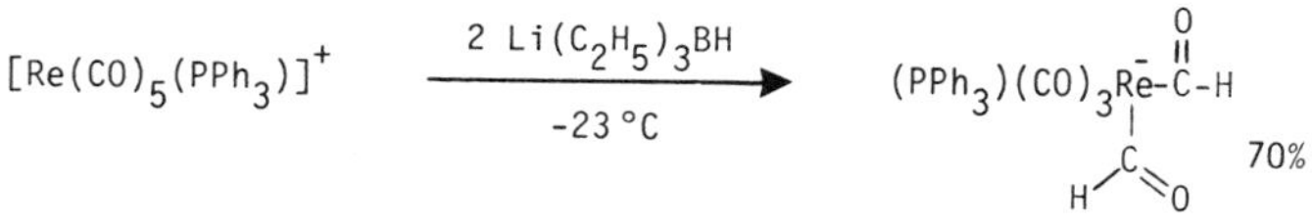

$$[Re(CO)_5(PPh_3)]^+ \xrightarrow[-23\,°C]{2\ Li(C_2H_5)_3BH} (PPh_3)(CO)_3\overset{-}{Re}\overset{\overset{O}{\|}}{-}C-H \quad 70\% \qquad (6)$$

$$\underset{H}{\overset{C}{\diagdown}}{}_O$$

20

$$\xrightarrow{Li(C_2H_5)_3BH} \qquad (7)$$

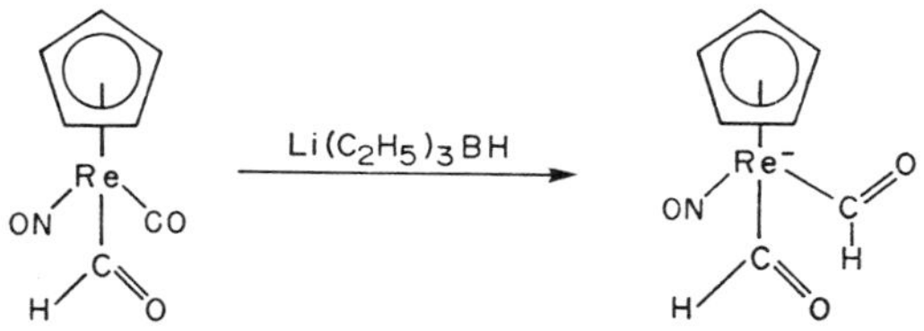

38   21

C. Neutral Formyl Complexes

All neutral transition metal formyl complexes described in the literature through the end of 1980 (52–68) are compiled in Table II. General comments made in the previous section regarding Table I apply.

The first neutral formyl complex, $Os(Cl)(CO)_2(PPh_3)_2(CHO)$ **(41)**, was reported by Collins and Roper in 1976 (52). As was the case with the first anionic formyl complex, the synthesis of **41** [Eq. (8)] was elegant but of limited generality. In a route belonging to approach C of Scheme 1, *thio*formyl **48** [Eq. (8)] was first S-methylated to a cationic thiocarbene complex. Subsequent hydrolysis afforded formyl **41**, which lost CO at room temperature. Homologs of **41** were prepared by identical routes (53); one of these, $Os(Cl)(CO)(CN-p-C_6H_4CH_3)(PPh_3)_2(CHO)$ **(43)**, proved

TABLE II

NEUTRAL TRANSITION METAL FORMYL COMPLEXES

Complex	Stability	Characterization	Reference
(34) $(\eta\text{-}C_5H_5)Mo(CO)_2(PPh_3)(CHO)$	Rapid decomp, $-41°C$	^1H NMR	57, 68
(35) $Mn(CO)_3(PPh_3)_2(CHO)$	Rapid decomp, $0°C$	$^1H, ^{13}C$ NMR	57, 68
(36) $(\eta\text{-}C_5H_5)Mn(NO)(CO)(CHO)$	Rapid decomp, $10°C$	$^1H, ^{13}C$ NMR	57, 68
(37) $Re(CO)_4(PPh_3)(CHO)$	Rapid decomp, $20°C$	$^1H, ^{13}C$ NMR	57, 68
(38) $(\eta\text{-}C_5H_5)Re(NO)(CO)(CHO)$	half-life 3–10 h, $25°C$, isolable as solid	$^1H, ^{13}C$ NMR, IR	56–58, 60, 62, 63, 68
(39) $(\eta\text{-}C_5H_5)Re(NO)(PPh_3)(CHO)$	Dec $\sim91°C$	$^1H, ^{13}C$ NMR, IR, anal., X ray	57, 59, 64–66
(40) $Ru(H)(solv)(PPh_3)_3(CHO)$	Dec $25°C$, isolable as impure solid	IR	55
(41) $Os(Cl)(CO)_2(PPh_3)_2(CHO)$	Evolves CO, $25°C$, isolable as impure solid	IR	52, 53
(42) $Os(Br)(CO)_2(PPh_3)_2(CHO)$	Isolable as impure solid	^1H NMR, IR	53
(43) $Os(Cl)(CO)(CN\text{-}p\text{-}C_6H_4CH_3)(PPh_3)_2(CHO)$	mp 151–153°C (darkens $\geqslant130°C$)	^1H NMR, IR, anal.	53
(44) $Os(H)(CO)_2(PPh_3)_2(CHO)$	Evolves $H_2 \geqslant40°C$	^1H NMR, IR	54
(45) $Ir(CO)_2(PPh_3)_2(CHO)$	Rapid decomp, $-30°C$	^1H NMR	57, 68
(46) $Ir(Cl)(H)[P(CH_3)_3]_3(CHO)$	mp 130°C, dec	^1H NMR, IR	67
(47) $Ir(CH_3)(H)[P(CH_3)_3]_3(CHO)$	mp 135°C, dec	^1H NMR, IR	67

stable to 130°C. Since only a very limited number of thioformyl and [metal=C(H)SR]$^+$ complexes have been synthesized, Eq. (8) does not provide a widely applicable route to neutral formyl complexes.

$$(CO)_2(PPh_3)_2(Cl)Os-C\overset{S}{\underset{H}{\diagdown}} \xrightarrow{CF_3SO_3CH_3} (CO)_2(PPh_3)_2(Cl)Os\overset{+}{=}C\overset{SCH_3}{\underset{H}{\diagdown}}$$

48

$$CF_3SO_3^-$$

$$\downarrow H_2O$$

$$(CO)_2(PPh_3)_2(Cl)Os-H \xleftarrow[-CO]{25°C} (CO)_2(PPh_3)_2(Cl)Os-C\overset{O}{\underset{H}{\diagdown}}$$

41

$$(8)$$

Roper has been able to isolate another osmium formyl by rearrangement of an η^2-formaldehyde complex, as shown in Eq. (9) (54). Because of the unavailability of such precursors, this reaction also does not provide a general entry into neutral formyl complexes. However, Eq. (9) does lend support to the claim that the related ruthenium formyl, Ru(H)(solv) (PPh$_3$)$_3$(CHO) (**40**), can be isolated as an impure solid, contaminated with substantial quantities of an η^2-formaldehyde precursor (55).

$$\begin{array}{c} OC\diagdown \overset{PPh_3}{\underset{|}{Os}} \diagup CH_2 \\ OC \diagup \overset{|}{\underset{PPh_3}{}} \diagdown O \end{array} \xrightarrow[75°C]{solid} \begin{array}{c} OC\diagdown \overset{PPh_3}{\underset{|}{Os}} \diagdown H \\ OC \diagup \overset{|}{\underset{PPh_3}{}} \diagdown \underset{O}{C} - H \end{array} \xrightarrow[\geq 40°C]{solution} Os(CO)_3(PPh_3)_2 + H_2 \quad (9)$$

With the exception of **46** and **47** [Eqs. (12) and (13)], the other neutral formyls in Table II have been prepared by hydride attack on metal carbonyl cation precursors. The most extensively studied of these have been $(\eta\text{-}C_5H_5)Re(NO)(CO)(CHO)$ (**38**) and $(\eta\text{-}C_5H_5)Re(NO)(PPh_3)(CHO)$ (**39**). The former was reported approximately simultaneously by Casey (56), Gladysz (57), and Graham (58). Conditions employed for its preparation are summarized in Eq. (10). Formyl **38** can be obtained as a solid, but the

$$\begin{array}{c} \overset{Re^+}{\underset{ON \diagup | \diagdown CO}{}} \\ CO \end{array} \xrightarrow[\text{Li}(C_2H_5)_3\text{BH, or}]{K(\underline{i}\text{-}C_3H_7O)_3 \text{ BH,}} \underset{\text{NaBH}_4 \text{ in THF/H}_2O}{} \begin{array}{c} \overset{Re}{\underset{ON \diagup | \diagdown CO}{}} \\ \underset{H \diagup}{C} \diagdown O \end{array} \quad (10)$$

38

isolation of analytically pure material has not yet been reported. However, the phosphine-substituted homolog **39**, synthesized by Tam, Wong, and Gladysz, can be obtained as a pure crystalline material by either of the routes shown in Eq. (11) (*57, 59, 64–66*).

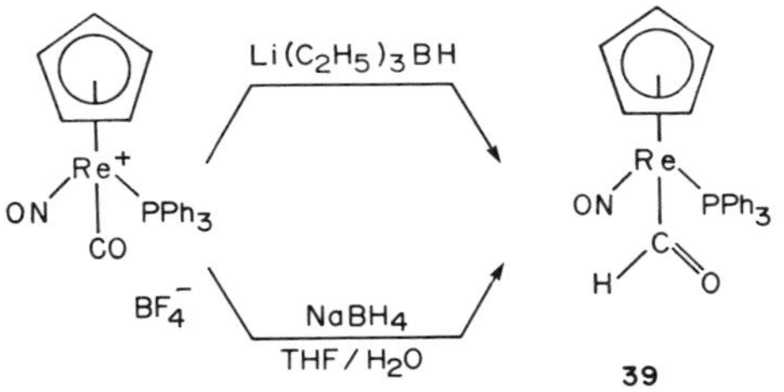

(11)

Recently, Thorn has found that several d^8 iridium compounds undergo (after ligand dissociation) oxidative addition to $H_2C{=}O$ (generated from paraformaldehyde). Thus obtained were the formyl hydrides shown in Eqs. (12) and (13) (*67*). These compounds are thermally robust (dec $\geqslant 130°C$) and will undoubtedly be the object of additional study.

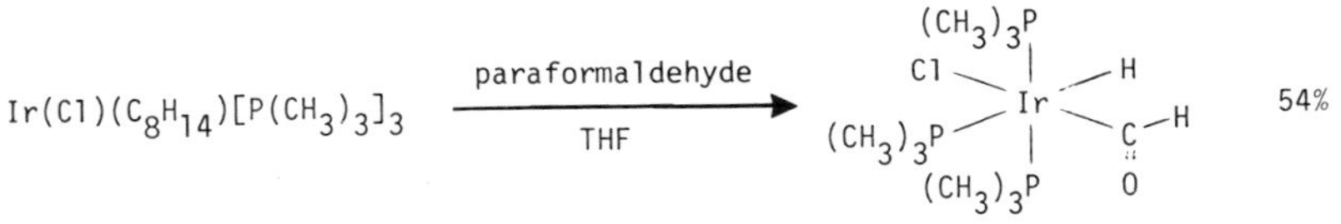

(12)

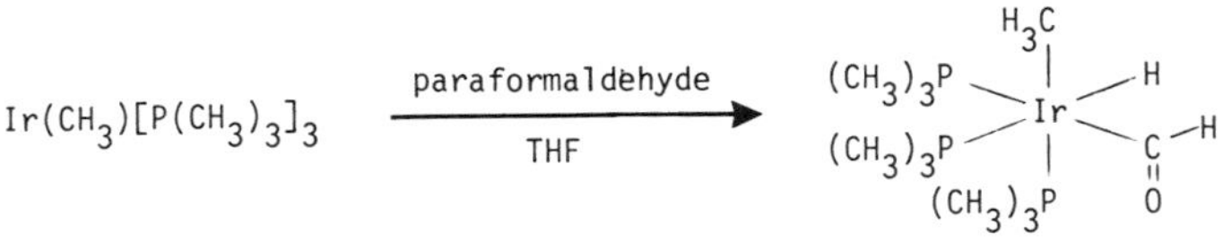

(13)

D. *Cationic Formyl Complexes*

The sole example of a cationic formyl complex was reported recently by Thorn (*67*). It was obtained by the oxidative addition of formaldehyde to a coordinatively unsaturated iridium cation, as shown in Eq. (14). Characterization included IR and ^1H, ^{13}C, and ^{31}P NMR. Formyl **49** is stable as a solid to 146°C.

$$[(CH_3)_3P]_4Ir^+ PF_6^- \xrightarrow[\text{THF}]{\text{paraformaldehyde}}$$

(CH₃)₃P structure **49** PF₆⁻ 60-80%

$$(14)$$

49

III

PHYSICAL PROPERTIES OF TRANSITION METAL FORMYL COMPLEXES

Formyl complexes are characterized by distinct spectroscopic features. The most frequently used diagnostic probe has been ^{1}H NMR. Chemical shifts in the δ 12–17 range (several ppm downfield from normal aldehyde or formamide ^{1}H-NMR resonances) (69) are observed. However, since chemical shifts measured in the presence of trialkylborane by-products can show marked temperature dependence (32, 38, 47, 66), absolute values must be treated with caution.

Formyl ^{13}C-NMR chemical shifts fall in the 240–310 ppm range, generally within a few ppm of where homologous metal acyl carbons would appear (27). The one-bond coupling constant $J_{^{13}C-^1H}$ can be used to distinguish a formyl from an acyl resonance. For Li$^+$ **19** this coupling has been determined to be 123 ± 1 Hz (38, 47, 66). For cationic formyl **49**, $J_{^{13}C-^1H}$ was found to be 150 Hz (67). Although these values are low compared to $J_{^{13}C-^1H}$ found for sp^2-hybridized carbons in organic molecules (70), they are typical for sp^2-hybridized carbons bound to coordinatively saturated metals (71). This may indicate slightly greater than normal p character in formyl C—H bonds. Collman has noted that under conditions that promote tight ion pairing, the formyl ^{13}C-NMR resonance of **22** shifts ~15 ppm to lower field (27, 28). However, CO ^{13}C-NMR chemical shifts in **22** are scarcely affected. This has been suggested as resulting from increased importance of tight ion pairs such as Na$^+$ **22b** below. In Na$^+$ **22b** the ligating carbon closely resembles the carbon of an oxygen-substituted "Fischer-carbene." These generally have ^{13}C-NMR chemical shifts downfield of 310 ppm (72).

$$(CO)_4\overset{-}{Fe}-C\overset{O}{\underset{H}{\diagup}}\ Na^+ \longleftrightarrow (CO)_4Fe=C\overset{O-Na}{\underset{H}{\diagup}}$$

(a)   Na$^+$ **22**   (b)

IR spectra of both anionic and neutral formyl complexes show $\nu_{C=O}$ between 1530 and 1630 cm^{-1} which are medium in intensity relative to

$\nu_{C\equiv O}$. Since these are much lower stretching frequencies than are encountered in esters (1720–1750 cm^{-1}) or amides ($\sim$1680 cm^{-1} in the absence of H bonding) (73), substantial contributions by resonance forms **50b** (anionic) and **51b** (neutral) to the ground state of formyl complexes are suggested.

(a)　　**50**　　(b)

(a)　　**51**　　(b)

Some typical $\nu_{C=O}$ for anionic formyl complexes include 1607 cm^{-1} for PPN$^+$(CO)$_4$Fe(CHO)$^-$ (PPN$^+$ **22**) (28), 1584 cm^{-1} for (C$_2$H$_5$)$_4$N$^+$*trans*-(CO)$_3$[(PhO)$_3$P]Fe(CHO)$^-$ [(C$_2$H$_5$)$_4$N$^+$ **24**] (29), and 1529 cm^{-1} for Li$^+$*cis*-(CO)$_5$ReRe(CO)$_4$(CHO)$^-$ (Li$^+$ **19**) (38, 47, 66). Some typical $\nu_{c=o}$ for neutral formyl complexes include 1610 cm^{-1} for Os(Cl)(CO)$_2$(PPh$_3$)$_2$(CHO) (**41**) (53), 1614–1635 cm^{-1} for (η-C$_5$H$_5$)Re(NO)(CO)(CHO) (**38**) (56–58), 1558–1566 cm^{-1} for (η-C$_5$H$_5$)Re(NO)(PPh$_3$)(CHO) (**39**) (57, 59), and 1589 cm^{-1} for Ir(CH$_3$)(H)[P(CH$_3$)$_3$]$_3$(CHO) (**47**) (67). For the cationic formyl **49** [Eq. (14)], $\nu_{c=o}$ was found to be 1600 cm^{-1} (67).

Most aldehydes also show two weak ν_{C-H} in the 2830–2695 cm^{-1} region (72). Similar absorptions have been observed in some formyl complexes (28, 29, 52–54, 67). However, these are not intense enough to be diagnostic, and confirmation of this assignment by study of the corresponding deuteroformyls (—CDO) would seem desirable.

The X-ray crystal structure of formyl (η-C$_5$H$_5$)Re(NO)(PPh$_3$)(CHO) (**39**) has been reported (59). Other structures have been completed, but are unpublished (67); that of (C$_2$H$_5$)$_4$N$^+$ *trans*-(CO)$_3$[(3,5-(CH$_3$)$_2$C$_6$H$_3$O)$_3$P]-Fe(CHO)$^-$ (**35**) has been briefly described in a review (63). The molecular structure of **39** is shown in Fig. 1. Bond angles indicate the formyl ligand to be reasonably trigonal: $\angle$Re–C(1)–O(1), 128.1 (8)°; $\angle$Re–C(1)–H(1), 119 (5)°. There is substantial uncertainty in the C—H bond distance [1.08(9) Å], but it is in the range expected for aldehydic C—H bonds (74). The formyl $\angle$Fe—C—O in **35** is reported to be 134°.

Several structural features of **39** suggest that resonance structure **51b** significantly contributes to its ground state. First, the formyl C=O distance, 1.220(12) Å, is as long or longer than the corresponding C=O bond distance in formamide (1.193 Å) (74). Second, the plane of the formyl

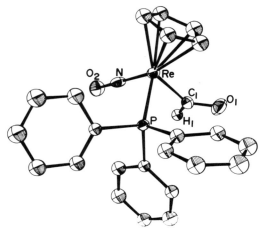

FIG. 1. Molecular structure of (η-C$_5$H$_5$)Re(NO)(PPh$_3$)(CHO) (**39**). Reprinted with permission from Wong *et al.* (*59*).

ligand virtually eclipses the C(1)–Re–N–O(2) plane (dihedral angle ~4°). This feature (which is not particularly well illustrated in Fig. 1) is important in the following context: several cationic alkylidene complexes of the formula [(η-C$_5$H$_5$)Re(NO)(PPh$_3$)(=CHR)]$^+$PF$_6^-$ have been synthesized (*75*). One of these (R = C$_6$H$_5$) has been subjected to an X-ray crystal structure determination (*76*). The alkylidene plane was found to similarly eclipse the C–Re–N–O plane. Hückel MO calculations indicate that this ligand geometry maximizes overlap with a donor orbital on the rhenium (*77, 78*). Thus the formyl ligand in **39** appears "alkylidene-like" (**51b**) in bonding geometry. Finally, the Re–C(1) distance in **39**, 2.055(10)Å, compares more closely with the rhenium–benzylidene bond distance in [(η-C$_5$H$_5$)Re(NO)(PPh$_3$)(=CHC$_6$H$_5$)]$^+$ PF$_6^-$ (1.949(6)Å) (*76*) than the rhenium–carbon bond distance in rhenium alkyls (i.e., 2.29 Å in (η-C$_5$H$_5$)-Re(CO)$_2$(H)(CH$_2$C$_6$H$_5$) (*79*); 2.215(4)Å in (η-C$_5$H$_5$)Re(NO)(PPh$_3$)(CH-(C$_6$H$_5$)CH$_2$C$_6$H$_5$) (*76*)).

IV

REACTIONS OF TRANSITION METAL FORMYL COMPLEXES

A. Reactions with Ketones and Aldehydes

Transition metal formyl complexes are capable of donating hydride to several classes of substrates, of which ketones and aldehydes are prototypical. Carbonyl group reduction was first noted with benzaldehyde and the manganese formyl **6**, as shown in Eq. (15) (*31*); by ^{1}H NMR, reaction was observed to be rapid at $-78°C$ (*32*). However, formyl **6** decomposes below room temperature and cannot be separated from the trialkylborane by-product formed in its preparation. Therefore, it is possible that the trialkylborane in some way mediates the hydride transfer.

$$(15)$$

96%

6

The involvement of trialkylboranes in these reactions was probed by use of the optically active trialkylborohydride **52**, shown in Eq. (16) (*39*). In previous work, **52** had been demonstrated to reduce the prochiral ketone acetophenone to 1-phenylethanol of 17% optical purity (*80*). Compound **52** was then used to generate the unstable anionic formyls **6**, **12**, and **26** (Table I); subsequently, acetophenone was added to these reaction mixtures. If Eq. (16) were reversible and **52** were the active hydride transfer agent, 1-phenylethanol of 17% optical purity would be expected. In practice, optical purities of 3.1–11.7% were obtained (*39*). This indicates some type of trialkylborane involvement in the hydride transfer (the exact role cannot be readily determined by experiment). Therefore, it became important to attempt similar reactions with isolable, purified formyl complexes.

$$(16)$$

52

As shown in Eqs. (17) and (18), the isolated formyls **19** and **24** are capable of reducing aldehydes and ketones (*37, 38, 42, 47, 66*). Thus there is no doubt that hydride transfer is an intrinsic chemical property of anionic formyl complexes. One reaction of a neutral formyl complex with an aldehyde has been reported: addition of benzaldehyde to $(\eta\text{-}C_5H_5)Re(NO)(CO)(CHO)$ (**38**) yields the alkoxycarbonyl complex $(\eta\text{-}C_5H_5)Re(NO)(CO)(CO_2CH_2C_6H_5)$ (*62*). This transformation, which appears to require catalysis by adventitious acid, can be viewed as occurring via attack of initially formed benzyl alcohol upon the intermediate carbonyl cation $[(\eta\text{-}C_5H_5)Re(NO)(CO)_2]^+$.

$$cis\text{-}(CO)_5Re\text{-}\overset{\mid}{\underset{\underset{H}{}}{Re}}(CO)_4\underset{\overset{C}{H}\diagdown^O}{} + C_6H_5\text{-}\overset{\overset{O}{\parallel}}{C}\text{-}H \xrightarrow{\ H_2O\ workup\ } (CO)_5Re\text{-}Re(CO)_5 + C_6H_5\text{-}\overset{\overset{O\text{-}H}{\mid}}{\underset{\underset{H}{\mid}}{C}}\text{-}H \tag{17}$$

19   100%   74%

$$trans\text{-}(CO)_3[(PhO)_3P]\overset{\mid}{\underset{\overset{C}{H}\diagdown^O}{Fe}} + \overset{\overset{O}{\parallel}}{\underset{}{C}}\diagdown\diagup \xrightarrow{\ H_2O\ workup\ } (CO)_4[(PhO)_3P]Fe + \overset{\overset{O\text{-}H}{\mid}}{\underset{\underset{H}{\mid}}{C}}\diagdown\diagup \tag{18}$$

24   95%

B. *Reactions with Metal Carbonyls*

A unique reaction of formyl complexes is "formyl transfer," in which the formyl ligand undergoes *apparent* migration from one metal to another. This transformation was first observed with the manganese formyl **12**, as shown in Eq. (19) (*35, 47*). However, **12** is unstable at room temperature and cannot be separated from trialkylborane by-product. Therefore, it is again important to establish that this type of reaction proceeds with pure formyl complexes.

$$(CO)_5Mn\text{-}\overset{\mid}{\underset{\overset{C}{H}\diagdown^O}{Mn}}(CO)_4 + (CO)_5Fe \longrightarrow (CO)_5Mn\text{-}Mn(CO)_5 + (CO)_4\overset{\mid}{\underset{\overset{C}{H}\diagdown^O}{Fe}} \tag{19}$$

12   79%   76%

                                                                                               22

Two "formyl transfer" reactions of isolated formyl complexes are shown in Eqs. (20) (*37, 42*) and (21) (*37, 38, 42, 47, 66*). Control experiments indicate that these reactions do not involve metal hydride intermediates (formed via decarbonylation). Straightforward intermolecular H^- transfer (rather than formyl ligand transfer) is believed to be taking place.

$$trans\text{-}(CO)_3[(PhO)_3P]\overset{\mid}{\underset{\overset{C}{H}\diagdown^O}{Fe}} + (CO)_5Re\text{-}Re(CO)_5 \longrightarrow$$

24

$$(CO)_4[(PhO)_3P]Fe + cis\text{-}(CO)_5Re\text{-}\overset{\mid}{\underset{\overset{C}{H}\diagdown^O}{Re}}(CO)_4 \tag{20}$$

                                              82%

                                                                          19

Hydride donor strengths can be summarized as follows: **24** > **19** > **12** > **22**. This type of reaction provides a useful alternative to formyl synthesis by borohydride reductants; neutral metal carbonyl by-products are produced instead of boron-containing species.

$$cis\text{-}(CO)_5Re\text{-}\overline{Re}(CO)_4 \; + \; (CO)_5\overline{Fe} \longrightarrow (CO)_5Re\text{-}Re(CO)_5 \; + \; (CO)_4\overline{Fe} \qquad (21)$$

19   90%   72%

22

"Formyl transfers" involving neutral formyl complexes, such as shown in Eq. (22), have recently been reported (*68*). By precipitation of the metal carbonyl cation by-product, quite pure solutions of kinetically labile neutral formyl complexes may be obtained.

$$(22)$$

39   **38**

C. Reactions with Alkylating Agents

Reactions of formyl complexes with alkylating agents can be more complex than the reductions in Eqs. (15–22). Some examples of simple hydride transfer exist. For instance, $(CO)_4Fe(CHO)^-$ (**22**) reduces octyl iodide to octane (75%) (*27, 28*); $(C_2H_5)_4N^+$ **25** (Table I) reacts with heptyl iodide (overnight, room temperature, THF) to give heptane (71%) and $(CO)_4[(ArO)_3P]Fe$ (*37, 42*); $cis\text{-}(CO)_5ReRe(CO)_4(CHO)^-$ (**19**) converts octyl iodide to octane (68%) (*47*).

Reactions of neutral formyl complexes with alkylating agents can follow different courses. Roper has observed the O-methylation reaction shown in Eq. (23) (*54*). Cationic methoxymethylidene complex **53** was obtained in excellent yield.

Attempted O-methylation of formyl **39** [Eq. (24)] did not yield appreciable quantities of the anticipated methoxymethylidene product $[(\eta\text{-}C_5H_5)Re(NO)(PPh_3)(=CHOCH_3)]^+ SO_3F^-$ (**54**) (*61, 64*). Rather, $(\eta\text{-}C_5H_5)Re(NO)(PPh_3)(CH_3)$ (**55**) and $[(\eta\text{-}C_5H_5)Re(NO)(PPh_3)(CO)]^+SO_3F^-$ (**56**) were formed, as shown in Eq. (24). When CD_3SO_3F was used in place of CH_3SO_3F, no CD_3 was incorporated into product. Therefore, **55** and **56** would appear to represent formyl reduction and oxidation products,

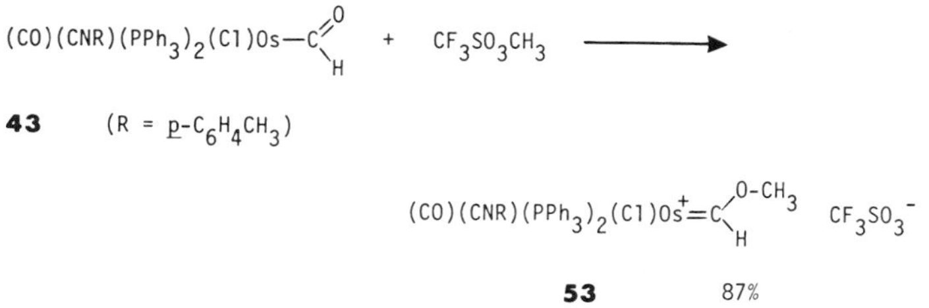

$$(CO)(CNR)(PPh_3)_2(Cl)Os—C\overset{O}{\underset{H}{\diagdown}} \quad + \quad CF_3SO_3CH_3 \quad \longrightarrow$$

43 $(R = \underline{p}\text{-}C_6H_4CH_3)$ (23)

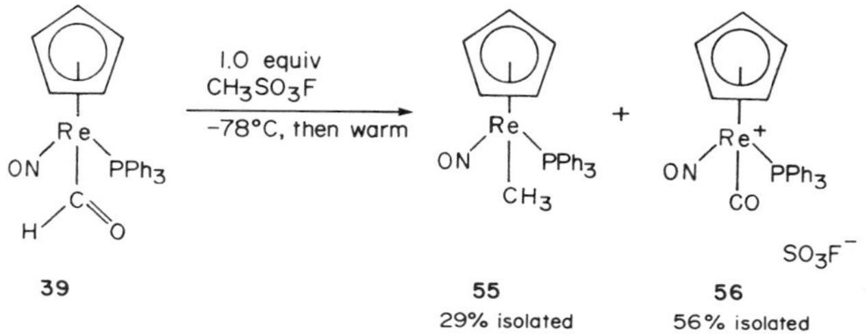

$$(CO)(CNR)(PPh_3)_2(Cl)Os\overset{+}{=}C\overset{O-CH_3}{\underset{H}{\diagdown}} \quad CF_3SO_3^-$$

53 87%

(24)

39   **55**   **56**

29% isolated   56% isolated

respectively. In ^{1}H-NMR monitored reactions, CH_3OCH_3 was observed to form about equimolar with **55**.

To aid in elucidating the mechanism of Eq. (24), independent syntheses of suspected intermediates **54**, $(\eta\text{-}C_5H_5)Re(NO)(PPh_3)(CH_2OCH_3)$ (**57**), and $[(\eta\text{-}C_5H_5)Re(NO)(PPh_3)(=CH_2)]^+X^-$ (**58**) were undertaken (*61, 64, 65*). By use of low temperature ^{1}H-NMR monitoring and conducting cross-checks on the reactivities of **54–58** [Eqs. (25) and (26)], the disproportionation mechanism shown in Scheme 2 was proposed.

Observations pertinent to Scheme 2 are as follows (*61, 64–66*): as followed by ^{1}H NMR, CH_3SO_3F reacts slowly with **39** at $-70°C$ but rapidly at $\geqslant -40°C$. However, under *no* conditions (including inverse addition, use of other methylating agents, etc.) can the formation of **54** from **39** be directly observed. Rather, ^{1}H NMR indicates that substantial quantities of methoxymethyl complex **57** build up. The carbonyl cation product **56** is also present. These data are readily explained if, subsequent to initial O-methylation (Scheme 2, step a), rapid hydride transfer from unreacted **39** to **54** occurs (step b). As shown in Eq. (25), this reaction (conducted with independently synthesized samples) is indeed rapid at $-70°C$.

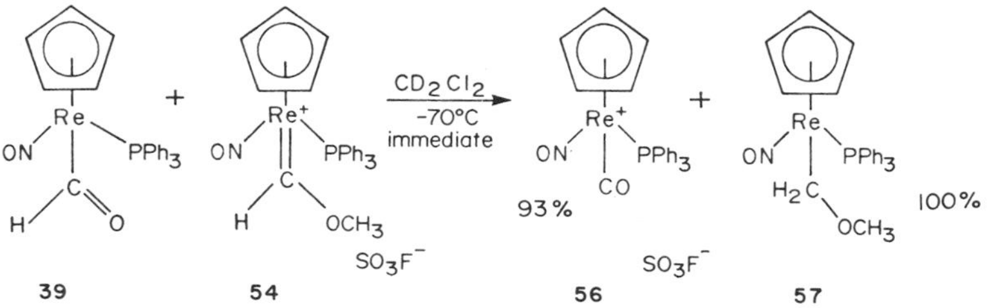

(25)

39   **54**   **56**   **57**

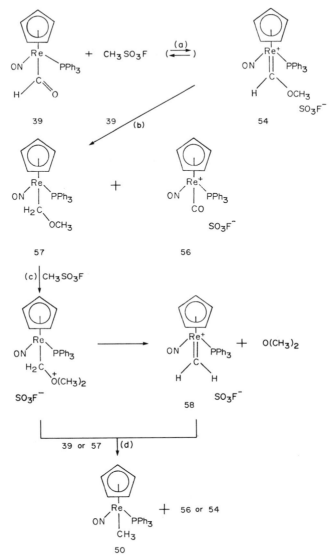

SCHEME 2. Proposed mechanism for the generation of **55** on treatment of **39** with CH$_3$SO$_3$F. Reprinted with permission from Wong *et al.*, *J. Am. Chem. Soc.* **101**, 3371 (1979). Copyright 1979 American Chemical Society.

On warming to $-20°C$, methoxymethyl **57** reacts with the remaining CH_3SO_3F. The rhenium methyl product **55** forms, but oxonium salt or methylidene (**58**) intermediates are not detected. This is understandable if the methylidene **58** (or the oxonium salt) is reduced more rapidly than it is formed (Scheme 2, step d). Depending upon reaction conditions, **39** and/or **57** might serve as hydride donors. Reactions of both of these species with **58** [shown in Eq. (26) for **39**] are rapid at $-70°C$. In slowly warmed reactions, **54** was usually detected in the final product mixture, indicating the participation of **57** as a hydride donor in the final step (d in Scheme 2). The precise distribution of final products obtained should reasonably be a sensitive function of reactant ratios and concentrations, order of reactant addition, and reaction temperature and time.

$$\text{(26)}$$

The reason for the differing behavior of formyls **43** [Eq. (23)] and **39** [Eq. (24) and Scheme 2] toward CH_3SO_3F is not clear. Perhaps the former is a less active hydride donor. There have been two reports of CH_4 formation on treatment of anionic formyl complexes (**6, 19**) with CH_3SO_3F (*35, 38*). In each case, good yields of the corresponding neutral metal carbonyls (H⁻ loss products) were obtained (*35, 38, 47*). However, the observation of CH_4 formation could not be reproduced (*47*). It seems possible that chemistry along the lines of Scheme 2 (O-methylation followed by hydride transfer) might be occurring. Thus anionic formyl complexes can give different reactions depending on whether they are treated with hard alkylating agents, such as CH_3SO_3F, or soft alkylating agents, such as octyl iodide.

D. Reactions with Protonating Agents

Formyl complexes show varying behavior when treated with protonating agents. Reaction of $(CO)_4Fe(CHO)^-$ (**22**) with acid gives formaldehyde in 13–20% yield (*27–29*). Homologous $(CO)_4Fe(COR)^-$ acyls afford excellent yields of aldehydes when protonated, presumably via

$(CO)_4Fe(H)(COR)$ intermediates (27, 28). Formaldehyde may be produced from a similar precursor.

Reaction of trans-$(CO)_3[(PhO)_3P]Fe(CHO)^-$ (24) with excess CF_3CO_2H results in the formation of CH_3OH (0.27 equiv) but not formaldehyde (<0.05 equiv) (29). There are several possible routes for CH_3OH formation. One would involve initial formaldehyde production as observed upon protonation of 22, followed by a rapid back-reaction with unprotonated 24 to give methoxide (Section IV,A). Another would involve initial O-protonation, followed by rapid hydride transfer from unprotonated 24 to give $(CO)_3[(PhO)_3P]Fe(CH_2OH)^-$; CH_3OH might then form from a $L_4Fe(H)(CH_2OH)$ intermediate. Either of these routes affords a maximum of 0.5 equiv of CH_3OH.

Two reports of H_2 formation upon acidification of anionic formyls 6 (31) and 19 (38) could not be reproduced (32, 47). Thus there are no documented examples of H_2 evolution upon protonation of anionic formyl complexes. It is clear, however, that rapid reactions ensue in all cases (32, 47, 66) and that good yields of neutral metal carbonyl (H^- loss) products are obtained.

When treated with CF_3CO_2H at $-70°C$, neutral formyl 39 undergoes O-protonation. Hydroxymethylidene 58 is obtained, as shown in Eq. (27) (57, 64). Although 58 is unstable above $-40°C$, it can be characterized by ^1H and ^{13}C NMR. On warming, a mixture of $(\eta\text{-}C_5H_5)Re(NO)(PPh_3)(CH_3)$ (55) and $[(\eta\text{-}C_5H_5)Re(NO)(PPh_3)(CO)]^+ \ CF_3CO_2^-$, similar to that observed in Eq. (24), is obtained. Thus Eq. (27) is believed to be reversible, enabling disproportionation chemistry analogous to Scheme 2 to occur. When formyl 39 is treated with the stronger acid CF_3SO_3H (64) the isolable (stable to $\geq 80°C$ in $CDCl_2CDCl_2$) hydroxymethylidene $[(\eta\text{-}C_5H_5)\text{-}Re(NO)(PPh_3)(CHOH)]^+ \ CF_3SO_3^-$ is obtained. These constitute the only complexes containing a $=CHOH$ ligand to be detected and/or isolated to date; such species have received consideration as intermediates in the Fischer–Tropsch process.

$$\text{(27)}$$

Recently, Geoffroy and Steinmetz reported that reaction of the unstable cluster formyl $Os_3(CO)_{11}(CHO)^-$ [Eq. (28)] with excess H_3PO_4 yielded the methylidene bridged cluster $Os_3(CO)_{11}CH_2$ (0.20–0.30 equiv) (45). When CF_3CO_2D was used as the acid, no deuterium was incorporated into $Os_3(CO)_{11}CH_2$; reaction of H_3PO_4 with $Os_3(CO)_{11}(CDO)^-$ yielded $Os_3(CO)_{11}CD_2$. In all cases the formation of about equimolar amounts of $Os_3(CO)_{12}$ was noted. These results suggest the reaction path shown in Eq. (28). When acidification was conducted with CF_3CO_2H, some methanol was also produced (43, 45).

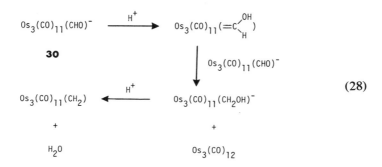

$$(28)$$

E. Reactions with Reducing Agents

In Sections IV,A–D, reactions which involved formyl oxidation or disproportionation were described. Reactions of formyl complexes with reducing agents will now be examined. At the outset, it can be stated that no well-defined reductions utilizing H_2 have been found (47, 62, 66). This is disappointing, since such reactions would probably have relevancy to homogeneous CO reduction catalyst pathways.

Anionic formyl complexes can be reduced when treated with the more potent trialkylborohydride nucleophiles. Thus K^+ **22** reacts with $K(sec\text{-}C_4H_9)_3BH$ in refluxing THF to yield $K_2Fe(CO)_4$, as shown in Eq. (29). After acidification, methanol (but not formaldehyde) can also be isolated (66); any formaldehyde would likely be reduced by $(sec\text{-}C_4H_9)_3B$ (81) (or $K(sec\text{-}C_4H_9)_3BH)$) under the reaction conditions. Similarly, K^+ $cis\text{-}(CO)_5\text{-}ReRe(CO)_4(CHO)^-$ reacts with $K(sec\text{-}C_4H_9)_3BH$ at room temperature to yield a new anion formulated as $K_2Re_2(CO)_9$ (38, 47, 66). Depending on conditions, either formaldehyde or methanol accompanies its formation. Although exact mechanisms are not known for these transformations, initial H^- attack upon a CO and/or formyl ligand is undoubtedly involved. Recently, the plausibility of bisformyl intermediates in these reductions has gained support: the ^{13}C-NMR monitored decomposition of a species believed to be $Ir_4(CO)_{10}(CDO)_2^{2-}$ indicated the formation of formaldehyde, methanol, and methoxide (46).

$$K^+ \ (CO)_4\overset{|}{\underset{\underset{H}{C}\diagdown_O}{Fe^-}} \quad \xrightarrow[\text{refluxing THF}]{K(\text{sec-}C_4H_9)_3BH} \quad K_2Fe(CO)_4 \tag{29}$$

$$K^+ \ \mathbf{22} \qquad\qquad\qquad 95\text{-}100\%$$

The reduction chemistry of $(\eta\text{-}C_5H_5)Re(NO)(CO)(CHO)$ (**38**) has been studied by Casey, Graham, and Gladysz (56–58, 60, 62). Some representative transformations are given in Scheme 3.

As previously noted, the nucleophilic hydride reductant $Li(C_2H_5)_3BH$ converts **38** to the bisformyl **21** (Scheme 3) (56, 57, 62). However, electrophilic reducing agents act on the formyl ligand. For instance, $BH_3 \cdot THF$ reduces the formyl ligand in **38** to a methyl ligand (**59**, Scheme 3) (56, 57, 62). Masters has previously observed the related $BH_3 \cdot THF$ reduction of metal carbonyl acyls to metal carbonyl alkyls (82). Significantly, N,N-dialkylamides ($RCONR'R''$) are similarly reduced to amines ($RCH_2NR'R''$) by $BH_3 \cdot THF$ (83); this reinforces the analogy drawn between amides and formyl complexes in Section III and indicates that the metal cannot be viewed as playing a unique activating role in these conversions.

Casey demonstrated that $HAl(i\text{-}C_4H_9)_2$, which has only one transferable hydride, could be used to reduce the formyl ligand in **38** to a hydroxymethyl ligand (**60**, Scheme 3) (62). Compound **60** was the first well-characterized metal complex with a $—CH_2OH$ ligand to be synthesized (60, 62). It could be more conveniently prepared by $NaAlH_2(C_2H_5)_2$ attack upon $[(\eta\text{-}C_5H_5)Re(NO)(CO)_2]^+BF_4^-$, which avoids work-up and handling of the sensitive formyl **38**. Graham found that under carefully controlled conditions, **38** could also be converted to hydroxymethyl **60** using $NaBH_4$ in THF/H_2O (58). Apparently, by-product BH_3 (or a product–BH_3 adduct) is hydrolyzed more rapidly than it can effect further reduction. Obviously, rapid hydrolysis of BH_3 (or a formyl–BH_3 adduct) (58) must also be invoked when formyls **38** and **39** are prepared using $NaBH_4$ in THF/H_2O [Eqs. (10) and (11)].

Other formyl complexes can be reduced with BH_3. Thus $(\eta\text{-}C_5H_5)$-$Re(NO)(PPh_3)(CHO)$ (**39**) is converted to $(\eta\text{-}C_5H_5)Re(NO)(PPh_3)(CH_3)$ (**55**) in 72% yield (59, 64); similarly, $[Ir(H)[P(CH_3)_3]_4(CHO)]^+ PF_6^-$ is reduced to $[Ir(H)[P(CH_3)_3]_4(CH_3)]^+ PF_6^-$ in $\sim$26% spectroscopic yield (67).

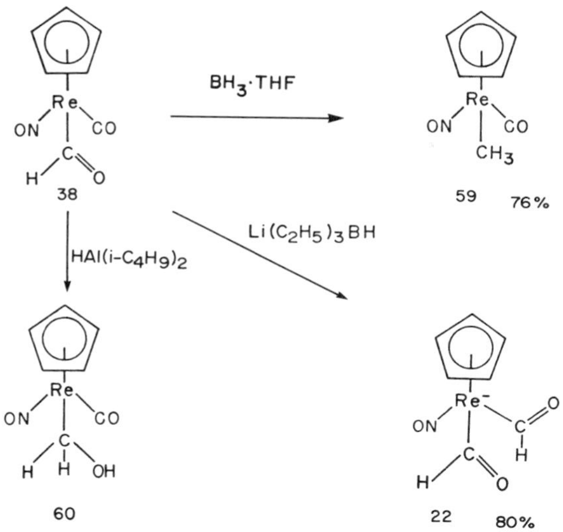

SCHEME 3.   Some reductions of the neutral formyl $(\eta\text{-}C_5H_5)Re(NO)(CO)(CHO)$ (38).

F. Other Reactions

Although the anionic formyl $(C_2H_5)_4N^+$ **19** does not undergo thermal decarbonylation at room temperature, photochemical CO loss occurs upon irradiation [Eq. (30)] (37, 47, 66). This reaction can even be effected under laboratory fluorescent lights.

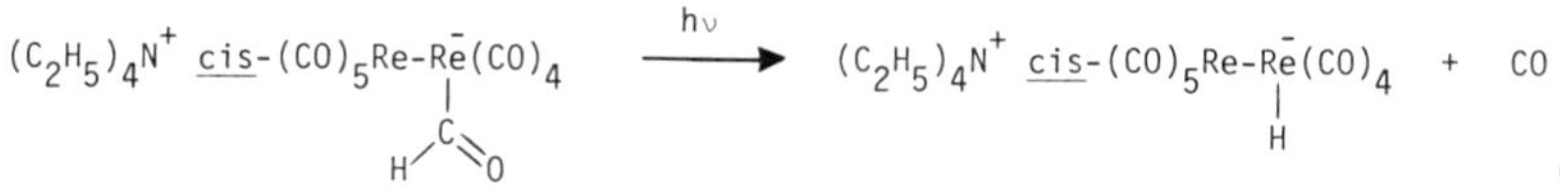

$$(30)$$

After exposure of $(CO)_4Fe(CHO)^-$ to O_2 and acidification, an explosion occurred (27, 28). The formation of performic acid was suggested.

Reaction of $(\eta\text{-}C_5H_5)Re(NO)(PPh_3)(CHO)$ (**39**) with $(CH_3)_3SiCl$ leads to formyl disproportionation similar to that observed in Scheme 2 (61, 66). Reaction of **39** with the more reactive silylating agent $CF_3SO_3Si(CH_3)_3$ affords quantitative spectroscopic yields of $[(\eta\text{-}C_5H_5)Re(NO)(PPh_3)\text{-}(=CHOSi(CH_3)_3)]^+$ $CF_3SO_3^-$, which decomposes upon attempted purification at room temperature (84).

V

TOPICS RELATING TO THE STABILITY
AND DECOMPOSITION OF TRANSITION METAL
FORMYL COMPLEXES

A. *Decomposition Chemistry*

Discussion of the decomposition chemistry of formyl complexes has been deferred until this stage because some of the reactivity modes described in Section IV can play important roles.

Iron formyls $(CO)_3LFe(CHO)^-$ typically decompose by loss of CO and/or L. For instance, Collman and Winter reported that $Na^+(CO)_4$-$Fe(CHO)^-$ underwent slow decarbonylation to $Na^+(CO)_4FeH^-$ ($t_{1/2}$, 25°C, $\geqslant 12$ days) (27, 28). Casey and Neumann have studied phosphite-substituted homologs (42). At 65°C in THF, $(C_2H_5)_4N^+(CO)_3[(PhO)_3P]$-$Fe(CHO)^-$ decomposed to a 4:1 mixture of $(CO)_4FeH^-$ (phosphite loss product) and $(CO)_3[(PhO)_3P]FeH^-$ (CO loss product). The closely related formyl $(C_2H_5)_4N^+$ **25** [Table I and Eq. (31)] afforded only the phosphite loss product $(CO)_4FeH^-$.

The rate of decomposition of $(C_2H_5)_4N^+$ **25** was found to be first order and independent of added phosphite. At 63°C, the following activation parameters were obtained: $\Delta H^{\ddagger} = 29.0 \pm 1.5$ kcal/mol; $\Delta S^{\ddagger} = 7.9 \pm 6.1$ eu. These data suggest that (provided the reaction is analogous to well-established metal acyl decarbonylation mechanisms) (4, 5) loss of phosphite is followed by rapid hydride migration to the metal, as shown in Eq. (31). None of the formyl could be detected to be in equilibrium with $(CO)_4FeH^-$, even in the presence of excess $(ArO)_3P$.

Other anionic formyl complexes decompose by more complex pathways. Unstable formyl **6** (Scheme 4) yielded approximately equimolar amounts of $(CO)_5Mn^-$, $(CO)_5Mn(COC_6H_5)$, and (after protonation) benzyl alcohol (31, 32). The rate of decomposition was first order, accelerated by

added $(C_2H_5)_3B$, and showed a kinetic deuterium isotope effect of ~3.3
(32). From k_{obs}, $\Delta S^{\ddagger}$ was calculated to be -8.8 ± 2 eu. In contrast to
Eq. (31), these data are not consistent with initial and rate-determining
CO loss.

A decomposition mechanism that satisfactorily accounts for the preced-
ing data is given in Scheme 4. The slow step would involve hydride
migration [assisted by $(C_2H_5)_3B$] from the formyl to the benzoyl ligand,
generating an alkoxide intermediate (step a). Rapid fragmentation to benzaldehyde and $(CO)_5Mn^-$ would follow (step b); when independently generated (85), the alkoxide does indeed fragment as indicated. However,
under the conditions of Scheme 4, intermediate benzaldehyde is rapidly
reduced by undecomposed formyl [see Eq. (15)]. Thus both formyl oxidation [$(CO)_5MnCOC_6H_5$] and reduction [$(CO)_5Mn^-$, $C_6H_5CH_2O^-$] products
are generated in Scheme 4. The overall decomposition can be termed
disproportionative.

The manganese formyl 12 also decomposes to oxidized [$(CO)_5MnMn-$
$(CO)_5$] and reduced [$(CO)_5Mn^-$] products, as shown in Eq. (32) (35, 39, 47,
66). Interestingly, decomposition is retarded by added $(C_2H_5)_3B$. Some
manganese–manganese bond cleavage obviously takes place [Eq. (32)].
This may occur via an intramolecular α-elimination of $(CO)_5Mn^-$ from 12.
This would be analogous to the recently demonstrated decomposition of
fac-$Li^+[(CO)_3Re(PPh_3)(COCH_3)(Br)]^-$ to $Li^+ Br^-$ and cis-$(CO)_4Re(PPh_3)$-
(CH_3) (86). When the homologous formyl cis-$(CO)_5ReRe(CO)_4(CHO)^-$
is decomposed at 90–100°C, $(CO)_5ReRe(CO)_5$ is isolated in 39–52% yields
(38, 47, 66); however, no well-defined reduction [i.e, $(CO)_5Re^-$] or
rhenium hydride products can be detected.

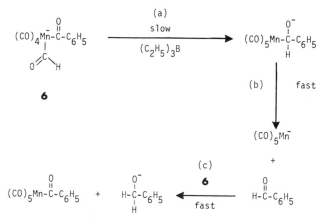

SCHEME 4.   Proposed mechanism for the decomposition of 6.

$$2 \text{ Li}^+ \ (CO)_5\text{Mn-Mn}(CO)_4 \quad \xrightarrow{>0 \ °C} \quad (CO)_5\text{Mn-Mn}(CO)_5 \ + \ 2 \text{ Li}^+ (CO)_5\text{Mn}^- \ + \ H_2$$

$$\underset{\underset{\text{Li}^+ \ \mathbf{12}}{\overset{|}{\underset{H}{\overset{\diagup C}{\diagdown}}}O}}{}$$

$$(32)$$

The unstable anionic cluster formyls (**29–33**) in Table I all appear to decompose, at least in part, by decarbonylation to metal hydrides (*43–46*). Enhanced stability has been noted for the corresponding deuteroformyls (*46*). Surprisingly, many anionic formyls that might be expected to decompose to metal hydrides give no sign of such products by ^{1}H NMR (*32*). The decomposition chemistry of anionic bisformyl complexes would seemingly be of great interest (*56, 57, 62, 66*) but is only beginning to receive detailed attention (*46*).

Neutral formyl complexes which contain ligating CO often decompose by decarbonylation; however, several exceptions exist. For instance, the osmium formyl hydride $Os(H)(CO)_2(PPh_3)_2(CHO)$ evolves H_2(*54*). Although the data are preliminary, the cationic iridium formyl hydride **49** [Eq. (14)] may also decompose by H_2 evolution (*67*). These reactions have some precedent in earlier studies by Norton (*87*), who obtained evidence for rapid alkane elimination from osmium acyl hydride intermediates $Os(H)(CO)_3(L)(COR)$ [L = PPh_3, $P(C_2H_5)_3$]. Additional neutral formyls which do not give detectable metal hydride decomposition products have been noted (*57, 68*); however, in certain cases this can be attributed to the instability of the anticipated hydride under the reaction conditions (H_2 loss or reaction with halogenated solvents).

The decomposition chemistry of neutral formyl **38** can vary depending upon conditions (*56, 57, 60, 62, 68*). Dilute solutions decompose principally to the rhenium hydride $(\eta\text{-}C_5H_5)Re(NO)(CO)(H)$ [Eq. (33)]. However, Casey has found that concentrated solutions (encountered upon attempted purification of **38**) decompose to the novel ester **61** (isolated as a 1 : 1 mixture of diastereomers) (*60, 62*). Two pathways have been suggested for the formation of **61**. In the presence of acid, initial O-protonation of formyl **38** (to give hydroxymethylidene $[(\eta\text{-}C_5H_5)\text{-}Re(NO)(CO)(=CHOH)]^+$) is believed to be followed by rapid hydride transfer from unprotonated formyl to give equal amounts of hydroxymethyl **60** (Scheme 3) and $[(\eta\text{-}C_5H_5)Re(NO)(CO)_2]^+$; subsequent combination and H^+ loss would yield **61**. In the absence of acid, initiation of hydride transfer has been proposed to occur via the attack of the carbonyl group of **38** upon the formyl oxygen of a second **38** (*60, 62*). These transformations have some features in common with the electrophile induced disproportionation of **39** in Scheme 2; Casey has noted parallels with the Cannizzaro and Tischenko reactions of aldehydes.

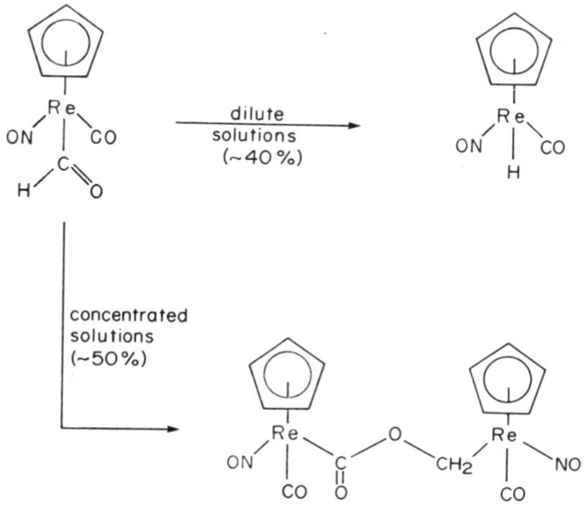

(33)

61

Decomposition of the homologous formyl $(\eta\text{-}C_5H_5)Re(NO)(PPh_3)(CHO)$ (**39**) also depends somewhat upon conditions (57, 64, 65). When **39** is heated at 105°C in toluene-d_8, a 1 : 1 mixture of $(\eta\text{-}C_5H_5)Re(NO)(CO)(H)$ and $(\eta\text{-}C_5H_5)Re(NO)(PPh_3)(H)$ forms cleanly ($t_{\frac{1}{2}} \cong 1$ h) (64). However, yields of these hydrides decrease markedly (to 1–15% each) when **39** is decomposed at lower temperatures in THF-d_8 (4–12 days, 50–60°C). Decomposition by-product PPh_3 does not significantly react with $(\eta\text{-}C_5H_5)Re(NO)(CO)(H)$ under the decomposition conditions; hence $(\eta\text{-}C_5H_5)Re(NO)(PPh_3)(H)$ appears to be a primary product. Thus **39** is the only formyl found to date which can decompose via loss of the formyl $C{=}O$.

B. Thermodynamic and Kinetic Stability

Some of the decomposition reactions in Section V,A above bear upon the thermochemistry of Eq. (2) (Section II). Since a metal hydride has never been observed to be in equilibrium with a measurable quantity of a metal formyl (in Eq. (31), 1% of **25** would have been detectable) (42), ΔG_{rxn} for Eq. (2) (forward direction) is commonly >3 kcal/mol (see, however, Addendum, p. 34). In contrast to Eq. (31), the acyl–alkyl equilibrium in Eq. (34) lies far to the left (42).

$$(CO)_3[(ArO)_3P]\overset{-}{Fe}{-}\overset{O}{\underset{CH_3}{C}} \quad \underset{\longleftarrow}{\longrightarrow} \quad (CO)_4\overset{-}{Fe}{-}CH_3 \quad + \quad (ArO)_3P \qquad (34)$$

The difference in ΔG_{rxn} for Eqs. (31) and (34) is most easily rationalized in terms of the bond energy comparisons shown in Scheme 5. According to current estimates, metal–hydrogen bonds (50–60 kcal/mol) are as much as 30 kcal/mol stronger than homologous metal–carbon bonds (20–30 kcal/mol) (*17, 88*). Thus on the basis of this comparison alone (Scheme 5, a), metal hydride carbonylation should be less favored than metal alkyl carbonylation. However, before a final conclusion is reached, bond energies in the formyl and acyl products must be considered. The acetaldehyde C—H bond dissociation energy is 87 kcal/mol, whereas the acetone C—C bond dissociation energy is 81 kcal/mol (*89*). The difference between these two values (6 kcal/mol) serves as a reasonable estimate for comparison (c) in Scheme 5. Since ketone C=O bond energies are only a few kilocalories stronger than aldehyde C=O bond energies, comparison (b) in Scheme 5 is small and in the opposite direction of (c) (*89*). This indicates the relative thermochemistry of the reactions in Scheme 5 to be essentially controlled by comparison (a), and that CO "insertion" into a M—H bond should be the more difficult reaction to achieve.

The above bond energy analysis is a relative one and does not address whether or not Eq. (2) *can* (despite the contrary examples above) be exothermic. Closely related hydride migrations have been observed to proceed. For instance, the thioformyl utilized by Roper in Eq. (8) was prepared as shown in Eq. (35) (*52, 53*). As indicated, hydride migration to coordinated CS occurs readily and in preference to coordinated CO. Since π-bonds between first- and second-row atoms (i.e., CS) are not as strong as those between first-row atoms (i.e., CO) hydride migration (which entails a π-bond order reduction) appears to follow the thermodynamically preferred course.

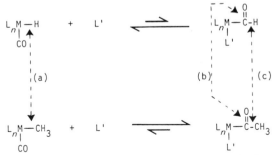

SCHEME 5.   Critical bond energy comparisons in "insertion" reactions.

$$(CO)(PPh_3)_2(Cl)Os-CS \quad \xrightarrow[40 \ psi]{CO} \quad (CO)_2(PPh_3)_2(Cl)Os-C\overset{S}{\underset{H}{\diagdown}} \tag{35}$$

48 99%

In the same vein, several examples of stoichiometric hydride migration to coordinated olefins have been observed (*90, 91*); corresponding alkyl migrations (which should be more favorable based upon a Scheme 5-type analysis) have not yet been noted to occur (*91*). Bercaw has reported the insertion of an isonitrile (subsequent to initial coordination) into a zirconium–hydrogen bond to yield the imino formyl derivative **62** [Eq. (36)] (*92*).

$$(\eta\text{-}C_5Me_5)_2Zr\overset{H}{\underset{H}{\diagdown}} \quad \xrightarrow[-65 \ ^\circ C]{CH_3N\equiv C} \quad (\eta\text{-}C_5Me_5)_2Zr\overset{\overset{H}{\underset{|}{C}}=N-CH_3}{\underset{H}{\diagup}} \tag{36}$$

90%

42

Hückel MO calculations have not revealed any intrinsic kinetic barrier to hydride migration to coordinated CO (*93*). Thus it is worthwhile to consider possibilities that might mask the occurrence of a metal hydride carbonylation reaction. For instance, metal hydrides have been observed to react rapidly with metal acyls; reduction products such as aldehydes or bridging —CHRO— species form (*94–96*). Therefore, it is possible that a formyl complex might react with a metal hydride precursor at a rate competitive with its formation. Such a reaction could also complicate the decomposition chemistry of formyl complexes. Preliminary studies have in fact shown that metal hydrides can react with formyl complexes (*35, 57*), but a complete product analysis has not yet been done.

In several cases, transient zirconium formyls have been suggested to react with their zirconium hydride precursors (*97, 98*). For instance, treatment of $(\eta\text{-}C_5H_5)_2ZrHCl$ with CO yields the biszirconium complex $(\eta\text{-}C_5H_5)_2Zr(Cl)CH_2O(Cl)Zr(\eta\text{-}C_5H_5)_2$ (*98*). An initially formed formyl, generated analogously to the imino formyl in Eq. (36), might subsequently react with starting $(\eta\text{-}C_5H_5)_2Zr(H)(Cl)$ to yield this product. Analogous complexes are formed when $(\eta\text{-}C_5H_5)_2Zr(H)(Cl)$ is reacted with $(\eta\text{-}C_5H_5)_2Zr(COR)(Cl)$ acyls (*96*). The possibility that similar steps might occur in reactions of $(\eta\text{-}C_5Me_5)_2ZrH_2$ with CO has been discussed (*97*).

Several trends in the kinetic stability of formyl complexes are evident in Tables I and II. First, formyl complexes of third row transition metals tend to be more stable than those of first row transition metals. It is well established that third row metals (e.g., Re) form much stronger metal–ligand and metal–metal bonds than first row metals (e.g., Mn) (88). Since most formyl decomposition mechanisms are initiated by ligand dissociation, third-row formyl complexes should have greater kinetic stability. Also, the substitution of good donor ligands such as PPh_3 for CO usually provides additional stability. This would increase $d\pi$–$p\pi$ backbonding to any remaining COs, decreasing their lability. Furthermore, the metal-formyl bond would be strengthened by the increased importance of resonance forms 50b or 51b. Formal charge on the metal is clearly a secondary stability consideration; when substituted with sufficiently good donor ligands, neutral and even cationic formyl complexes prove isolable. From these generalizations and the data in Tables I and II, it is clear that many additional examples of stable formyl complexes should exist.

C. *Other Stoichiometric Reactions That Are Likely to Involve Formyl Intermediates and/or Bridging CHO Species*

In addition to the zirconium hydride/CO chemistry described in the previous section, other reactions have been observed that are likely to involve unstable formyl intermediates. For instance, $Na^+(CO)_5Mn^-$ reacts with ^{13}C-labeled formic acetic anhydride to give ^{13}CO-substituted $(CO)_5MnH$ as product (99). This is consistent with the intermediacy of $(CO)_5Mn^{13}CHO$, which (by comparison to 35 in Table II) would be expected to be extremely unstable and decompose by loss of a terminal ^{12}CO (4, 5). The same formyl has also been suggested to be a labile intermediate in CO exchange reactions of $(CO)_5MnH$ (100). Oxidative addition of formic acetic anhydride to trans-$[IrCl(CO)L_2]$ (L = PPh_3 or $PPh(CH_3)_2$) has been studied (101); although only iridium hydride products were obtained, these have been rationalized as being formed from formyl intermediates.

Cutler found that treatment of iron alkoxymethylidenes of the formula 63 with I^- (0.5 equiv) resulted in the product mixture shown in Eq. (37) (102). The initial step of this reaction is almost certainly I^- attack upon the alkyl group of 63 to yield the neutral formyl $(\eta\text{-}C_5H_5)Fe(CO)(L)(CHO)$ (Scheme 1, approach C-ii). Subsequent rapid hydride donation from $(\eta\text{-}C_5H_5)Fe(CO)(L)(CHO)$ to unreacted starting material [see Eq. (26)] would give the observed product distribution.

$$2 \; [\text{Cp-Fe}^+(CO)(L)=\overset{H}{\underset{OR}{C}}] \; + \; I^- \longrightarrow [\text{Cp-Fe}^+(CO)(L)(CO)] \; + \; [\text{Cp-Fe}(CO)(L)\overset{H_2C}{\underset{OR}{}}] \; + \; RI \tag{37}$$

63

L = CO, PPh$_3$

R = −CH$_3$, −CH$_2$CH$_3$

Bercaw has prepared complexes containing bridging ligands of formula CHO, a typical example of which is shown in Eq. (38) (*97*). Such species may be unstable intermediates in other reactions (*103, 104*). For instance, a transient ^{1}H-NMR resonance at δ 14.3 can be detected during the course of Eq. (39) (*104*). Since early transition metal hydrides of the types utilized in Eqs. (38) and (39) can transfer H$^-$ to aldehydes and ketones (*104*), the initial step in these transformations is probably related to the R$_3$BH$^-$ formyl synthesis [Eq. (3)]. Schrock has reported that η^2 bridging formyls are obtained when dimeric tantalum hydrides such as (η-C$_5$Me$_4$Et)$_2$Ta$_2$Cl$_4$H$_2$ are reacted with CO (*103*). Although bridging CHO ligands might plausibly form on certain CO/H$_2$ reduction catalysts, their chemistry (which has not yet been extensively developed) can be expected to differ significantly from that of the simple terminal formyls that are the focus of this chapter.

$$(\eta\text{-}C_5H_5)_2W\text{—CO} \; + \; (\eta\text{-}C_5Me_5)_2Zr\overset{H}{\underset{H}{\diagup}} \longrightarrow (\eta\text{-}C_5H_5)_2W=C\overset{H}{\underset{O}{\diagdown}}\underset{H}{\overset{}{Zr(\eta\text{-}C_5Me_5)_2}} \tag{38}$$

95%

$$(\eta\text{-}C_5H_5)_2Nb\overset{H}{\underset{H}{\diagup}} \; + \; Fe(CO)_5 \longrightarrow (\eta\text{-}C_5H_5)_2Nb\overset{H}{\underset{CO}{|}}\text{—}Fe(CO)_4 \; + \; H_2 \tag{39}$$

VI

CONCLUSION

It is evident that transition metal formyl complexes possess a unique and rich chemistry. There are numerous surprising contrasts to reactions of transition metal acyl complexes. Since synthetic routes to formyl complexes are now well established, research will likely seek applications for their chemical properties.

With regard to the role of formyl intermediates in catalytic reactions of CO and H_2, the reluctance of transition metal formyl complexes to form from CO and metal hydrides suggests that the corresponding step in any catalytic cycle would also be difficult. Although formyl complexes have not yet been observed to undergo well-defined reactions with H_2, they can be reduced by hydride reagents. These transformations probably have little bearing upon the fate of catalyst-bound formyls; however, the disproportionative tendencies of formyl complexes (both thermally and in the presence of added electrophiles) may be of greater significance. CO/H_2 reduction catalysts often contain electrophilic components, and Shriver has demonstrated that Lewis acid electrophiles can promote the migration of metal alkyls to CO, both in solution and on surfaces (105, 106). Thus electrophiles could conceivably facilitate both the formation and disproportionation of catalyst-bound formyls. Clearly, however, this is only one speculative possibility. Many major mechanistic questions regarding formyl complexes remain, and these will undoubtedly continue to be probed by the full armament of methods available to chemists.

ADDENDUM

Several articles describing formyl chemistry appeared after the body of this chapter was written. In an exceedingly important communication (107), Wayland reported the reversible carbonylation of Rh(OEP)(H) (OEP = octaethylporphyrin) to an isolable neutral formyl Rh(OEP)(CHO) (^{1}H NMR: δ 2.90, d, $J_{103Rh-1H} = 1.75$ Hz; ^{13}C NMR: 194.4 ppm, $J_{13C-1H} = 200$ Hz; IR (cm^{-1}, nujol): $\nu_{C=O}$ 1700 cm^{-1}). An X-ray crystal structure has confirmed the Rh(OEP)(CHO) structure. The porphyrin ligand clearly imparts special properties (stability, ^{1}H NMR, IR) to the formyl ligand. However, the reason why the carbonylation of Rh(OEP)(H) should be more facile than the carbonylation of other metal hydrides is not yet apparent.

Marks has reported (108) the reversible carbonylation of thorium hydrides $(\eta$-$C_5Me_5)_2Th(H)(OR)$ [R = t-C_4H_9, 2,6-$(t$-$C_4H_9)_2C_6H_3$] to spectroscopically observable η^2-formyls, $(\eta$-$C_5Me_5)_2Th(\eta^2$-CHO)(OR). Bulky alkoxide ligands must be present to retard subsequent dimerization to $L_nThO(H)C\!=\!C(H)OThL_n$ species. This carbonylation is clearly driven by the oxygenophilicity of thorium. The product is probably more accurately formulated as an oxycarbene, $Th\!\leftarrow\!C(H)O$.

An account by Casey (109) describes, in addition to material cited elsewhere in this chapter, the detection of the neutral formyl

$(\eta\text{-}C_5H_5)Ru(CO)_2(CHO)$ following the reaction of $(\eta\text{-}C_5H_5)Ru(CO)_3^+$ with $(i\text{-}C_3H_7O)_3BH^-$ at $-90°C$; rapid decomposition subsequently occurred. An account by Schoening, Vidal, and Fiato (110) mainly describes chemistry cited elsewhere in this chapter $(43, 46, 99)$.

ACKNOWLEDGMENTS

The preparation of this review was aided by grants from the Department of Energy, the Alfred P. Sloan Foundation, and the Camille and Henry Dreyfus Foundation (Teacher–Scholar Grant).

REFERENCES

1. C. Masters, *Adv. Organomet. Chem.* **17**, 61 (1979), and references therein.
2. E. L. Muetterties and J. Stein, *Chem. Rev.* **79**, 479 (1979), and references therein.
2a. R. L. Pruett, *Science* **211**, 11 (1981).
3. H. Pichler and H. Schulz, *Chem. Ing. Tech.* **42**, 1162 (1970).
4. A. Wojcicki, *Adv. Organomet. Chem.* **11**, 87 (1973).
5. F. Calderazzo, *Angew. Chem. Int. Ed. Engl.* **16**, 299 (1977).
6. I. Wender, *Catal. Rev. Sci. Eng.*, **14**, 97 (1976).
7. G. Henrici-Olivé and S. Olivé, *Angew. Chem. Int. Ed. Engl.* **15**, 136 (1976).
8. M. Araki and V. Ponec, *J. Catal.* **44**, 439 (1976).
9. A. Jones and B. D. McNicol, *J. Catal.* **47**, 384 (1977).
10. R. W. Joyner, *J. Catal.* **50**, 176 (1977).
11. G. G. Low and A. T. Bell, *J. Catal.* **57**, 397 (1979).
12. P. Biloen, J. N. Helle, and W. M. H. Sachtler, *J. Catal.* **58**, 95 (1979).
13. R. C. Brady, III, and R. Pettit, *J. Am. Chem. Soc.* **102**, 6181 (1980).
14. F. Fischer and H. Tropsch, *Brennst. Chem.* **7**, 97 (1926).
15. A. Takeuchi and J. R. Katzer, *J. Phys. Chem.* **85**, 937 (1981).
16. R. L. Pruett, *Ann. N.Y. Acad. Sci.* **295**, 239 (1977).
17. J. W. Rathke and H. M. Feder, *J. Am. Chem. Soc.* **100**, 3623 (1978).
18. H. M. Feder and J. W. Rathke, *Ann. N.Y. Acad. Sci.* **33**, 45 (1980); D. R. Fahey, *J. Am. Chem. Soc.* **103**, 136 (1981).
19. J. S. Bradley, *J. Am. Chem. Soc.* **101**, 7419 (1979); W. Keim, M. Berger, and J. Schlupp, *J. Catal.* **61**, 359 (1980).
20. B. D. Dombek, *J. Am. Chem. Soc.* **102**, 6855 (1980); J. F. Knifton, *J. Chem. Soc. Chem. Commun.* p. 188 (1981).
21. R. J. Daroda, J. R. Blackborow, and G. Wilkinson, *J. Chem. Soc. Chem. Commun.* pp. 1098, 1101 (1980); T. E. Paxson, C. A. Reilly, and D. R. Holeck, ibid., p. 618 (1981).
22. C. P. Casey and C. A. Bunnell, *J. Am. Chem. Soc.* **98**, 436 (1976), and references therein.
23. C. M. Lukehart, *Acc. Chem. Res.* **14**, 109 (1981), and references therein.
24. J. A. Gladysz, G. M. Williams, W. Tam, D. L. Johnson, D. W. Parker, and J. C. Selover, *Inorg. Chem.* **18**, 553 (1979).
25. J. W. Suggs, *J. Am. Chem. Soc.* **100**, 640 (1978), and references therein.
26. A. Davison and D. L. Reger, *J. Am. Chem. Soc.* **94**, 9237 (1972).
27. J. P. Collman and S. R. Winter, *J. Am. Chem. Soc.* **95**, 4089 (1973).
28. S. R. Winter, Ph.D. Thesis, Stanford Univ. (1973).

29. C. P. Casey and S. M. Neumann, *J. Am. Chem. Soc.* **98**, 5395 (1976).
30. J. A. Gladysz and J. C. Selover, *Am. Chem. Soc. Natl. Meet., 172nd, San Francisco, Calif.* Abstr. INOR 41 (1976).
31. J. A. Gladysz and J. C. Selover, *Tetrahedron Lett.* p. 319 (1978).
32. J. C. Selover, M. Marsi, D. W. Parker, and J. A. Gladysz, *J. Organomet. Chem.* **206**, 317 (1981).
33. S. R. Winter, *in* "20th Annual Report on Research Under Sponsorship of the Petroleum Research Fund," p. 116. Am. Chem. Soc., Washington, D.C., 1976.
34. S. R. Winter, G. W. Cornett, and E. A. Thompson, *J. Organomet. Chem.* **133**, 339 (1977).
35. J. A. Gladysz, G. M. Williams, W. Tam, and D. L. Johnson, *J. Organomet. Chem.* **140**, C1 (1977).
36. J. A. Gladysz and W. Tam, *J. Org. Chem.* **43**, 2279 (1978).
37. C. P. Casey and S. M. Neumann, *J. Am. Chem. Soc.* **100**, 2544 (1978).
38. J. A. Gladysz and W. Tam, *J. Am. Chem. Soc.* **100**, 2545 (1978).
39. J. A. Gladysz and J. H. Merrifield, *Inorg. Chim. Acta* **30**, L317 (1978).
40. K. P. Darst and C. M. Lukehart, *J. Organomet. Chem.* **171**, 65 (1979).
41. J. A. Gladysz, *Aldrichimica Acta* **12**, 13 (1979).
42. C. P. Casey and S. M. Neumann, *Adv. Chem. Ser.* No. 173, p. 131 (1979).
43. R. L. Pruett, R. C. Schoening, J. L. Vidal, and R. A. Fiato, *J. Organomet. Chem.* **182**, C57 (1979).
44. B. F. G. Johnson, R. L. Kelly, J. Lewis, and J. R. Thornback, *J. Organomet. Chem.* **190**, C91 (1980).
45. G. R. Steinmetz and G. L. Geoffroy, *J. Am. Chem. Soc.* **103**, 1278 (1981).
46. R. C. Schoening, J. L. Vidal, and R. A. Fiato, *J. Organomet. Chem.* **206**, C43 (1981).
47. W. Tam, M. Marsi, and J. A. Gladysz, submitted for publication.
48. G. A. Olah, Y. D. Vankar, M. Arvanaghi, and J. Sommer, *Angew. Chem. Int. Ed. Engl.* **18**, 614 (1979).
49. J. E. Ellis, *J. Organomet. Chem.* **86**, 1 (1975).
50. H. D. Kaesz and R. B. Saillant, *Chem. Rev.* **72**, 231 (1972).
51. K. Inkrott, R. Goetze, and S. G. Shore, *J. Organomet. Chem.* **154**, 337 (1978).
52. T. J. Collins and W. R. Roper, *J. Chem. Soc. Chem. Commun.* p. 1044 (1976).
53. T. J. Collins and W. R. Roper, *J. Organomet. Chem.* **159**, 73 (1978).
54. K. L. Brown, G. R. Clark, C. E. L. Headford, K. Marsden, and W. R. Roper, *J. Am. Chem. Soc.* **101**, 503 (1979).
55. B. N. Chaudret, D. J. Cole-Hamilton, R. S. Nohr, and G. Wilkinson, *J. Chem. Soc. Dalton Trans.* p. 1546 (1977).
56. C. P. Casey, M. A. Andrews, and J. E. Rinz, *J. Am. Chem. Soc.* **101**, 741 (1979).
57. W. Tam, W.-K. Wong, and J. A. Gladysz, *J. Am. Chem. Soc.* **101**, 1589 (1979).
58. J. R. Sweet and W. A. G. Graham, *J. Organomet. Chem.* **173**, C9 (1979).
59. W. K. Wong, W. Tam, C. E. Strouse, and J. A. Gladysz, *J. Chem. Soc. Chem. Commun.* p. 530 (1979).
60. C. P. Casey, M. A. Andrews, and D. R. McAlister, *J. Am. Chem. Soc.* **101**, 3371 (1979).
61. W.-K. Wong, W. Tam, and J. A. Gladysz, *J. Am. Chem. Soc.* **101**, 5440 (1979).
62. C. P. Casey, M. A. Andrews, D. R. McAlister, and J. E. Rinz, *J. Am. Chem. Soc.* **102**, 1927 (1980).
63. C. P. Casey, S. M. Neumann, M. A. Andrews, and D. R. McAlister, *Pure Appl. Chem.* **52**, 625 (1980).

64. W. Tam, G. Y. Lin, W.-K. Wong, W. A. Kiel, V. K. Wong, and J. A. Gladysz, *J. Am. Chem. Soc.*, **104**, 141 (1982).
65. J. A. Gladysz, W. A. Kiel, G.-Y. Lin, W. K. Wong, and W. Wam, *Adv. Chem. Ser.* No. 152, p. 147 (1981).
66. W. Tam, Ph.D. Thesis, Univ. of California at Los Angeles (1979).
67. D. L. Thorn, *J. Am. Chem. Soc.* **102**, 7109 (1980); *Organometallics* **1**, 197 (1982).
68. W. Tam, G.-Y. Lin, and J. A. Gladysz, *Organometallics* **1**, in press (1982).
69. R. M. Silverstein, G. C. Bassler, and T. C. Morrill, "Spectrometric Identification of Organic Compounds," 3rd ed., Chap. 4. Wiley, New York, 1974.
70. F. W. Wehrli and T. Wirthlin, "The Interpretation of Carbon-13 NMR Spectra," pp. 48–53. Heyden, London, 1978.
71. R. R. Schrock, *Acc. Chem. Res.* **12**, 98 (1979).
72. M. H. Chisholm and S. Godleski, *Prog. Inorg. Chem.* **20**, 299 (1976).
73. R. M. Silverstein, G. C. Bassler, and T. C. Morrill, "Spectrometric Identification of Organic Compounds," 3rd ed., Chap. 3. Wiley, New York, 1974.
74. A. Streitwieser and C. H. Heathcock, "Introduction to Organic Chemistry," pp. 348, 452–453. Macmillan, New York, 1976.
75. W. A. Kiel, G.-Y. Lin, and J. A. Gladysz, *J. Am. Chem. Soc.* **102**, 3329 (1980).
76. W. A. Kiel, unpublished observations.
77. B. E. R. Schilling, R. Hoffmann, and J. W. Faller, *J. Am. Chem. Soc.* **101**, 592 (1979).
78. O. Eisenstein and R. Hoffmann, unpublished calculations.
79. E. O. Fischer and A. Frank, *Chem. Ber.* **111**, 3740 (1978).
80. S. Krishnamurthy, F. Vogel, and H. C. Brown, *J. Org. Chem.* **42**, 2534 (1977).
81. H. C. Brown, M. M. Midland, P. Jacob, III, N. Miyaura, M. Itoh, and A. Suzuki, *J. Am. Chem. Soc.* **94**, 6549 (1972).
82. J. A. Van Doorn, C. Masters, and H. C. Volger, *J. Organomet. Chem.* **105**, 245 (1976).
83. H. O. House, "Modern Synthetic Reactions," 2nd ed., p. 80. Benjamin, New York, 1972.
84. G-Y. Lin, unpublished observations.
85. J. A. Gladysz, J. C. Selover, and C. E. Strouse, *J. Am. Chem. Soc.* **100**, 6766 (1978).
86. D. W. Parker, M. Marsi, and J. A. Gladysz, *J. Organomet. Chem.* **194**, C1 (1980).
87. J. R. Norton, *Acc. Chem. Res.* **12**, 139 (1979).
88. J. A. Conner, *Top. Curr. Chem.* **71**, 71 (1977).
89. S. W. Benson, "Thermochemical Kinetics," 2nd ed., p. 309. Wiley, New York, 1976.
90. J. W. Byrne, H. U. Blaser, and J. A. Osborn, *J. Am. Chem. Soc.* **97**, 3871 (1975).
91. H.-F. Klein, R. Hammer, J. Gross, and U. Schubert, *Angew. Chem. Int. Ed. Engl.* **19**, 809 (1980), and references therein.
92. P. T. Wolczanski and J. E. Bercaw, *J. Am. Chem. Soc.* **101**, 6450 (1979).
93. H. Berke and R. Hoffmann, *J. Am. Chem. Soc.* **100**, 7224 (1978).
94. J. Halpern, *Pure Appl. Chem.* **51**, 2171 (1979).
95. J. A. Gladysz, W. Tam, G. M. Williams, D. L. Johnson, and D. W. Parker, *Inorg. Chem.* **18**, 1163 (1979).
96. K. I. Gell and J. Schwartz, *J. Organomet. Chem.* **162**, C11 (1978).
97. P. T. Wolczanski and J. E. Bercaw, *Acc. Chem. Res.* **13**, 121 (1980).
98. G. Fachinetti, C. Floriani, A. Roselli, and S. Pucci, *J. Chem. Soc. Chem. Commun.* p. 269 (1978).
99. R. A. Fiato, J. L. Vidal, and R. L. Pruett, *J. Organomet. Chem.* **172**, C4 (1979).
100. F. Basolo and R. G. Pearson, "Mechanisms of Inorganic Reactions," 2nd ed., Chap. 7. Wiley, New York, 1978; B. H. Byers and T. L. Brown, *J. Organomet. Chem.* **127**,

181 (1977); R. G. Pearson, H. W. Walker, H. Mauermann, and P. C. Ford, *Inorg. Chem.* **20,** 2741 (1981).

101. J. A. van Doorn, C. Masters, and C. van der Woude, *J. Organomet. Chem.* **141,** 231 (1977).
102. A. R. Cutler, *J. Am. Chem. Soc.* **101,** 604 (1979).
103. P. Belmonte, R. R. Schrock, M. R. Churchill, and W. J. Youngs, *J. Am. Chem. Soc.* **102,** 2858 (1980).
104. J. A. Labinger, K. S. Wong, and W. R. Scheidt, *J. Am. Chem. Soc.* **100,** 3254 (1978).
105. S. B. Butts, S. H. Strauss, E. M. Holt, R. E. Stimson, N. W. Alcock, and D. F. Shriver, *J. Am. Chem. Soc.* **102,** 5093 (1980).
106. F. Correa, R. Nakamura, R. E. Stimson, R. L. Burwell, Jr., and D. F. Shriver, *J. Am. Chem. Soc.* **102,** 5112 (1980).
107. B. B. Wayland and B. A. Woods, *J. Chem. Soc., Chem. Commun.,* p. 700 (1981).
108. P. J. Fagan, K. G. Moloy, and T. J. Marks, *J. Am. Chem. Soc.* **103,** 6959 (1981).
109. C. P. Casey, M. A. Andrews, D. R. McAlister, W. D. Jones, and S. G. Harsy, *J. Mol. Cat.* **13,** 43 (1981).
110. R. C. Schoening, J. L. Vidal, and R. A. Fiato, *J. Mol. Cat.,* **13,** 83 (1981).

The Organic Chemistry of Gold

GORDON K. ANDERSON*

Department of Chemistry
University of Guelph
Guelph, Ontario, Canada

* Present address: Department of Chemistry, University of Missouri-St. Louis, St. Louis, Missouri 63121.

I

INTRODUCTION

Since the earliest times elemental gold has held a special fascination for man, as evidenced by the alchemists' desire to convert base materials into the metal; although the days of alchemy are now long past, gold is still prized for its beauty.

The study of gold compounds, on the other hand, was to a considerable extent neglected over the years, at least by comparison with the efforts expended in investigating the compounds of neighboring metals. However, there has been a significant growth of interest in gold chemistry recently, with the use of gold–phosphine complexes in the treatment of arthritis being a topical field of study (1).

The area of organogold chemistry has also blossomed in the past ten years. In a 1926 review article (2) it was stated that "Gold has no affinity for carbon. Complex organic compounds of gold in which the gold is directly attached to carbon are incapable of existence, or at least cannot be isolated." In fact, only a few workers were active in this field for many years, and the past decade has produced about 75% of the presently existing contributions to the subject.

Organogold complexes and gold compounds in general predominantly contain the metal in one of two oxidation states, namely gold(I) or gold(III). The former has a d^{10} electronic configuration and is necessarily diamagnetic. Gold(I) compounds are commonly two-coordinate, linear, 14-electron species, in contrast to univalent copper and silver complexes, which tend to adopt four-coordinate structures. Higher coordination number gold(I) compounds are known, but they readily undergo ligand dissociation.

Gold(III) has a d^8 electronic configuration, and its complexes are usually four coordinate. These 16-electron species are square planar, and hence diamagnetic, and are quite amenable to study by NMR techniques.

The first ionization energy of gold (890 kJ mol^{-1}) is higher than that of copper or silver because of the inadequately shielded higher nuclear charge, but the third ionization energy is considerably lower (2943 kJ mol^{-1}) and may account for the large number of trivalent gold complexes. Gold(III) is a fairly "hard" acid and is more prone, therefore, to formation of compounds with oxygen and nitrogen donors than the univalent metal is. Further background information on general gold chemistry is available in a recent monograph (3).

Organogold chemistry was reviewed by Armer and Schmidbaur (4) in 1970 and has been discussed only briefly (3, 5) during the past ten years, and a more specific review of arylgold complexes has appeared (6). The present work includes all reports, up to mid-1980, of compounds containing a gold–carbon bond, with the exceptions of cyanides, cyanates, and thiocyanates, which are mentioned only briefly. Where the same work has been published in English and another language, only the former is referenced. Assignment of gold(I) and gold(III) compounds to separate sections is largely achieved, but in certain cases, as in the oxidative addition of halogens to organogold(I) complexes, it is deemed more pertinent to discuss in detail the preparation of gold(III) complexes under the reactions of their gold(I) analogs. In such cases a cross-reference is made to the appropriate section.

II
ORGANOGOLD(I) COMPLEXES: PREPARATION AND CHARACTERIZATION

A. Dialkyl- and Diarylaurates

Complexes of gold(I) containing organic groups only as ligands have been prepared recently for the first time. The dimethylaurate(I) anion was produced by treatment of $[AuMe(PPh_3)]$ in diethyl ether solution with methyllithium (7, 8). The ^1H-NMR spectrum contained a singlet at δ -0.19, and in the ^{31}P-NMR spectrum a signal at δ P -5.8, due to free PPh_3, was observed. It was suggested that the dimethylaurate(I) anion existed as an ether complex; treatment with pyridine resulted in formation of the $[AuMe_2py_2]^-$ anion. When the dialkylaurate(I) anion was treated with an alkyl halide, the corresponding trialkylgold(III) complex was produced. This proved to be a useful route to mixed alkyl complexes [Eq. (1)].

$$[AuR^1L] + R^2Li \xrightarrow{-L} [AuR^1R^2]^- Li^+ \xrightarrow{R^3X} [AuR^1R^2R^3L] + LiX \qquad (1)$$

The dimethylaurate(I) anion was more readily isolated (9) when the lithium ion was complexed with pentamethyldiethylenetriamine (PMDT) [Eq. (2)]. The greater thermal stability of these complexes compared with their phosphine analogs was explained in terms of less-ready ligand dissociation and complexation of the lithium ion, preventing its attack at the gold center.

$$[AuMe(PPh_3)] + MeLi + PMDT \rightarrow [Li(PMDT)][AuMe_2] + PPh_3 \qquad (2)$$

Hydrolysis of dimethylaurate(I) salts in the presence of PPh$_3$ gave [AuMe(PPh$_3$)], whereas the complex [Li(PMDT)][AuMe$_2$] produced ethane and metallic gold (10). Rapid decomposition took place in the presence of oxygen to give the above products, and the intermediacy of paramagnetic gold(II) species has been postulated. The possible further reaction pathways of the latter have been the subject of a theoretical treatment (10).

Bis(polyfluorophenyl)aurate(I) complexes have been similarly prepared and isolated as their tetra-n-butylammonium salts ($11-13$). Treatment of [AuCl(tht)] (tht = tetrahydrothiophene) with excess LiC$_6$F$_5$, followed by n-Bu$_4$N$^+$Br$^-$, gave the diarylaurate(I) complex. A similar reaction with 1 mol equiv LiC$_6$F$_5$, followed by treatment with acid, gave the corresponding n-Bu$_4$N[AuX(C$_6$F$_5$)] complex [Eq. (3)]. The bis(pentachlorophenyl)aurate(I) analogs were also prepared (12), and treatment of [Au(C$_6$F$_5$)(tht)] with LiC$_6$H$_2$F$_3$ gave (13) the mixed anion [Au(C$_6$F$_5$)(C$_6$H$_2$F$_3$)]$^-$. The diarylaurates reacted with bis(triphenylphosphine)gold(I) cations to yield neutral products [Eq. (4)].

$$[\text{AuCl(tht)}] + \text{LiC}_6\text{F}_5 \longrightarrow [\text{Au(C}_6\text{F}_5\text{)(tht)}] \xrightarrow[n\text{-Bu}_4\text{N}^+]{\text{HX}} n\text{-Bu}_4\text{N[AuX(C}_6\text{F}_5\text{)]} \quad (3)$$

$$[\text{Au(PPh}_3\text{)}_2]\text{X} + n\text{-Bu}_4\text{N[Au(C}_6\text{F}_5\text{)}_2] \longrightarrow 2\,[\text{Au(C}_6\text{F}_5\text{)(PPh}_3\text{)}] + n\text{-Bu}_4\text{NX} \quad (4)$$

The reactions of Li[AuMe$_2$] with methyl iodide and methyl tosylate have been investigated (14), but the relatively high reactivity toward the latter remains unexplained.

B. Complexes of the Type [AuRL] (R = Alkyl or Aryl)

The most common route to alkyl or aryl complexes of the type [AuRL] is by the treatment of a halide complex with an alkyl- or aryllithium reagent. The first reactions of this type were performed (15) in 1959 [Eq. (5)], and the methyl and phenyl compounds were found to have chemical and thermal stabilities intermediate between those of the previously known organopalladium and -platinum complexes.

$$[\text{AuCl(PR}_3'\text{)}] + \text{RLi} \rightarrow [\text{AuR(PR}_3'\text{)}] + \text{LiCl} \quad (5)$$

R = Me or Ph; PR$_3'$ = PEt$_3$ or PPh$_3$

The ^{1}H-NMR spectrum of [AuMe(PMe$_3$)] exhibited two doublet resonances, with ^2J(P—H) 8.9 Hz and ^3J(P—H) 8.7 Hz (16). Addition of 10^{-3}–

10^{-2} mol equiv PMe_3 caused collapse of the CH_3Au doublet due to phosphine exchange. Further PMe_3 addition caused a reduction in $^2J(P—H)$, which became equal to zero at a $Au:PMe_3$ ratio of $1:3.4$ and eventually became identical to that in free PMe_3, the resonance being the average of the values for complexed and free PMe_3. The intermediate zero value indicated that the values of $^2J(P—H)$ in the free and complexed trimethylphosphine were of opposite sign.

It was shown (17) that phosphine exchange occurred in $[AuMe(PR_3)]$ complexes, but species with a coordination number greater than two were not observed, even with bidentate phosphorus ligands. Although such species were not observed, an associative mechanism was thought to be operative (16, 18). With trimethylphosphite also, only the two-coordinate complex $[AuMe\{P(OMe)_3\}]$ was detected in solution (19).

The trimethylsilylmethyl complex $[Au(CH_2SiMe_3)(PPh_3)]$ was prepared (20), and it was suggested that the α-silyl group lowered the thermal stability of the complex because the PMe_3 and $AsPh_3$ analogs could not be isolated. The following year, however, the neopentyl complex was also prepared (21) and was found to be more thermally unstable than the complex containing the α-silyl group. This was attributed to possible $d\pi-d\pi$ overlap with the metal in the latter, which was reflected in ^1H-NMR chemical shift changes of the methylene group in this and analogous platinum compounds.

Bidentate phosphorus ligands, such as bis(diphenylphosphino)methane (dpm), form bridges between gold centers, and these binuclear complexes may also be readily alkylated [Eq. (6)] (22). Arylgold complexes of this type have also been prepared [Eq. (7)] (23). The complex $[AuCl(AsPh_3)]$ reacted with 2-pyridyllithium at $-40°C$ to give an arylgold(I) complex (24), which reacted by arsine dissociation to yield a cyclic trimer [Eq. (8)]. On heating, coupling of the 2-pyridyl groups occurred. 8-Quinolylgold(I) complexes were found to exist as dimers (25), which, along with other quinolyl compounds, also underwent coupling reactions on heating.

$$
\begin{array}{ccc}
\overset{\displaystyle\frown}{Ph_2P\quad PPh_2} & & \overset{\displaystyle\frown}{Ph_2P\quad PPh_2} \\
|\qquad\ | & & |\qquad\ | \\
Au\quad Au + 2\,MeLi \longrightarrow & & Au\quad Au + 2\,LiCl \\
|\qquad\ | & & |\qquad\ | \\
Cl\quad\ Cl & & Me\quad Me
\end{array}
\qquad (6)
$$

$$[ClAu\{Ph_2P(CH_2)_nPPh_2\}AuCl] + 2\,p\text{-}MeC_6H_4Li \xrightarrow{-2LiCl}$$
$$[p\text{-}MeC_6H_4Au\{Ph_2P(CH_2)_nPPh_2\}AuC_6H_4Me\text{-}p]$$

$n = 2, 3, \text{or } 4$

$$(7)$$

$$\text{(8)}$$

Reactions of aryllithium reagents containing potential coordination sites in the ortho position with [AuClL] produced no evidence for three-coordination. Values of $\nu(C{=}C)$ in the complexes $[Au(C_6H_4CH{=}CH_2\text{-}o)\text{-}(PPh_3)]$ and $[Au(C_6H_4CH_2CH{=}CH_2\text{-}o)(PPh_3)]$ confirmed that the olefinic groups were not coordinated (26), and there was no N-coordination in $[Au(C_6H_4CH_2NMe_2\text{-}o)(PPh_3)]$ (27). The reaction of $[AuCl(PEt_3)]$ with $o\text{-}Ph_2PCH_2C_6H_4Li$ produced a binuclear complex (28), triethylphosphine being displaced by the chelating ligand [Eq. (9)].

$$\text{(9)}$$

Polyfluorophenylgold(I) complexes have been prepared with mono- and bidentate phosphine ligands (29). With $[AuR_f(PCy_3)]$ ($R_f = C_6H_3F_2$ or $C_6H_2F_3$) an equilibrium was established [Eq. (10)] involving the anionic diarylaurates described previously (Section II,A), which depended on the nature of the aryl group.

$$\text{(10)}$$

Alkyl and aryl complexes of the type [AuRL] may also be prepared by reaction of a halide complex with a Grignard reagent. Thus have been prepared the methyl complexes containing triphenylphosphine (30) and diphenyl(ferrocenyl)phosphine (31), as well as the first benzylgold complex (32). The m- and p-fluorophenylgold(I) complexes containing phosphine or isonitrile ligands were also produced by this route (33), and the phosphine complexes were found to be less stable than their platinum analogs (34). From ^{19}F-NMR chemical shift data, it was found that the gold center withdraws electron density from the isonitrile in [Au(C$_6$H$_4$F)-(CNC$_6$H$_4$F)] (33), but is a weak donor into the fluorophenyl ring. Chemical shift data for [Au(C$_6$H$_4$F)(PR$_3$)] complexes have also been used (35) to compile a π-acceptor series for various phosphorus ligands, although variations in σ-donor ability have been neglected. Pentafluorophenyl- and pentabromophenylgold(I) complexes have also been prepared by the Grignard route (36, 37).

Thallium(III) compounds usually react to cause oxidation of gold(I), but in certain cases arylgold(I) species have been isolated [Eqs. (11) and (12)] (38, 39). Oxidative addition was prevented in the former due to the inability of X to form the necessary bridges in the intermediate species (38).

$$[AuXL] + TlBr(C_6F_5)_2 \rightarrow [Au(C_6F_5)L] + TlBrX(C_6F_5) \tag{11}$$
L = PPh$_3$ or AsPh$_3$; X = Ph, NO$_3$, O$_2$CMe, or SCN)

$$[AuCl(PPh_3)] + Tl(C_6HCl_4)_3 \rightarrow [Au(C_6HCl_4)(PPh_3)] + TlCl(C_6HCl_4)_2 \tag{12}$$

Alkylgold(I) complexes have also been obtained from reactions of tris(triphenylphosphinegold)oxonium salts with activated organic reagents, such as ketones (40), vinyl ethers (41, 42), and esters (42). With methyl ketones, β-ketonylgold(I) complexes were always obtained [Eq. (13)]. Treatment of a vinylgold(I) complex with KMnO$_4$ in acetone also produced [Au(CH$_2$COMe)(PPh$_3$)] (43).

$$(Ph_3PAu)_3O^+MnO_4^- + CH_3COR \rightarrow [Au(CH_2COR)(PPh_3)] \tag{13}$$
R = Me, Et, or Ph

A few, less general, routes to [AuRL] complexes have been discovered. Reaction of [AuCl(PPh$_3$)] with diazomethane resulted (44) in methylene insertion into the Au—Cl bond [Eq. (14)], but decomposition to [AuCl(PPh$_3$)] occurred on standing. Benzylgold(I) complexes were prepared by aromatization auration (45) using an olefinic precursor [Eq. (15)], the ratio of the products obtained upon cleavage with acid of the Au—C bond suggesting that aromatization auration may have occurred more readily for the more conjugated triene.

$$[AuCl(PPh_3)] + CH_2N_2 \rightarrow [Au(CH_2Cl)(PPh_3)] + N_2 \tag{14}$$

$$(15)$$

L = PPh$_3$

The trifluoromethyl complex [Au(CF$_3$)(PPh$_3$)] was produced (46) when the methyl analog was treated with CF$_3$I. This reaction was peculiar to the triphenylphosphine complex, and to a lesser extent its PMePh$_2$ analog, as the species [AuMeL] (L = PMe$_3$ or PMe$_2$Ph) reacted to give gold(III) complexes. In the trifluoromethylgold(I) complexes, [Au(CF$_3$)L] (L = PMe$_3$ or PMe$_2$Ph), the values of ^2J(P—H) and ^3J(P—F) were found to be of opposite sign (47).

C. Ylide Complexes

Gold(I) ylide complexes of various types have been prepared (48). Reactions of [AuClL] complexes with ylides led initially to monoylide species (49), but displacement of L also occurred with excess ylide [Eq. (16)]. Analogous compounds with one or two α-silyl groups were also obtained.

$$[AuCl(PMe_3)] + Me_3P{=}CH_2 \rightarrow [Me_3PAuCH_2\overset{+}{P}Me_3]Cl^-$$
$$\downarrow Me_3P{=}CH_2, \ -PMe_3 \qquad (16)$$
$$[Me_3PCH_2AuCH_2PMe_3]^+Cl^-$$

The triphenylphosphine analog, [Ph$_3$PAuCH$_2\overset{+}{P}$Ph$_3$]Cl$^-$, was prepared similarly (50), and its ^{1}H-NMR spectra in the temperature range $-50°$C to 25°C showed that dissociation of PPh$_3$ occurred in solution. Again, in the presence of excess Ph$_3$P=CH$_2$, further reaction took place to give the bisylide complex.

Mono- and bis(ylide)gold(I) complexes were also formed with Ph$_3$PCHCOPh (51). With [AuMe(PMe$_3$)], however, displacement of the methyl group could not occur, and a neutral complex was produced, [Eq. (17)] (52, 53). The analogous species containing α-trimethylsilyl groups in the ylide or alkyl moieties were similarly prepared. The first tris(dimethylamino)methylenephosphorane complex of a transition metal was obtained recently (54) from [AuCl(PPh$_3$)] [Eq. (18)], and the ^{1}H- and ^{13}C-NMR data for the methylene groups were similar to those in [Au(CH$_2$-PPh$_3$)$_2$]Cl.

$$[AuMe(PMe_3)] + Me_3P{=\!\!=}CH_2 \rightarrow [Me\overset{-}{Au}CH_2\overset{+}{P}Me_3] + PMe_3 \qquad (17)$$

$$[AuCl(PPh_3)] + 2\,(Me_2N)_3P{=\!\!=}CH_2 \xrightarrow{-PPh_3} [(Me_2N)_3\overset{+}{P}CH_2\overset{-}{Au}CH_2\overset{+}{P}(NMe_2)_3]Cl^- \qquad (18)$$

Six gold(I) ylide complexes of the types described above have been tested for use in arthritis therapy (55), but only one showed an activity comparable to standard chrysotherapeutic agents. The separation between toxic and therapeutic doses was too narrow, but this is clearly an area where further study could be of immense value.

The majority of gold(I) ylide complexes prepared have been of a binuclear constitution. The first compound of this type was obtained (56) when [Au(CH$_2$PMe$_3$)$_2$]Cl was allowed to stand in the presence of the ylide for seven days [Eq. (19)]. The ^{31}P-NMR spectrum showed only one signal, indicating that the two phosphorus centers were in identical environments, and the ^{1}H-NMR spectrum exhibited two doublets with ^2J(P—CH$_3$) and ^2J(P—CH$_2$) having the same sign. Thus the symmetrical structure depicted in Eq. (19) was invoked, and the presence of two onium centers adjacent to the Au—C bonds was believed to stabilize this unlikely species.

$$[Me_3PCH_2AuCH_2PMe_3]Cl \xrightarrow[-\,(Me_4P)Cl]{Me_3PCH_2\ cat.}
\begin{array}{c}
Me \quad\ CH_2{-}Au{-}CH_2 \quad\ Me \\
\diagdown\ \diagup \qquad\qquad\ \diagdown\ \diagup \\
P \qquad\qquad\qquad\ P \\
\diagup\ \diagdown \qquad\qquad\ \diagup\ \diagdown \\
Me \quad\ CH_2{-}Au{-}CH_2 \quad\ Me
\end{array} \qquad (19)$$

Since this discovery a number of binuclear complexes of this type have been isolated, including the arsenic analog of the above (57) and compounds prepared from more exotic ylide precursors (58–60). A polymeric gold(I) complex was obtained by reaction of [AuCl(PMe$_3$)] with the bidentate ylide CH$_2$=PMe$_2$(CH$_2$)$_6$PMe$_2$=CH$_2$ (60).

$$2\,[AuMe(PMe_3)] + Me_3P{=\!\!=}C{=\!\!=}PMe_3 \xrightarrow{-2PMe_3}
\begin{array}{c}
Me_3P \quad\ Au{-}Me \\
\diagdown\ \diagup \\
C \\
\diagup\ \diagdown \\
Me_3P \quad\ Au{-}Me
\end{array} \qquad (20)$$

Other types of binuclear compounds have been prepared from bis(trimethylphosphoranylidene)methane [Eq. (20)] (61), and from the reaction of $Et_2MeP=CH_2$ with the polymeric species $[Au\{CH(PPh_2)_2\}]_n$ [Eq. (21)] (62).

$$2/n\ [Au\{CH(PPh_2)_2\}]_n + Et_2MeP=CH_2 \xrightarrow{-CH_2(PPh_2)_2}
\begin{array}{c}
\underset{Ph_2}{P}-Au-CH_2 \\
HC \diagup \qquad \diagdown PEt_2 \\
\underset{Ph_2}{P}-Au-CH_2
\end{array} \quad (21)$$

D. Vinyl and Acetylide Complexes

Gold(I) acetylide complexes have been prepared from alkynyllithium compounds [Eq. (22)] (63), and a dimeric complex [(PhC≡C)Au(μ-dpe)-Au(C≡CPh)] [dpe = bis(diphenylphosphino)ethane] was obtained analogously. The simplest acetylide complexes are perhaps those containing only acetylide ligands. The compounds $[Au(C≡CPh)]_x$ and $[Au(C≡C$-t-Bu)$]_4$ (1) contain σ- and π-bonded acetylide groups, the values of

$$
\begin{array}{c}
t\text{-Bu} \\
C \\
t\text{-BuC}\equiv\text{C}-\text{Au}-||| \\
| \qquad\qquad\quad C \\
\text{Au} \qquad\qquad | \\
| \qquad\qquad\quad \text{Au} \\
C \qquad\qquad\quad | \\
|||-\text{Au}-\text{C}\equiv\text{C}-t\text{-Bu} \\
C \\
t\text{-Bu}
\end{array}
$$

(1)

$\nu(C≡C)$ being 100–150 cm^{-1} lower than in the corresponding [Au(C≡CR)L] complexes (63). The latter were obtained by reaction of the clusters with neutral ligands, and the donor strength in [Au(C≡CPh)L] complexes was found to decrease in the order $PR_3 > P(OR)_3 > RNC > AsR_3 > SbR_3 >$ amines. The crystal structure of [Au(C≡CPh)(NH$_x$$i$-Pr)] showed (64) the solid state structure to involve infinite zigzag chains of gold atoms, and the probability of hydrogen bonding between the amino hydrogens and the alkynyl groups.

$$[AuBr(PEt_3)] + LiC≡CPh \rightarrow [Au(C≡CPh)(PEt_3)] + LiBr \quad (22)$$

Acetylide complexes were also produced in liquid ammonia in the reaction of gold(I) iodide with potassium acetylides (65). With excess K(C≡

CPh), the complex $K[Au(C{\equiv}CPh)_2]$ was obtained, whereas the presence of excess $K(C{\equiv}CH)$ yielded $K_2[HC{\equiv}CAuC{\equiv}CAuC{\equiv}CH]$ and $K[Au(C{\equiv}CH)_2]$. The addition of $HC{\equiv}CCH_2CR_2CN$ (R = Me or Ph) to gold(I) iodide in liquid ammonia gave (66) the gold(I) acetylide, whose infrared spectrum indicated a $d\pi–p\pi^*$ interaction between the gold atom and the alkynyl group.

Gold(I) ketenide was prepared from $[AuCl(2,6\text{-}Me_2C_5H_3N)]$ (67), and this unusual compound reacted with phenylacetylene in the presence of phenylamine to give an acetylide cluster [Eq. (23)].

$$[Au_2C_2O] + PhC{\equiv}CH + PhNH_2 \rightarrow [Au(C{\equiv}CPh)]_x + MeCONHPh \qquad (23)$$

Vinyl compounds have been prepared (68, 69) from gold(I) halide complexes and Grignard reagents [Eq. (24)]. In the case of styryl(triphenylphosphine)gold two isomers were obtained, having cis and trans configurations about the carbon–carbon double bond. Vinylgold(I) compounds have also been prepared via their mercury analogs [Eq. (25)] (70), the metathetical replacement of mercury by gold occurring with retention of configuration. The reactions of these compounds will be discussed later (Section III).

$$PhCH{=}CHMgBr + [AuX(PPh_3)] \rightarrow [PhCH{=}CHAu(PPh_3)] + MgBrX \qquad (24)$$

$$\left(\begin{array}{c} Ph \\ \diagdown \\ \quad C{=}C \\ \diagup \quad \diagdown \\ H \qquad CO_2Me \end{array} \right)_2 Hg + 2\,[AuMeL] \xrightarrow{HBF_4} 2 \quad \begin{array}{c} Ph \qquad AuL \\ \diagdown \quad \diagup \\ C{=}C \\ \diagup \quad \diagdown \\ H \qquad CO_2Me \end{array} \qquad (25)$$

$L = PPh_3$ or $P(C_5H_4FeC_5H_5)_3$

The reaction of $[AuMe(PPh_3)]$ with hexafluorobut-2-yne ultimately gave the binuclear vinylgold(I) complex $[(Ph_3P)AuC(CF_3){=}C(CF_3)Au(PPh_3)]$ (71); the crystal structure of this compound has been determined (72) and shows a cis configuration about the double bond.

E. *Olefin and Acetylene Complexes*

Several gold(I) olefin complexes have been prepared, although the exact constitution was often in some doubt. The first olefin complex was reported (73) in 1964, when white crystals were obtained from the reaction of cycloocta-1,5-diene (cod) with AuCl or $HAuCl_4$. These products were believed to be a dimeric aurous complex $[Au_2Cl_2(cod)]$. Subsequent reports, however, showed that although gold(I) chloride reacted (74) to give a species analyzing as $[Au_2Cl_2(cod)]$, the gold(III) precursor gave (75) a yellow, polymeric compound $[AuCl(cod)]_n$.

With monoolefins apparently simple complexes of the type [AuCl-(olefin)] were obtained (74–76), whereas di- and triolefins gave dimeric species. With norbornadiene both monomeric and dimeric compounds were isolated (74). A gold(I) complex was also obtained from the reaction of auric chloride with hexamethyl Dewarbenzene (77).

Mixed gold(I)–gold(III) olefin complexes have been prepared by reacting $AuCl_3$ with norbornadiene (nbd) (78). Complexes of the type $[Au_2Cl_4(nbd)_n]$ (n = 1, 2, or 3) were isolated, for which NMR measurements indicated the presence of π-bonded olefins and ESR spectra showed the nonexistence of gold(II) centers. They were thus formulated as mixed complexes, having Au(I) and Au(III) centers in a 1 : 1 ratio (Fig. 1). Reaction of [AuCl(olefin)] with auric chloride also gave rise to the mixed complex $[Au_2Cl_4(olefin)]$ (79). The cyclooctatetraene (cot) analog was prepared in a stepwise fashion (80) at low temperature from the olefin and aurous and auric chlorides [Eq. (26)].

$$AuCl + cot \xrightarrow[-45°C]{CH_2Cl_2} [AuCl(cot)] \xrightarrow[SO_2, -50°C]{AuCl_3} [Au_2Cl_4(cot)] \qquad (26)$$

Molecular weight and conductivity measurements on the simple gold(I) complexes [AuCl(olefin)] showed (81) them to be monomeric and undissociated in solution. The mixed complexes, on the other hand, were found to be 1 : 1 electrolytes in polarizing solvents with $AuCl_4^-$ as anion and, presumably, a bis- or tris(olefin)gold(I) species as cation. Of the mixed valence complexes only $[Au_2Cl_4(nbd)]$ was undissociated, due to the necessity to retain a linear, two-coordinate gold(I) center.

Tris(olefin)gold(I) cations were mentioned above (81), but only one such species has been isolated. Whereas cis-cyclooctene formed only an unstable species [AuCl(cis-C_8H_{14})], its trans analog produced (82) a stable complex, which by comparison with its copper analog was assumed to be a chloride-bridged dimer, $[Au_2(\mu-Cl)_2(trans-C_8H_{14})_2]$. Treatment of this material with $AgSO_3CF_3$ or $MeSO_3F$ in the presence of trans-cyclooctene

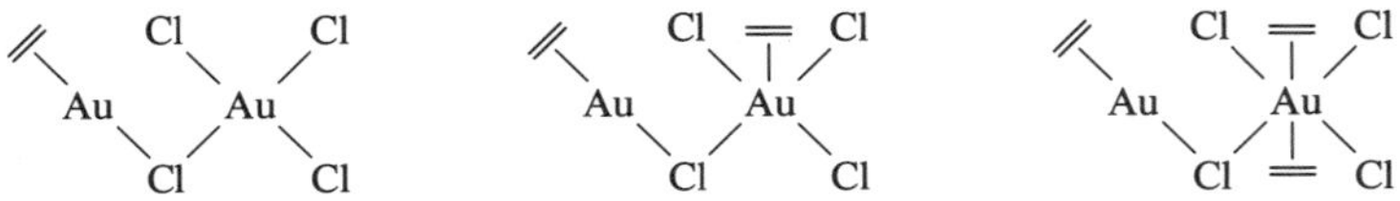

FIG. 1.   Structures of mixed gold(I)–gold(III) olefin complexes.

gave the tris(*trans*-cyclooctene)gold(I) cation. Its ^{1}H-NMR spectrum indicated the equivalence of the cyclooctene ligands, and a structure similar to the copper (*83*) and silver (*82*) analogs was considered likely.

The infrared and Raman spectra of a number of gold(I) and gold(III) diolefin complexes have been recorded (*84*), and on this basis the gold(I) complexes [Au$_2$X$_2$(cod)] (X = Cl or Br) were thought to be polymeric with halide and perhaps diolefin bridges, with the latter adopting the chair form.

The simplest gold–olefin compound, monoethylenegold(0), has been synthesized (*85*) in argon matrices at 8–10 K from gold atoms and ethylene. A theoretical study of the ethylene–metal bond in gold(I) compounds, among others, has appeared (*86*), suggesting that π-backdonation is relatively unimportant in gold(I) compounds.

Few gold(I) acetylene complexes are known. The simple compounds [AuCl(RC≡CR)] (R = Me, Et, or Ph) have been prepared from aurous chloride (*87*), whereas reactions with AuCl$_3$ gave the mixed valence species, [Au$_2$Cl$_4$(RC≡CR)$_n$] (*n* = 1 or 2). The ^{197}Au Mössbauer spectrum of [Au$_2$Cl$_4$(MeC≡CMe)$_2$] showed the presence of gold(I) and gold(III) centers, and the compound has been formulated as [Au(MeC≡CMe)$_2$]-[AuCl$_4$]. With excess but-2-yne, half the auric chloride was reduced to gold(I) with simultaneous formation of *trans*-3,4-dichlorotetramethyl-cyclobutene.

The only other report (*88*) of a gold(I) acetylene complex concerns [AuBr(cyclooctyne)$_2$], which was prepared from aurous bromide and cyclooctyne.

F. β-Diketonate Complexes

Several β-diketonate complexes of gold(I) have been prepared, and since the latter is a "soft" acid, these are invariably bonded through the carbon atom. This contrasts with the behavior of copper(I) and silver(I), which form O-bonded complexes. Thus [Ag(acac)(PPh$_3$)] exhibited (*89*) four infrared bands in the 1500–1700 cm^{-1} region, due to coupling of ν(CO) and ν(CC) when the complex was O-bonded, whereas only three were observed for the gold(I) analog. The acetylacetonate complexes have generally been prepared (*89–92*) by reacting [AuClL] with Tl(acac). They exhibit at ~1650 cm^{-1} two strong absorptions in their infrared spectra attributed to ν(CO). The complex [Au(acac)(PPh$_3$)] was also prepared (*93*) by treating [AuBr(PPh$_3$)] with acetylacetone in the presence of Ag$_2$O.

G. *Cyclopentadienyl Complexes and Aryl Complexes Containing 3-Center 2-Electron Bonds*

Gold(I) forms "simple" cyclopentadienyl compounds as well as more complex species involving 3-center 2-electron (3c–2e) bonds, and these together with analogous 3c–2e arylgold(I) compounds are discussed here.

Cyclopentadienylgold(I), formed by reacting [AuCl(olefin)] complexes with NaC_5H_5, was found (*94*) to be unstable, but [Au(C_5H_5)(PPh$_3$)], which was prepared analogously, was more amenable to study. Its electronic spectrum is similar to those of [Cu(σ-C_5H_5)(PEt$_3$)] and [Hg(σ-C_5H_5)$_2$] (*95*), and its ^{1}H-NMR shows a singlet at δ 6.3. On cooling below $-50°$C, this resonance splits into a doublet due to phosphorus coupling (*96, 97*). With excess PPh$_3$, a singlet was observed (*96*) even at $-100°$C, and the phosphine exchange was thought to be associative (*97*).

The cyclopentadienyl ring in [Au(C_5H_5)(PPh$_3$)] (**2**) is σ-bonded and fluxional. In the methylcyclopentadienyl analog, three ^{1}H-NMR signals were observed which was taken (*94*) to indicate that the gold was bonded to the carbon atom bearing the methyl group. It was shown (*96*), however, that although two ring-proton resonances were observed at ambient temperatures, the high-field resonance was resolved into a doublet at $-90°$C, whereas the low field resonance remained unaffected. The high-field signal was assigned to the β-hydrogens, and hence it was suggested that the structure shown was the most stable one, with the β-hydrogens having the greater probability of occupying the aliphatic site.

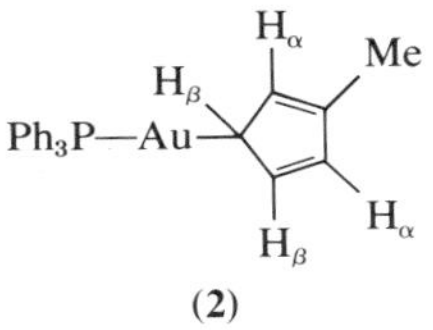

(**2**)

Ferrocenylgold(I) complexes were also found to contain σ-bonded cyclopentadienyl ligands. Treatment of [AuCl(PPh$_3$)] with LiC$_5$H$_4$FeC$_5$H$_5$ gave the parent compound (*98*), while substituted ferrocenyllithium reagents reacted similarly (*99*) [Eq. (27)]. The substituted complexes were of greater stability, and it was suggested there might be some interaction between the gold center and the X substituent. Substitution in both cyclopentadienyl rings was achieved [Eq. (28)], although the product decomposed readily. A similar reaction was observed (*100*) between [AuCl(PPh$_3$)] and LiC$_5$H$_4$Mn(CO)$_3$. The reactions of these complexes with halogens or acids resulted (*98–102*) in cleavage of the gold–cyclopentadienyl bonds.

$$X = Cl, \ OMe, \ or \ CH_2NMe_2$$

Treatment of $[Au(C_5H_4FeC_5H_5)(PPh_3)]$ with HBF_4 led (*103*) to a species of formula $[C_5H_5FeC_5H_4(AuPPh_3)_2]^+BF_4^-$, where the positive charge was thought to be delocalized over the two gold centers. The crystal structure of this complex was determined (*104*), and the cation was found to contain a novel Fe—Au—Au chain, the two gold centers being nearly symmetrically bonded to the cyclopentadienyl ring with gold–carbon distances of 2.13(4) and 2.27(4) Å. The use of $MeCO^+BF_4^-$ or $NO_2^+BF_4^-$ instead of tetrafluoroboric acid afforded the same products (*103, 105*).

Direct auration of ferrocene with $(Ph_3PAu)_3O^+BF_4^-$ also led (*106*) to the complex $[C_5H_5FeC_5H_4(AuPPh_3)_2]^+BF_4^-$. With substituted ferrocenes auration occurred primarily in position 2 [Eq. (29)], irrespective of the electron-withdrawing or -releasing nature of the substituent X.

Reaction of $[AuMe(PPh_3)]$ with ferrocenylmercury precursors caused (*107, 108*) gold for mercury exchange, giving analogous 3c–2e complexes [Eq. (30)]. Treatment of these compounds with aqueous sodium chloride produced $[Au(C_5H_4FeC_5H_4X)(PPh_3)]$ and $[AuCl(PPh_3)]$. When

$[F_c(AuPPh_3)_2]BF_4$ ($F_c = C_5H_4FeC_5H_5$) was treated with $[F_cSAu(PPh_3)]$ an exchange reaction took place (109) to give $[F_cAu(PPh_3)]$ and the 3c–2e bonded compound $[F_cS(AuPPh_3)_2]BF_4$.

The reaction of organogold complexes with HBF_4, and of $[AuMe(PPh_3)]$ with organomercury compounds in the presence of HBF_4, was extended (110) to aryl and vinyl complexes [Eq. (31)]. It was suggested (111, 112) that reaction with HBF_4 generated the reactive species $[(Ph_3P)Au]^+BF_4^-$, which readily attacked another organogold molecule to yield the observed product.

$$[RAu(PPh_3)] + HBF_4 \rightarrow [R(AuPPh_3)_2]^+BF_4^- \tag{31}$$

R = C_6H_5, C_6H_4Me-p, $CH_2{=}CH$, $C_5H_5FeC_5H_4$, $C_5H_5FeC_5H_3Cl$, $C_5H_5FeC_5H_3OMe$, or $C_5H_5FeC_5H_3CH_2NMe_2$

The ^{31}P-NMR spectrum of $[C_5H_5FeC_5H_4(AuPPh_3)_2]^+BF_4^-$ exhibited (111) two resonances, indicating nonequivalence of the gold centers with respect to the cyclopentadienyl ring. This nonequivalence was not found, however, for the complex $[p\text{-}MeC_6H_4(AuPPh_3)_2]^+BF_4^-$. These 3c–2e complexes act as $Au(PPh_3)^+$ transfer agents, giving $[AuR(PPh_3)]$ and a variety of aurated products (111, 113), the donor strength toward $Au(PPh_3)^+$ lying in the following order (L = PPh_3) (111):

$C_6H_5AuL < p\text{-}MeC_6H_4AuL < O\!\!\bigcirc\!\!NH <$ [ferrocenyl-Cl–AuL] $<$ [ferrocenyl-OMe–AuL] $\sim$

[ferrocenyl–AuL] $<$ [ferrocenyl-CH_2NMe_2–AuL] $<$ [ferrocenyl–SAuL] $\sim$ $PPh_3 < Cl^- < I^-$

The complexes $[L(CO)_2MnC_5H_4(AuPPh_3)_2]^+BF_4^-$ (L = CO or PPh_3) have been prepared (114) and react with $[Au(C_6H_4Me)(PPh_3)]$ or $[Au(C_5H_4FeC_5H_5)(PPh_3)]$ by transfer of the $Au(PPh_3)^+$ moiety, simultaneously regenerating the complex $[AuC_5H_4Mn(CO)_2L(PPh_3)]$.

The ^{19}F-NMR data for $[p\text{-}FC_6H_4(AuPPh_3)_2]^+BF_4^-$ indicated (115) a symmetrical structure, and the stability of the complexes $[R(AuPPh_3)_2]^+BF_4^-$ was found to decrease in the following order:

R = $p\text{-}NH_2C_6H_4 > C_5H_5FeC_5H_4 > p\text{-}MeOC_6H_4 > p\text{-}MeC_6H_4 > p\text{-}FC_6H_4 > C_6H_5$

The chemistry of these aryl- and ferrocenylgold(I) complexes has been summarized in a review (*116*).

When diphenylzinc was reacted with gold(III) chloride, a [Ph₃ZnAu] species was obtained (*117, 118*). Molecular weight measurements showed this product to be dimeric, and the ¹H-NMR spectrum indicated the presence of two types of aryl groups in a 1 : 2 ratio, whereas the ¹³C-NMR data showed these to be due to zinc-bonded and bridging aryls, respectively. Structure **3** was postulated. The same complex was obtained when Ph₂Zn was reacted with [AuCl(CO)]; the *p*-tolyl analog was also prepared.

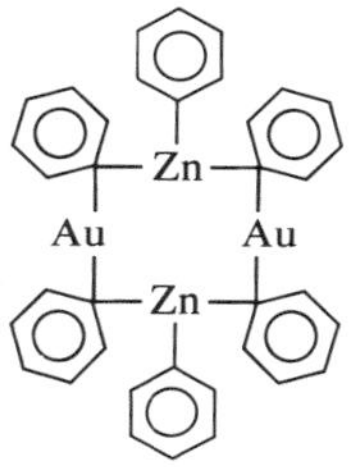

(3)

Reaction of [Ph₃ZnAu]₂ with PPh₃ produced (*118*) [AuPh(PPh₃)] and diphenylzinc, whereas with bipyridine a species with considerable charge separation was obtained. This was suggested to be of the form Ph(bipy)Zn⁺ · · · AuPh₂⁻, its formation being promoted by the strong affinity of zinc for nitrogen donors.

When [AuCl(CO)] and Ph₂Zn were reacted in equimolar amounts a species of constitution [Ph₂ZnAuCl] was formed (*117*), whose ¹³C-NMR spectrum showed only the presence of bridging phenyl groups. It was later found (*119*) that addition of [AuCl(CO)] to [Ar₃ZnAu] (Ar = C₆H₅ or *p*-MeC₆H₄) also produced this species, whereas the reactions of Ar₂Zn (Ar = *o*-tolyl, *o*-vinylphenyl, or 2,6-dimethoxyphenyl) and Ph₂M (M = Cd or Hg) produced the chloride complexes [Ar₂AuMCl] only. It was suggested that these were polymeric species **4**.

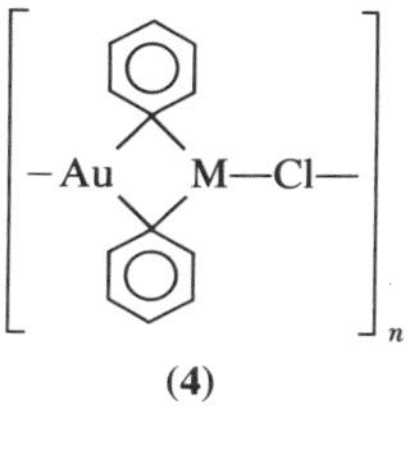

(4)

Although it is well known that aryllithium reagents react with gold(I) halide complexes to give the arylgold(I) analogs, it was found that when the aryl group contained an ortho substituent capable of coordination, a different reaction takes place. Addition of o-$Me_2NCH_2C_6H_4Li$ to [$AuBr(PPh_3)$] gave (*120*) a species of the form [$Ar_4Au_2Li_2$] (Ar = $C_6H_4CH_2NMe_2$-o) [Eq. (32)] where coordination of the NMe_2 groups gave rise to tetragonally coordinated lithium. By analogy with the copper (*121*) and silver (*122*) complexes, a structure involving 3c–2e bonds was suggested. Reaction with CuBr gave the [$Ar_4Au_2Cu_2$] complex and lithium bromide.

$$4\ o\text{-}Me_2NCH_2C_6H_4Li + 2\ [AuBr(PPh_3)]$$
$$\rightarrow [(o\text{-}Me_2NCH_2C_6H_4)_4Au_2Li_2] + 2\ PPh_3 + 2\ LiBr \quad (32)$$

In the complexes [$Ar_4M_2Li_2$] (Ar = $C_6H_4CH_2NMe_2$-o; M = Cu, Ag, or Au) (**5**), C-1 was recognized as a chiral center, and rotation about the C-1–C-4 axis would cause continuous inversion of configuration at C-1 (*123*). Monitoring of the prochiral CH_2 group allowed study of the configuration at C-1, and the barrier to rotation of the 3c–2e system was found to depend on the size and coordinating ability of the ortho substituent. For example, the barrier to rotation was greater for Ar = $C_6H_4CHMeNMe_2$-o (*124*). In the ^{1}H-NMR spectrum, the CH_2 and NMe_2 signals for [(o-$Me_2NCH_2C_6H_4)_4M_2Li_2$] coalesced at different rates, and the methyl signals moved downfield, indicating that Li—N dissociation occurred and that rotation of the aryl group about the C-1–C-4 axis was indeed involved.

(5)

Hexanuclear aryl complexes of the type [$Ar_4M_4Au_2(SO_3CF_3)_2$] (Ar = $C_6H_4NMe_2$-o; M = Cu or Ag) have been obtained (*125, 126*) according to Eqs. (33) and (34). These involve 3c–2e bonds, with the gold atoms occupying the apical positions in the M_4Au_2 octahedra and the SO_3CF_3

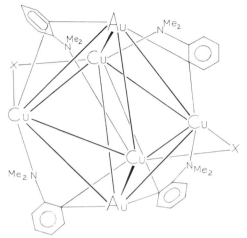

FIG. 2. Structure of [Ar$_4$Cu$_4$Au$_2$(SO$_3$CF$_3$)$_2$] (Ar = C$_6$H$_4$NMe$_2$-o). X = SO$_3$CF$_3$. Reprinted with permission from G. van Koten *et al.*, *Inorg. Chem.*, **16**, 1782 (1977). Copyright 1977 American Chemical Society.

groups each bridging two equatorial metal centers (Fig. 2).

$$Ar_4Au_2Li_2 + 4\ MSO_3CF_3 \xrightarrow{-2\ LiSO_3CF_3} Ar_4M_4Au_2(SO_3CF_3)_2 \qquad (33)$$
$$M = Cu\ or\ Ag$$

$$(4/n)\ (ArAu)_n + 4\ CuSO_3CF_3 \xrightarrow{-2\ AuSO_3CF_3} Ar_4Cu_4Au_2(SO_3CF_3)_2 \qquad (34)$$

The reaction of [Ar$_4$Au$_2$Li$_2$] (Ar = C$_6$H$_4$CH$_2$NMe$_2$-o) with [Rh$_2$Cl$_4$-(CO)$_2$] involves (*127*) transfer of an aryl group to rhodium and the formation of the polymeric species [Ar—Au]$_n$. Such species are also obtained (*128*) in the reaction of Li[AuAr$_2$] with SnMe$_3$Br (Ar = C$_6$H$_4$NMe$_2$-o, C$_6$H$_4$CH$_2$NMe$_2$-o, or C$_6$H$_4$CHMeNMe$_2$-o). (*S*)-2-Me$_2$NCHMeC$_6$H$_4$Au was found to be dimeric, probably involving intermolecular Au—N bonding to give a ten-membered ring, but the structures of (2-Me$_2$NCH$_2$C$_6$H$_4$Au)$_n$ and (2-Me$_2$NC$_6$H$_4$Au)$_n$ are less clear.

Bi- and tetranuclear, three-center two-electron gold(I) complexes have been prepared containing polyfluorophenyl groups (*129*). The binuclear complexes are of the type discussed previously, [Ar(AuL)$_2$]X and [Ar{Au$_2$(dpe)}]BF$_4$, whereas the diarylaurates reacted to give tetranuclear compounds [Eq. (35)] suggested to be of the same type as the mixed metal complexes discussed above.

An early review of these aryl clusters of the group IB metals has appeared (*130*).

$$n\text{-}Bu_4N[Au(2,4,6\text{-}C_6H_2F_3)_2] + HPF_6 \rightarrow$$

(35)

H. Isonitrile, Carbene, and Imino Complexes

Gold(I) isonitrile complexes were first reported (131) in 1956, when it was found that they were generally less stable than the isonitrile complexes of other transition metals. This was attributed to the low tendency of gold to form bonds with double bond character. A number of isonitrile complexes have been made, however, and they have been used extensively in the preparation of carbene and imino complexes of gold(I).

Treatment of $HAuCl_4$, $NaAuCl_4$, or $AuBr_3$ with isonitriles gives (132, 133) complexes of the type [AuX(CNR)]. Such complexes have also been prepared by displacement of dimethyl sulfide from [AuX(SMe_2)] (134). In the presence of free isonitrile, and with a noncoordinating anion, ionic complexes of the type $[Au(CNR)_2]^+X^-$ are obtained (132, 133).

The infrared spectra of the complexes [AuX(CNMe)] exhibited (133) $\nu(CN)$ bands at ~ 2270 cm^{-1}, slightly higher in frequency than those in the free ligand. Their ^{1}H-NMR spectra consisted of a $1:1:1$ triplet due to coupling of the methyl hydrogens with nitrogen-14, whereas the ionic species $[Au(CNMe)_2]^+X^-$ produced only a broad resonance. Coupling to nitrogen-14 was only observed where the rate of relaxation was slow enough, and this was related to the degree of π-bonding in the complexes (133).

The complexes [AuCl(CNR)] (R = Ph, p-MeOC_6H_4, or C_6H_{11}), among others, have been studied by ^{35}Cl-NQR spectroscopy (135), and it was concluded that the ^{35}Cl-NQR frequencies were sensitive to the nature of the trans ligand and that σ-bonding effects were dominant.

The cyano complex [Au(CN)(CNMe)] has been prepared by treating [Au(CN)$_2$]$^-$ with methyl iodide (136), and its near-linear structure was elucidated from X-ray crystallographic studies.

The preparations of all of the gold(I) carbene complexes to date have utilized isonitrile precursors, either directly or as intermediates. Treatment of [AuX(CNR)] with alcohols (134, 137, 138) or amines (139) affords the complexes [AuX{C(Y)NHR}] (Y = OR' or NHR') [Eq. (36)]. The use of excess isonitrile and a noncoordinating anion gives ionic bis(carbene)gold(I) complexes [Eq. (37)] (138). Organogold(I) isonitrile complexes react to give [Au(C$_6$X$_5$){C(Y)NHC$_6$H$_4$Me-p}] (X = F, Cl, or Br; Y = OR, NH$_2$, NHR, or NR$_2$) (140).

$$(RNC)AuX + MeOH \longrightarrow \begin{array}{c} RNH \\ \diagdown \\ C—Au—X \\ \diagup \\ MeO \end{array} \qquad (36)$$

$$(RNC)AuX + RNC + MeOH + BF_4^- \xrightarrow{-X^-} \left(\begin{array}{c} RNH \\ \diagup \diagdown \\ \ \ \ \ C— \\ \diagdown \diagup \\ MeO \end{array} Au \right)_2^+ BF_4^- \qquad (37)$$

Direct reaction of the bis(isonitrile)gold(I) cations with alcohols or amines also gave the biscarbene complexes [Eq. (38)] (132, 141, 142). With bulky C$_6$H$_{11}$NC, only one carbene moiety was formed (141, 142) under the conditions employed in the preparation of the other biscarbene compounds.

$$[Au(CNR)_2]^+ClO_4^- + HY \rightarrow [Au\{C(Y)NHR\}_2]^+ClO_4^- \qquad (38)$$
Y = OR' or NHR'

The biscarbene complexes were found (142) from their ^{1}H-NMR spectra to be mixtures of geometrical isomers, due to hindered rotation about the C—N or C—O bonds, and the isomers have been separated by fractional crystallization in the case of [Au{C(OEt)NHC$_6$H$_4$Me}$_2$]$^+$ClO$_4^-$. Treatment of these species with triphenylphosphine gave mixed ligand complexes and formamidines [Eq. (39)]. They also underwent oxidative addition of iodine to yield the first gold(III) carbene complexes.

$$[Au\{C(NHC_6H_4Me\text{-}p)_2\}_2]^+ + PPh_3$$
$$\rightarrow [Au\{C(NHC_6H_4Me)_2\}(PPh_3)]^+ + MeC_6H_4NHCH{=}NC_6H_4Me \qquad (39)$$

Reactions of tetrachloroaurates with isonitriles and amines constitutes a convenient *in situ* preparation of bis(carbene)gold(I) complexes (143,

144). These were found to produce formamidines on treatment with isonitriles or cyanide ion. The cleavage of electron-rich olefins in the presence of [AuCl(PPh$_3$)] also led (*145, 146*) directly to biscarbene complexes [Eq. (40)].

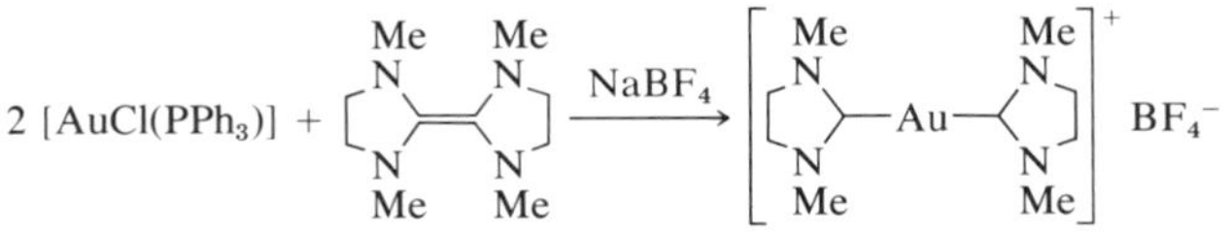

$$2 \ [\text{AuCl(PPh}_3)] + \left[\begin{array}{c} \text{Me} \quad \text{Me} \\ \text{N} \quad \text{N} \\ \diagdown \diagup \\ \text{N} \quad \text{N} \\ \text{Me} \quad \text{Me} \end{array} \right] \xrightarrow{\text{NaBF}_4} \left[\begin{array}{c} \text{Me} \qquad \text{Me} \\ \text{N} \qquad \text{N} \\ \diagdown \text{Au} \diagup \\ \text{N} \qquad \text{N} \\ \text{Me} \qquad \text{Me} \end{array} \right]^{+} \text{BF}_4^{-} \qquad (40)$$

When an isonitrile ligand containing a suitable hydroxyl group was reacted with AuCl$_4^-$, it underwent isomerization (*147*) via a carbene complex (Scheme 1). With CN(CH$_2$)$_3$OH, in the presence of a noncoordinating anion, the bis(carbene)gold(I) complex [Au{$\overline{\text{CO(CH}_2)_3\text{NH}}$}$_2$]$^+BPh_4^-$ was formed.

When gold(I) carbene complexes are treated with hydroxide ion, or if the corresponding isonitrile complexes are reacted with an alcohol and hydroxide ion, iminomethylgold(I) species result (*134, 138, 139, 148–150*). Molecular weight measurements showed these to be trimeric in nature, (*148*) when they did not contain other coordinated ligands. Thus treatment of [AuCl(SMe$_2$)] with isonitrile, alcohol, and base readily gives complexes of the type [Au{C(OR')=NR}]$_3$ (**6**), whereas with [AuCl(PPh$_3$)] the monomeric compounds [Au{C(OR')=NR}(PPh$_3$)] are obtained. The crystal structure of tris-μ-[(ethoxy)(*N*-*p*-tolylimino)methyl-*N,C*]trigold has been determined (*151*), and the compound was thus shown to contain a

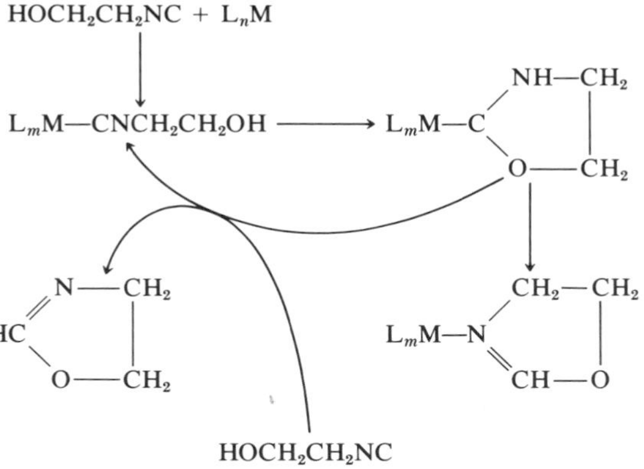

HOCH$_2$CH$_2$NC + L$_n$M

$$\downarrow$$

L$_m$M—CNCH$_2$CH$_2$OH $\longrightarrow$ L$_m$M—C
$$\begin{array}{c} \text{NH—CH}_2 \\ \diagup \qquad \quad | \\ \text{O——CH}_2 \end{array}$$

$$\begin{array}{c} \text{N——CH}_2 \\ \text{HC} \diagup \qquad \quad | \\ \diagdown \\ \text{O——CH}_2 \end{array}$$

L$_m$M—N
$$\begin{array}{c} \text{CH}_2\text{—CH}_2 \\ \diagup \qquad \quad | \\ \diagdown \\ \text{CH——O} \end{array}$$

HOCH$_2$CH$_2$NC

L$_n$M = PdI$_2$, PtCl$_2$, or AuCl$_4^-$

SCHEME 1

$$\begin{array}{c}
\text{R} \qquad\qquad \text{OR'} \\
\diagdown \qquad\qquad \diagup \\
\text{N} \rightarrow \text{Au} - \text{C} \\
\diagup\diagup \qquad\qquad \diagdown\diagdown \\
\text{R'O} - \text{C} \qquad\qquad \text{N} - \text{R} \\
\diagdown \qquad\qquad \diagdown \\
\text{Au} \qquad\qquad \text{Au} \\
\diagdown \qquad \diagup \\
\text{N} = \text{C} \\
\diagup \qquad \diagdown \\
\text{R} \qquad\quad \text{OR'}
\end{array}$$

(6)

somewhat irregular nine-membered ring exhibiting near-linear geometry at the gold(I) centers.

The silver analogs of these cyclic trimers have also been readily prepared and found (152, 153) to transfer the iminomethyl group to other metals, including gold, to give the complexes $[Au\{C(OR')=NR\}]_3$ and $[Au\{C(OR')=NR\}(PPh_3)]$.

The trimeric compounds have been cleaved with a variety of neutral ligands [Eq. (41)] (154). With PPh_3 the reaction proceeds smoothly, the rate and extent of reaction being dependent on the nature of R (the process occurs more readily when R is aromatic), but with isonitriles species of indeterminate constitution are obtained in addition to $\lfloor Au\{C(OR')=NR\}(CNR'')\rfloor$. The cyclic complexes also underwent stepwise oxidative addition of bromine or iodine to yield (155) mixed gold(I)–gold(III) species, and finally the analogous gold(III) trimers.

$$\tfrac{1}{3}[Au\{C(OR')=NR\}]_3 + L \rightarrow [L\text{-}Au\text{-}C(OR')=NR] \qquad (41)$$

The complexes $[Au\{C(OMe)=NR\}(PPh_3)]$ (R $= p\text{-}MeC_6H_4$ or $p\text{-}NO_2C_6H_4$) have been shown (156, 157) to act as monodentate ligands toward other metal centers. They were found to coordinate through nitrogen, and coordination was accompanied by a decrease in the value of $\nu(CN)$ in the infrared spectrum.

I. Other Complexes Containing Gold–Carbon Bonds

The complex $[AuCl(CO)]$ has been much used as a starting material in the preparation of organogold(I) compounds, but since it does contain a gold–carbon bond itself it will be considered briefly here. The complex has a long history, having been first reported (158) in 1925. It was prepared by heating auric chloride in the presence of carbon monoxide, or by passing CO into a benzene solution of anhydrous aurous chloride (159). Its isolation from the reaction of $HAuCl_4$ with carbon monoxide in thionyl chloride has more recently been reported (160, 161) as a more convenient synthesis.

It has been suggested (*162*) that there exists only negligible
π-backbonding in [AuCl(CO)], and a number of displacement reactions
have been described (*162, 163*). Vibrational and NMR spectroscopic studies have been made of this complex (*164*), and the results have been
compared with those for carbonyl complexes of palladium, platinum,
rhodium, and iridium.

Gold(0) species containing carbonyl ligands have been identified (*165,
166*) in matrix-isolation studies involving gold atoms with carbon
monoxide and dioxygen.

When metallic gold was evaporated into an ethanol solution of
bis(diphenylphosphino)methane and ammonium nitrate, the complex
$[Au_5(Ph_2PCH_2PPh_2)_3(Ph_2PCHPPh_2)](NO_3)_2$ was obtained (*167*). The X-ray
structure of this cluster (Fig. 3) showed it to contain a gold–carbon linkage,
and it was suggested that the compound might best be described as a tetranuclear cluster incorporating a linear gold(I) entity.

Treatment of the cluster complex $[Au_5(PPh_3)_4Cl \cdot 3.5\ H_2O]$ with potassium cyanide produces (*168*) a trinuclear species $[Au_3(CN)(PPh_3)_2]$, whose
infrared spectrum suggests the existence of a covalently bonded cyanide
group. The anion $[Au(CNO)_2]^-$ has been prepared (*169*) and stabilized by
large cations. The crystal structure of $[Ph_4As]^+[Au(CNO)_2]^-$ has been
determined (*170*), and the seven atoms of the anion found to be exactly
colinear, with an average gold–carbon distance of 2.01 Å.

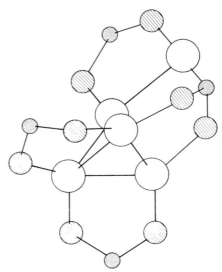

Fig. 3.   Structure of the $[Au_5(Ph_2PCH_2PPh_2)_3(Ph_2PCHPPh_2)]^{2+}$ cation; the phenyl groups
have been omitted. ○ = Au, ◍ = P, ⊛ = C. From van der Velden *et al.* (*167*).

III
REACTIONS OF ORGANOGOLD(I) COMPLEXES

A. *Thermal and Photochemical Decomposition*

When the complexes [AuR(PR$_3'$)] are heated either in solution or in the solid state, they decompose to metallic gold, free tertiary phosphine, and an organic fragment (*15, 30, 171*). The methyl and ethyl complexes produce ethane and *n*-butane, respectively, almost quantitatively by coupling of the organic groups [Eq. (42)]. The isopropyl and *t*-butyl analogs, on the other hand, undergo disproportionation reactions to yield alkanes and alkenes [Eqs. (43) and (44)]. First-order kinetics were obtained for the decomposition of [AuMe(PPh$_3$)], and excess triphenylphosphine was found to retard the reaction, so it was suggested (*30*) that initial PPh$_3$ dissociation was involved, and the mechanism of Eq. (45) was proposed.

$$2 \ [RAu(PPh_3)] \rightarrow R\text{-}R + 2 \ Au + 2 \ PPh_3 \tag{42}$$

$$i\text{-}PrAuL \rightarrow \underset{(28\%)}{Me_2CH_2} + \underset{(34\%)}{MeCH\!\!=\!\!CH_2} + \underset{(trace)}{C_6H_{14}} \tag{43}$$

$$t\text{-}BuAuL \rightarrow \underset{(47\%)}{Me_2CH\!\!=\!\!CH_2} + \underset{(13\%)}{Me_3CH} + C_8 \ \text{hydrocarbons} \tag{44}$$

$$L = PPh_3$$

$$MeAu(PPh_3) \overset{k_1}{\rightleftarrows} MeAu + PPh_3$$
$$MeAu + MeAu(PPh_3) \rightarrow Me\!\!-\!\!Me + 2 \ Au + PPh_3 \tag{45}$$

2-Pyridylgold(I) (*24*) and its quinolyl analogs (*25*) also decompose by coupling of the organic species. The complexes **7–9**, on heating, eliminate Me$_3$SiCl with concomitant rearrangement (*172*) to give polymeric species containing gold–carbon bonds; further reaction results in coupling of the organic groups.

(7)   (8)   (9)

Many organogold complexes are light sensitive, and it has proved advantageous to store them in the dark. Thus even α-silylmethyl complexes, which have been shown to be more stable in general than their purely organic counterparts, decompose to metallic gold under irradiation (*173*).

The photolysis of [AuMe(PPh$_3$)] in chloroform solution has been investigated (174) and found to involve the generation of singlet and triplet excited states [Eq. (46)]. The latter react with CDCl$_3$ to produce [AuCl(PPh$_3$)] and methyl radicals, which produce several species including CH$_3$CDCl$_2$, CH$_3$CCl$_3$, C$_2$H$_6$, and CH$_3$D.

$$[AuMe(PPh_3)] \rightarrow [AuMe(PPh_3)]^S \rightarrow [AuMe(PPh_3)]^T \qquad (46)$$

B. Oxidative Addition Reactions

Organogold(I) complexes readily undergo oxidative addition, but the products have not always proved to be the expected gold(III) species. Thus, although [AuXL] complexes react with halogens to give the corresponding gold(III) compounds (175, 176) and although [AuBr(PEt$_3$)] in particular reacts with chlorine to produce all six species of the type [AuBr$_{3-a}$Cl$_a$(PEt$_3$)] (a = 0,1,2,3) (177), organogold(III) complexes have been produced only transiently in analogous reactions of [AuR(PPh$_3$)] (R = alkyl or aryl) compounds (26, 173). The ultimate products were the gold(I) halide complexes.

Only with polyhaloaryl complexes have reactions with halogens consistently produced the expected organogold(III) compounds. The reaction of [Au(C$_6$F$_5$)(PPh$_3$)] with bromine at 0°C yields [AuBr$_2$(C$_6$F$_5$)(PPh$_3$)] (33), although some [AuBr(PPh$_3$)] is also obtained. Treatment of the analogous isonitrile (178) and tetrahydrothiophene (179) complexes with X$_2$ gives the organogold(III) species [AuX$_2$(C$_6$F$_5$)L] (X = Cl, Br, or I).

The pentabromophenyl complex [Au(C$_6$Br$_5$)(PPh$_3$)] undergoes oxidative addition of halogen to give [AuX$_2$(C$_6$Br$_5$)(PPh$_3$)] (36), the chloride complex having a trans geometry. The analogous pentachlorophenyl compounds [Au(C$_6$Cl$_5$)L] (L = PPh$_3$, PEt$_3$, or AsPh$_3$) also react (180–182) to give gold(III) products, whereas the reactions of pentafluorophenylgold(I) complexes are dependent on the nature of the halogen (180–183).

Oxidative addition of X$_2$ to anionic organogold(I) species affords the corresponding gold(III) compounds n-Bu$_4$N[AuBr$_3$(C$_6$F$_5$)] (11) and n-Bu$_4$N[AuX$_2$(C$_6$F$_5$)$_2$] (X = Cl, Br, or I) (13).

Cationic bis(carbene)gold(I) species undergo oxidative addition of halogens (141, 144, 184), and the crystal structure of [AuI$_2$\{C(NHC$_6$-H$_4$Me-p)$_2$\}$_2$]ClO$_4$ has been determined (185). It was discovered (184), however, that, although reactions in chloroform produced gold(III) carbene complexes, when [Au\{C(Y)NHR\}$_2$]$^+$ (Y = OEt or NHR) ions were treated with iodine in acetone solution, the products were [AuI\{C(Y)NHR\}] and RNHCOY. In the presence of water, trans-[AuI$_2$-\{C(NHC$_6$H$_4$Me-p)$_2$\}$_2$]$^+$ClO$_4^-$ decomposes to yield the N,N'-disubstituted urea (p-MeC$_6$H$_4$NH)$_2$CO (186).

Addition of Cl_2 to organogold(I) complexes has also been achieved by reaction with thallium(III) chloride [Eq. (47)] (13, 36, 180–182), the cis isomers being produced by this method. Reactions with X_2 gave the trans isomers (179).

$$[Au(C_6X_5)(PEt_3)] + TlCl_3 \rightarrow cis\text{-}[AuCl_2(C_6X_5)(PEt_3)] + TlCl$$

X = F, Cl, or Br   (47)

Other oxidative addition reactions involving thallium(III) reagents have resulted in transfer of two polyfluorophenyl groups to the gold center. Thus [AuClL] (L = PPh$_3$, AsPh$_3$, or CNPh) reacts (38, 178) with TlBr(C$_6$F$_5$)$_2$ to yield bis(pentafluorophenyl)gold(III) species [Eq. (48)], and the bridged complex [ClAu(μ-dpe)AuCl] reacts similarly (183). The ionic species [AuBr(C$_6$F$_5$)]$^-$ reacts with TlBr(C$_6$F$_5$)$_2$ to give the bromo-tris(pentafluorophenyl)gold(III) anion (11), whereas treatment of [Au(C$_6$F$_5$)(tht)] with TlClR$_2$ (R = C$_6$F$_4$H, C$_6$F$_3$H$_2$, 4-C$_6$H$_4$F, or 3-C$_6$H$_4$CF$_3$) yields mixed arylgold(III) complexes (187).

$$[AuClL] + TlBr(C_6F_5)_2 \rightarrow [AuCl(C_6F_5)_2L] + TlBr \qquad (48)$$

Stepwise addition of halogens to the trimeric complex, [Au{C(OMe)═NMe}]$_3$ produced (155) three complexes [Au$_3${C(OMe)═NMe}$_3$X$_n$] (n = 2,4,6) by consecutive oxidation of the three gold(I) centers. Binuclear gold(I) ylide complexes, however, underwent (56) consecutive oxidative addition of X_2 to give formal gold(II) and finally gold(III) products [Eq. (49)]. Crystal structure (188) and spectroscopic (56, 189, 190) data have

(49)

shown that a gold(II) species was indeed produced. Addition of $(Et_2NCS_2)_2$ to an analogous binuclear ylide complex also yielded an organogold(II) species [Eq. (50)] (191).

$$(50)$$

Organogold(III) complexes are produced when an alkyl halide is added to a dialkylaurate(I) in the presence of triphenylphosphine [Eq. (51)] (7, 8). Mixed complexes trans-$[AuMe_2R(PPh_3)]$ are obtained with other alkyl halides, whereas iodobenzene gives cis-$[AuMe_2Ph(PPh_3)]$. Addition of bromine to dimethylaurate(I) yields cis-$[AuBrMe_2(PPh_3)]$.

$$Me_2Au(PPh_3)^-Li^+ + MeI \rightarrow AuMe_3(PPh_3) + LiI \qquad (51)$$

Addition of methyl iodide to [AuMeL] (L = PMe_3, PMe_2Ph, $PMePh_2$, or PPh_3) proceeds by initial oxidative addition, but further reaction occurs to give $[AuMe_3L]$ and $[AuIL]$ (192–195). With CF_3I, the complex $[AuMe(PPh_3)]$ gave $[Au(CF_3)(PPh_3)]$ (46), but the PMe_3 and PMe_2Ph analogs produced $[AuMe_2(CF_3)L]$ and $[AuIL]$. The reaction with methyl iodide was suggested to be a two-stage process, as depicted by Eq. (52).

$$[AuMeL] + MeI \rightarrow [AuIMe_2L]$$
$$[AuIMe_2L] + [AuMeL] \rightarrow [AuMe_3L] + [AuIL] \qquad (52)$$

C. Insertion Reactions of Unsaturated Molecules

Organogold(I) complexes undergo a number of insertion reactions with unsaturated molecules, such as olefins, acetylenes, and sulfur dioxide. Insertion of carbon monoxide or carbon dioxide has not been achieved, although the reverse reaction has been observed with CO_2 (71).

Treatment of [AuMe(PPh$_3$)] with fluoroolefins under the influence of ultraviolet irradiation led (*71, 196*) to insertion of the olefin into the methyl–gold bond [Eq. (53)], but no thermal reaction was observed. Reaction of the product with Br$_2$ or C$_6$F$_5$CO$_2$H led to cleavage of the gold–carbon bond, and in the latter case heating above 120°C resulted in decarboxylation to yield [Au(C$_6$F$_5$)(PPh$_3$)].

$$[Au(CH_3)(PPh_3)] + F_2C{=}CF_2 \rightarrow [Au(CF_2CF_2CH_3)(PPh_3)] \qquad (53)$$

It was suggested (*196*) that the C$_2$F$_4$ insertion reactions probably proceed by a radical mechanism, although the isolation of a 1 : 1 adduct from the reaction of (NC)$_2$C=C(CN)$_2$ with [AuMe(PPh$_3$)] (*197, 198*) suggests that such an intermediate might be involved in the former reaction.

The reactions with acetylenes were somewhat more complex, the thermal and photochemical reactions of [AuMe(PPh$_3$)] with hexafluorobut-2-yne producing (*71, 196*) a binuclear complex, [(Ph$_3$P)AuC(CF$_3$)=C(CF$_3$)-Au(PPh$_3$)]. The crystal structure of this complex was determined (*72*), and it showed a cis arrangement of the gold centers about the olefinic moiety. It was suggested that the use of smaller tertiary phosphines might permit isolation of an intermediate species, and indeed with dimethylphenylphosphine a species of the constitution [{MeAu(PMe$_2$Ph)}$_2$C$_4$F$_6$] was obtained (*197*). This exhibited a complex ^{19}F-NMR spectrum, indicating the nonequivalence of the CF$_3$ groups, and two further reaction pathways were found, depending on the nature of the solvent [Eq. (54)]. The

$$(\text{MeAuL})_2\text{C}_4\text{F}_6 \begin{array}{l} \xrightarrow[\text{acetone}]{\text{C}_4\text{F}_6} \text{CF}_3(\text{AuL})\text{C}{=}\text{C}(\text{AuL})\text{CF}_3 + \text{C}_2\text{H}_6 \\ \\ \xrightarrow[\text{ether}]{} \quad \text{L—Au—C} \begin{array}{l} \overset{\text{Me}}{\underset{|}{\text{C}}}{-}\text{CF}_3 \\ {\parallel} \\ \text{CF}_3 \end{array} \end{array} \qquad (54)$$

L = PMe$_2$Ph

^{197}Au Mössbauer spectrum indicated (*199*) the existence of one Au(I) and one Au(III) center in this complex and in the analogous PMe$_3$ system. An X-ray diffraction study was made on the trimethylphosphine complex (**10**). The PMe$_2$Ph complex has an analogous structure, and in fact two isomers were obtained for [(Me$_3$P)Au(C$_4$F$_6$)AuMe$_2$(PMe$_3$)], the second having a cis configuration at the gold(III) center. It was suggested (*200*) that formation of these mixed Au(I)/Au(III) species may proceed by a free radical mechanism.

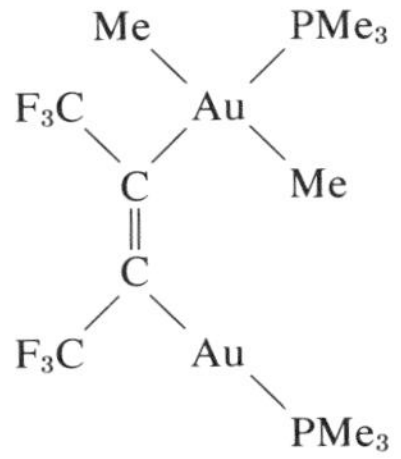

$$
\begin{array}{c}
\text{Me} \qquad \text{PMe}_3 \\
\diagdown \quad \diagup \\
\text{F}_3\text{C} \diagdown \text{Au} \\
\qquad \text{C} \qquad \text{Me} \\
\qquad \parallel \\
\qquad \text{C} \\
\text{F}_3\text{C} \qquad \text{Au} \\
\qquad \diagdown \\
\qquad \text{PMe}_3
\end{array}
$$

(10)

The ultimate products from these mixed species depend on the nature of the solvent and the tertiary phosphine (*197, 199, 200*) and are formed by reductive elimination of two organic fragments from the gold(III) center. A mechanism involving phosphine dissociation has been proposed (*200*) (Scheme 2).

Reaction of [AuMe(PPh₃)] with dimethylacetylene dicarboxylate (DMA) resulted (*198*) in insertion and formation of a small amount of a dimethylgold(III) complex [Eq. (55)], whereas but-2-yne reacted to give [(Ph₃P)AuC(CH₃)=C(CH₃)Au(PPh₃)]. With terminal acetylenes, however, the corresponding gold(I) acetylide complexes were produced, and methane was eliminated (*196, 198*). With perfluoroacetone, [AuMe(PPh₃)] yielded (*71*) a binuclear complex [(Ph₃P)AuC(CF₃)₂OAu(PPh₃)] from which the bridging ketone was displaced by the corresponding imine to produce [(Ph₃P)AuC(CF₃)₂N(H)Au(PPh₃)].

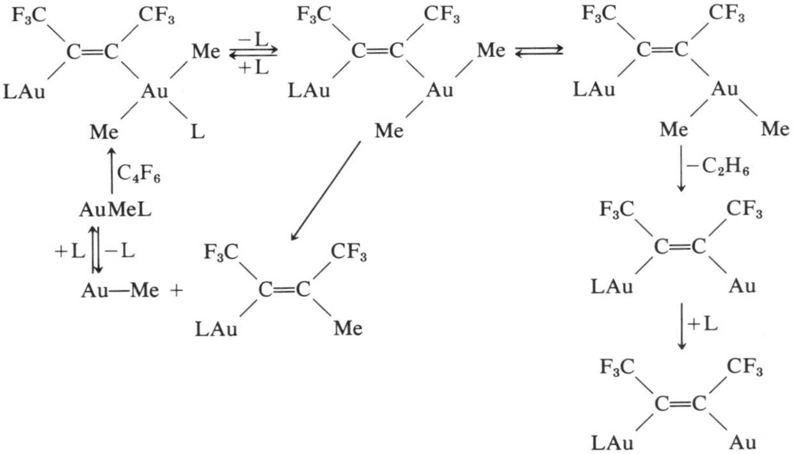

Scheme 2

$$[AuMe(PPh_3)] + MeO_2CC{\equiv}CCO_2Me$$
$$\rightarrow [Au\{C(CO_2Me){=}CMeCO_2Me\}(PPh_3)]$$
$$cis\text{-}[AuMe_2\{C(CO_2Me){=}CMeCO_2Me\}(PPh_3)] \quad (55)$$
$$\text{(trace)}$$

Organogold(I) compounds undergo insertion reactions (26, 198, 201, 202) with sulfur dioxide [Eq. (56)], and in each case the sulfinate ligand is S-bonded. The same complex (R = Ph) is obtained (201, 202) by treating [AuCl(PPh_3)] with AgO_2SPh, and the analogous sulfonates are prepared using $AgOSO_2Ph$.

$$[AuR(PPh_3)] + SO_2 \rightarrow [Au(SO_2R)(PPh_3)] \quad (56)$$
$$R = Me, Ph, o\text{-}CH_2{=}CHC_6H_4, \text{ or } o\text{-}CH_2{=}CHCH_2C_6H_4$$

D. Ligand-Substitution Reactions

Gold(I) complexes containing gold–carbon bonds have been prepared by displacement of a poorly nucleophilic ligand, and organogold(I) complexes themselves undergo neutral ligand-substitution reactions to yield new organogold species.

It has previously been noted (Section II,H) that several isonitrile compounds have been prepared by displacement of weakly bound ligands, such as dimethyl sulfide (33, 134, 138) or tetrahydrothiophene (140, 178). In particular, the pentahaloaryl complexes [Au(C_6X_5)(tht)] (X = F, Cl, or Br) react with isonitriles to give [Au(C_6X_5)(CNR)] species. Displacement of carbene ligands by tertiary phosphine (132, 142) and cleavage of the gold–nitrogen bonds in [Au{C(OR)=NR'}]_3 species (138, 154) have also been described.

The complex [AuCl(CO)] undergoes displacement of carbon monoxide with isonitriles to give [AuCl(CNR)] complexes (162, 163), or with other ligands to give species no longer containing a gold–carbon bond.

E. Reactions Involving Cleavage of the Gold–Carbon Bond

Many reagents cleave the gold–carbon bonds in σ-bonded organogold(I) complexes. Although the effect of halogens is usually to add oxidatively to gold(I) compounds, it was pointed out briefly (Section III,B) that in certain cases the corresponding gold(I) halide species is obtained. Thus whereas in some instances mixtures of gold(I) and gold(III) species are produced (33, 180, 181), for certain complexes quantitative gold–carbon bond cleavage occurs.

When [AuR(PPh$_3$)] (R = (Me$_3$Si)$_3$C or (Me$_3$Si)$_2$CH) is treated with iodine the products are [AuI(PPh$_3$)] and the alkyl iodide (173), whereas with bromine some oxidative addition takes place to yield [AuBr$_3$(PPh$_3$)]. Similarly, the reaction of [Au(CF$_2$CF$_2$CH$_3$)(PPh$_3$)], prepared by insertion of C$_2$F$_4$ into the gold–methyl bond, with bromine gave [AuBr(PPh$_3$)] (71, 196).

Under certain conditions aryl complexes are also prone to cleavage of the gold–carbon bond by the action of halogens. The o-vinylphenyl and o-allylphenyl complexes of the type [AuR(PPh$_3$)] react (26) with bromine to produce [AuBr(PPh$_3$)] and the bromo-substituted aryls. Also, the bridged complex [(C$_6$F$_5$)Au(μ-dpe)Au(C$_6$F$_5$)] reacts with chlorine or bromine to give the corresponding gold(III) species (183), but the effect of iodine is to cleave the gold–carbon bond [Eq. (57)].

$$[(C_6F_5)Au(\mu\text{-dpe})Au(C_6F_5)] + 2\ I_2 \rightarrow [IAu(\mu\text{-dpe})AuI] + 2\ C_6F_5I \qquad (57)$$

In the same manner, a number of alkyl (43), vinyl (69, 203), aryl (204), and cyclopentadienyl (98, 100–102) complexes of gold(I) are cleaved by halogens to yield the corresponding gold(I) halides and the halo-substituted organic species.

Treatment of [AuMe(PEt$_3$)] with HCl causes cleavage (15) of the gold–methyl bond [Eq. (58)], whereas similar treatment of the phenyl analog resulted in no reaction. More recently, it has been claimed (204) that [AuPh(PPh$_3$)] does react with HX (X = Cl, Br, or I) to yield the corresponding halide complex, and the ortho-substituted aryl complexes [AuR(PPh$_3$)] (R = o-CH$_2$=CHC$_6$H$_4$ or o-CH$_2$=CHCH$_2$C$_6$H$_4$) also react (26) to produce [AuCl(PPh$_3$)]. Vinyl (69, 203) and cyclopentadienyl (98, 100) compounds behave similarly with protic acids with cleavage of the gold–carbon linkage.

$$[AuMe(PEt_3)] + HCl \rightarrow [AuCl(PEt_3)] + CH_4 \qquad (58)$$

Addition of HCl to the gold(I) imine complex (149) [Au{C(OMe)=NAr}(PPh$_3$)] in aqueous media produces [AuCl(PPh$_3$)] [Eq. (59)]. A less common example of a reaction of this type involves addition (205) to [AuMe-(P—t-Bu$_3$)] of di-tert-butylphosphine oxide to yield a di-tert-butylphosphinite complex of gold(I) [Eq. (60)].

$$\left[Ph_3P-Au-C\overset{\displaystyle OMe}{\underset{\displaystyle NAr}{\big<}} \right] \xrightarrow[2\ H_2O]{2\ HCl} [AuCl(PPh_3)] + ArNH_3^+Cl^- + MeOH + HCO_2H \qquad (59)$$

$$[AuMe(P{-}t\text{-}Bu_3)] + t\text{-}Bu_2P\overset{O}{\underset{H}{\diagup\diagdown}} \longrightarrow \left[t\text{-}Bu_3P{-}Au{-}P\overset{t\text{-}Bu_2}{\underset{O}{\diagup\diagdown}} \right] + CH_4 \qquad (60)$$

Although organogold(I) complexes react with alkyl halides by initial oxidative addition, it has already been pointed out (Section III,B) that subsequent reductive elimination usually occurs. Thus addition of methyl iodide, for example, to alkylgold(I) complexes of the type [AuR'(PR₃)] provides a route to the corresponding [AuI(PR₃)] complexes (*173, 192–194, 206*). The reaction of [AuMeL] (L = PMe₃ or PMe₂Ph) with CF₃I was analogous to that with methyl iodide, but with [AuMe(PPh₃)] exchange (*46*) of the organic groups occurs [Eq. (61)]. With L = PMePh₂ both reaction pathways were followed.

$$[AuMe(PPh_3)] + CF_3I \rightarrow [Au(CF_3)(PPh_3)] + MeI \qquad (61)$$

Treatment of organogold(I) complexes with acyl halides (*43, 204*), carbon tetrachloride (*26*), or (SCN)₂ (*203*) all lead to cleavage of the gold–carbon bond. Radical pathways were found (*207–209*) to be involved in the reactions of [AuMeL] complexes with benzenethiol and -selenol, removal of the organic group again being the outcome [Eq. (62)]. It was shown (*210*) from a CIDNP study of the reaction of [AuMe(PPh₃)] with dibenzoyl peroxide that organogold(I) complexes of this type readily underwent such homolytic substitution reactions.

$$[AuMeL] + PhSH \rightarrow [Au(SPh)L] + CH_4 \qquad (62)$$
$$L = \text{tertiary phosphine}$$

Metal halide complexes also cause rupture of gold–carbon bonds in organogold(I) compounds. In most cases, treatment of organogold(I) complexes with TlBr(C₆F₅)₂ results in oxidation of the gold center, but with [AuPhL] (L = PPh₃ or AsPh₃) exchange (*38*) of aryl groups occurs to yield the pentafluorophenyl complex [Au(C₆F₅)L]. In general, mercuric halides cause cleavage of gold–carbon bonds to give the corresponding gold(I) halide complexes (*43, 69, 203, 204, 206, 211, 212*); the reaction of [AuMe(PPh₃)] with organomercuric halides produces the same result (*211*).

An equilibrium is set up when alkylgold(I) complexes are mixed with halide-containing organoplatinum(II) species [Eq. (63)] (*213*). The extent

$$[AuMe(PMe_2Ph)] + \textit{trans}\text{-}[PtXMe(PMe_2Ph)_2]$$
$$\rightleftharpoons [AuX(PMe_2Ph)] + \textit{cis}\text{-}[PtMe_2(PMe_2Ph)_2] \qquad (63)$$

of [AuX(PMe$_2$Ph)] formation was found to increase in the order Cl < Br < I. The analogous reactions with *cis*-[MCl$_2$(PMe$_2$Ph)$_2$] (M = Pd or Pt) were quantitative [Eq. (64)], the exchange occurring within 10 min for palladium but over three days for the platinum complex. Also, treatment with PtCl$_2$ of the complex [Au(C$_6$H$_4$CH$_2$CH=CH$_2$-*o*)(PPh$_3$)] produced (*26*) [AuCl(PPh$_3$)] and a diorganoplatinum(II) species.

$$[AuMeL] + cis\text{-}[MCl_2L_2] \rightarrow [AuClL] + trans\text{-}[MClMeL_2] \qquad (64)$$
$$M = Pd \text{ or } Pt; \ L = PMe_2Ph$$

Similar reactions with halide-containing gold(III) complexes also resulted (*213, 214*) in halide for organic group exchange [Eqs. (65) and (66)]. It was suggested (*214*) that such exchange proceeded via species containing bridging halide and methyl groups. It has previously been shown (Section III,B) that a process analogous to that depicted in Eq. (66) was believed to be implicated in the reactions of [AuMeL] complexes with methyl iodide (*193*).

$$2 \text{ [AuMe(PMe}_2\text{Ph)]} + \text{[AuBr}_3\text{(PMe}_2\text{Ph)]}$$
$$\rightarrow 2 \text{ [AuBr(PMe}_2\text{Ph)]} + cis\text{-[AuBrMe}_2\text{(PMe}_2\text{Ph)]} \qquad (65)$$

$$[AuMe(PPh_3)] + cis\text{-}[AuIMe_2PPh_3] \rightarrow [AuI(PPh_3)] + [AuMe_3(PPh_3)] \qquad (66)$$

Cleavage of the gold–methyl bond occurred in the reaction of [AuMe(PPh$_3$)] with [H$_2$Os$_3$(CO)$_{10}$] (*215*), where methane elimination was accompanied by formation of the mixed-metal cluster [HAuOs(CO)$_{10}$(PPh$_3$)].

Finally, hydride for methyl group exchange may have been involved (*216*) in a reaction with *trans*-[PtIH(PMe$_3$)$_2$] [Eq. (67)]. The hydridogold(I) species, if that was indeed the product obtained, underwent further reaction to yield a probable gold cluster compound.

$$[AuMe(PMe_3)] + trans\text{-}[PtIH(PMe_3)_2] \rightarrow trans\text{-}[PtIMe(PMe_3)_2] + \{[AuH(PMe_3)]\} \qquad (67)$$

IV

ORGANOGOLD(III) COMPLEXES: PREPARATION AND CHARACTERIZATION

A. *Tetraalkyl- and Tetraarylaurate Complexes*

The first compound containing four gold–carbon bonds was claimed (*217*) to have been prepared in 1933, but a reformulation of the obtained products was later published (218). Much more recently, the tetramethylaurate(III) anion was prepared (9, 219, 220) by the reaction of [AuMe$_3$(PPh$_3$)] with methyllithium [Eq. (68)]. Raman and ^{1}H-NMR data were consistent (219) with solvent-separated ion pairs, and the coordinatively saturated [AuMe$_4$]$^-$ anion showed no interaction with PPh$_3$. The declining significance of dπ–dπ backbonding was proposed as an explanation for the high stability of these electron-rich systems. As for the dialkylaurates(I), the lack of a readily dissociated ligand prevented decomposition (9).

$$[AuMe_3(PPh_3)] + MeLi \rightarrow Li^+[AuMe_4]^- + PPh_3 \qquad (68)$$

The analogous tetrakis(pentafluorophenyl)gold(III) anion has been prepared (11, 12), somewhat less cleanly, by the reaction of [AuCl$_3$(tht)] with pentafluorophenyllithium [Eq. (69)]. The tetra-n-butylammonium cation was used to stabilize the product, but even this did not prevent some reduction to the diarylaurate(I) species.

$$[AuCl_3(tht)] + C_6F_5Li + n\text{-}Bu_4NBr \rightarrow n\text{-}Bu_4N[Au(C_6F_5)_4] + n\text{-}Bu_4N[Au(C_6F_5)_2] \quad (69)$$

The use of ylides as ligands has allowed a number of gold(III) complexes containing four organic groups to be isolated. A review of ylide complexes of several transition metals, including gold, has appeared (48), and such species were featured briefly therein.

When [AuMe$_3$(PR$_3$)] complexes are treated with Me$_3$P=CH$_2$ displacement of tertiary phosphine occurs (221) to yield the zwitterionic species [Me$_3$AuCH$_2$PMe$_3$], which is stable up to a temperature of 185°C. Its ^{1}H-NMR spectrum exhibited two gold–methyl resonances in the ratio 2 : 1 as expected. Treatment of this complex with HCl gave cis-[AuClMe$_2$-(CH$_2$PMe$_3$)]. When dimethylgold(III) bromide is reacted with Me$_3$P=CH$_2$ an ionic bisylide complex is obtained (Scheme 3), which when treated with HCl yields cis-[AuClMe(CH$_2$PMe$_3$)$_2$]$^+$Br$^-$.

Sulfoxonium and sulfonium ylide complexes have also been prepared (222) from [AuMe$_3$(PPh$_3$)] [Eqs. (70) and (71)]. These complexes react

$$[AuMe_3(PPh_3)] + CH_2S(O)Me_2 \xrightarrow{\text{THF}} [Me_3Au-CH_2S(O)Me_2] + PPh_3 \qquad (70)$$

$$[AuMe_3(PPh_3)] + Me_3S^+Cl^- + NaH \xrightarrow[-H_2]{\text{THF}} [Me_3Au-CH_2SMe_2] + PPh_3 + NaCl \qquad (71)$$

with protic acids to give methane and cis-[AuXMe$_2$(CH$_2$SR)] species. The sulfoxonium ylide complex underwent reaction with PMe$_2$Ph in the presence of benzophenone, involving transfer of the methylene group [Eq.

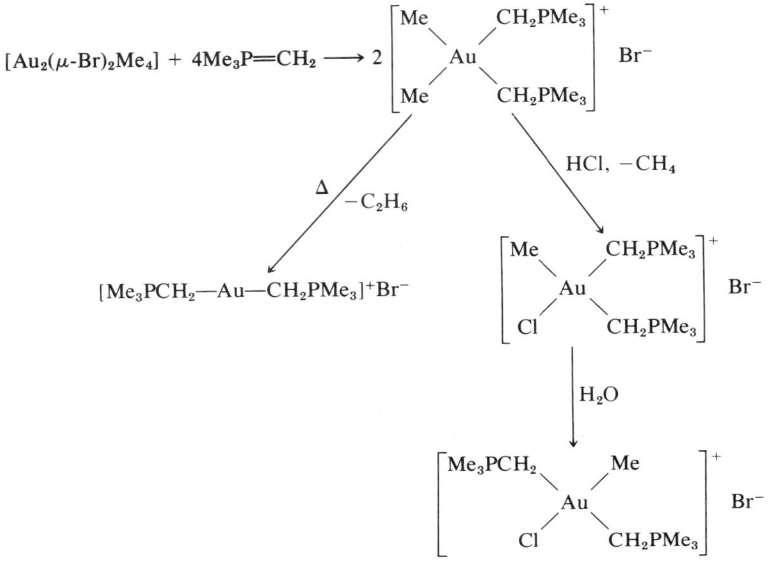

$$[Au_2(\mu\text{-Br})_2Me_4] + 4Me_3P{=}CH_2 \longrightarrow 2 \begin{bmatrix} Me & CH_2PMe_3 \\ & Au \\ Me & CH_2PMe_3 \end{bmatrix}^+ Br^-$$

$$\Delta \diagdown -C_2H_6$$

$$HCl, -CH_4$$

$$[Me_3PCH_2{-}Au{-}CH_2PMe_3]^+Br^-$$

$$\begin{bmatrix} Me & CH_2PMe_3 \\ & Au \\ Cl & CH_2PMe_3 \end{bmatrix}^+ Br^-$$

$$H_2O$$

$$\begin{bmatrix} Me_3PCH_2 & Me \\ & Au \\ Cl & CH_2PMe_3 \end{bmatrix}^+ Br^-$$

SCHEME 3

(72)]. Treatment of the sulfonium ylide complex with PMe_3 caused simple displacement of the ylide.

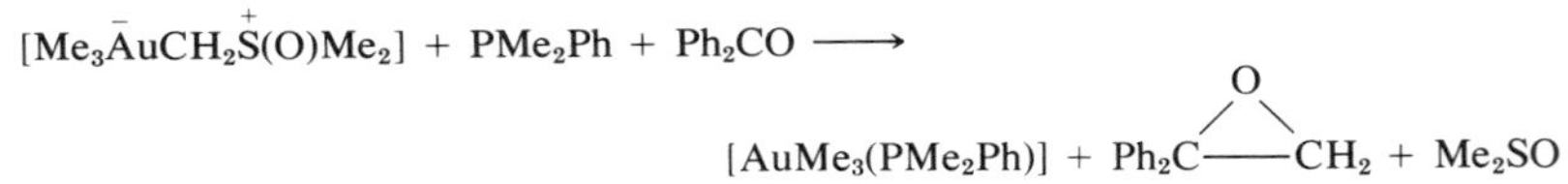

$$[Me_3\bar{A}uCH_2\overset{+}{S}(O)Me_2] + PMe_2Ph + Ph_2CO \longrightarrow$$

$$[AuMe_3(PMe_2Ph)] + Ph_2C\overset{O}{\overset{\diagup\diagdown}{-\!-\!-}}CH_2 + Me_2SO \quad (72)$$

Double ylide complexes have been obtained from the reactions of dimethylgold(III) chloride with $Me_3P{=}C{=}PMe_3$ (223) and $Me_3P{=}N{-}PMe_2{=}CH_2$ (224) [Eqs. (73) and (74)], and X-ray diffraction studies showed (225) the products to be nearly isostructural.

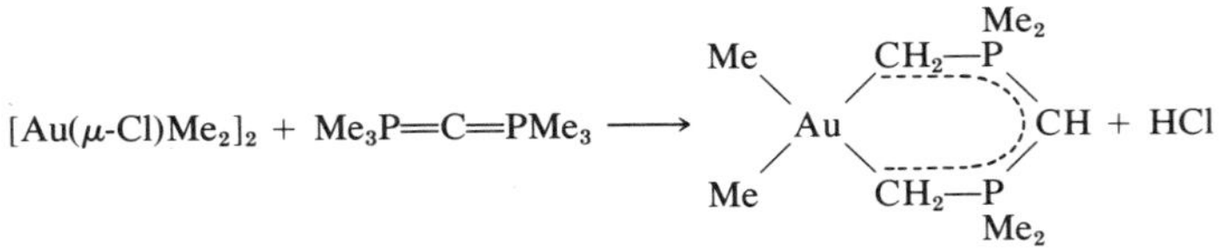

$$[Au(\mu\text{-Cl})Me_2]_2 + Me_3P{=}C{=}PMe_3 \longrightarrow \begin{matrix} Me & CH_2{-}P\overset{Me_2}{} \\ & Au & \quad CH \\ Me & CH_2{-}P\underset{Me_2}{} \end{matrix} + HCl \quad (73)$$

$$[Au(\mu\text{-Cl})Me_2]_2 + Me_3P{=}N{-}PMe_2{=}CH_2 \longrightarrow \begin{matrix} Me & CH_2{-}P\overset{Me_2}{} \\ & Au & \quad N \\ Me & CH_2{-}P\underset{Me_2}{} \end{matrix} + HCl \quad (74)$$

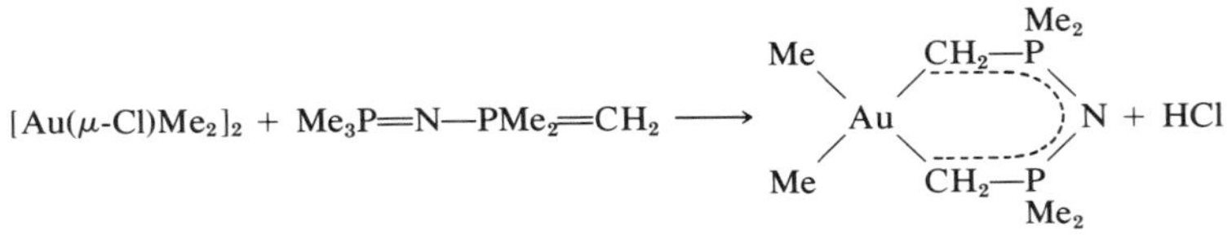

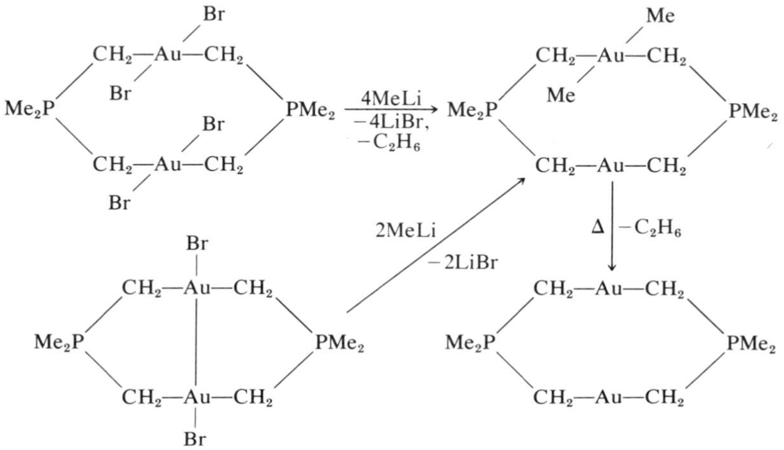

SCHEME 4

Another example of a complex with a tetraorganogold(III) center was obtained from the treatment (56) with methyllithium of the binuclear ylide complexes $[AuBr_n\{(CH_2)_2PMe_2\}]_2$ ($n = 1$ or 2) (Scheme 4). It was suggested that steric factors prevented identification of the tetramethyl analog because the tetraiodide complex could not be prepared either.

B. Triorganogold Complexes

Triorganogold(III) compounds containing no other ligands have been prepared, but were found to be thermally unstable. Thus trimethylgold(III) was prepared (226, 227) by treating gold(III) bromide or dimethylgold(III) bromide with methyllithium at −65°C, but it decomposed above −40°C. The Me_3Au moiety was stabilized by complexation with amines. Trimethylgold(III) itself reacted with protic acids or thiols by elimination of methane (227) and underwent disproportionation with gold(III) bromide [Eq. (75)]. Tris(pentafluorophenyl)gold(III) has been prepared (37) from auric chloride and the Grignard reagent and isolated as its PPh_3 adduct. Attempts to obtain $[Au(C_6F_5)_3]$ itself resulted in decomposition to give biaryls and metallic gold.

$$2\ Me_3Au + AuBr_3 \rightarrow 3\ Me_2AuBr \tag{75}$$

The complex $[AuMe_3(PPh_3)]$ has been prepared by addition (171) of triphenylphosphine at −65°C to an ether solution of trimethylgold(III) or by treating (228) $[Au(\mu\text{-}I)Me_2]_2$ with PPh_3 and then with methyllithium.

The complex exhibited two infrared bands assigned (*171*) to $\nu(\text{AuC})$ stretching frequencies. The ^{1}H-NMR spectrum of [AuMe$_3$(PMe$_3$)] contained (*16*) two doublets for the gold methyls, and in contrast to [AuMe(PMe$_3$)] (Section II,B), addition of excess PMe$_3$ did not affect the ^{1}H-NMR spectrum. Treatment with PEt$_3$, however, after a long reaction time did produce [AuMe$_3$(PEt$_3$)]. In the complexes *cis-* and *trans*-[AuMe$_2$(CF$_3$)(PMe$_3$)] it was found (*47*) that the value of $^3J(\text{P,H})$ was more negative with a trans coupling path, whereas the converse was true for $^3J(\text{P,F})$; furthermore, the geometry of the complex could be assigned on this basis. It was suggested that this difference was due to the large differences in the relative energies of the hydrogen and fluorine valence s electrons and to the availability of suitable electronic excitations involving the gold p orbitals for the trans relationship only. In the complex [AuMe$_3$(PMe$_3$)] (*229*) the ratio of the three-bond phosphorus–proton couplings, $^3J(\text{P,H})_{\text{trans}}/^3J(\text{P,H})_{\text{cis}}$, was found to be 1.3 ± 0.1.

It has already been mentioned (Section III,B) that [AuMe$_3$(PR$_3$)] complexes are isolated by reacting [AuMe(PR$_3$)] with methyl iodide (*192–195*), whereas the mixed alkyl analog is obtained (*46*) by addition of CF$_3$I to [AuMeL] (L = PMe$_3$ or PMe$_2$Ph). The *cis*-[AuMe$_2$(CD$_3$)(PMe$_3$)] complex has been prepared by treating the corresponding iodo complex with CD$_3$Li (*229*).

Complexes containing dissimilar organic groups have also been prepared from dialkylaurates(I) and organic halides. Thus treatment of AuMe$_2^-$ with iodobenzene in the presence of PPh$_3$ produced (*8*), among other products, *cis*-[AuMe$_2$Ph(PPh$_3$)]. Similarly, addition of *tert*-butyllithium to [AuMe(PPh$_3$)] followed by MeI led (*7, 230*) to mixed alkyl compounds, whereas addition of MeI or CD$_3$I to dimethylaurate(I) produced (*7*) [AuMe$_2$R(PPh$_3$)] (R = Me or CD$_3$). The cis and trans isomers of [AuMe$_2$Et(PPh$_3$)] were obtained (*231*) by treating *cis*-[AuIMe$_2$(PPh$_3$)] with an ethyl Grignard [Eq. (76)] and by addition of ethyl iodide to dimethylaurate(I) [Eq. (77)], respectively. The *n*-propyl and isopropyl complexes were prepared analogously. Reaction of cyclopentadienyl sodium with *cis*-[AuIMe$_2$(PPh$_3$)] at $-10°$C gave (*228*) *cis*-[AuMe$_2$(C$_5$H$_5$)(PPh$_3$)], for which Raman and ^{1}H-NMR spectra indicated a fluxional, σ-bonded cyclopentadienyl ring similar to that in [Au(C$_5$H$_5$)(PPh$_3$)] (*94*).

$$\text{cis-[AuIMe}_2\text{(PPh}_3\text{)] + EtMgI} \rightarrow \text{cis-[AuMe}_2\text{Et(PPh}_3\text{)] + MgI}_2 \tag{76}$$

$$\text{Li}^+[\text{AuMe}_2\text{(PPh}_3\text{)]}^- + \text{EtI} \rightarrow \text{trans-[AuMe}_2\text{Et(PPh}_3\text{)] + LiI} \tag{77}$$

Mixed triarylgold(III) complexes have been obtained by oxidation (*187*) of [Au(C$_6$F$_5$)(tht)] with thallium(III) compounds [Eq. (78)], whereas the reaction of cis-$[AuCl_2(C_6Br_5)(PPh_3)]$ with excess C_6F_5MgBr yielded (36) the cis-$[Au(C_6F_5)_2(C_6Br_5)(PPh_3)]$ complex.

$$[Au(C_6F_5)(tht)] + TlClR_2 \rightarrow cis\text{-}[Au(C_6F_5)R_2(tht)] + TlCl \qquad (78)$$
$$R = C_6F_5,\ C_6F_4H,\ C_6F_3H_2,\ p\text{-}C_6H_4F,\ \text{or}\ m\text{-}C_6H_4CF_3$$

An interesting example of a preparation of a triorganogold(III) species was found in the reaction of $[AuMe_2(acac)]$ with tertiary phosphine (232). Whereas the softer acid gold(I) formed only C-bonded species, the complex $[AuMe_2(acac)]$ contained an O-bonded acetylacetonate ligand; phosphine addition, however, caused conversion to a C-bonded moiety in cis-$[AuMe_2(acac)(PR_3)]$. The ability to promote this rearrangement was found to decrease in the order $PMe_2Ph > PMePh_2 > PPh_3$, whereas with PCy_3 a four-coordinate O-bonded species was observed, suggesting that steric constraints were significant in the rearrangement process. Use of the unsymmetrical benzoylacetonate ligand allowed two dynamic processes to be identified, namely, associative phosphine exchange via a five-coordinate species and a slower rearrangement to the C-bonded isomer (Scheme 5).

Anionic gold(III) species containing three gold–carbon bonds have also been prepared, by the action of $TlBr(C_6F_5)_2$ on the bromo(pentafluorophenyl)aurate(I) anion $(11, 12)$, [Eq. (79)]. An inner sphere mechanism has been proposed (13) for the reaction of $TlClR_2$ species with diarylaurates(I), where the anionic products may be of the types $[AuR_2]^-$, $[AuR_4]^-$, or $[AuClR_3]^-$.

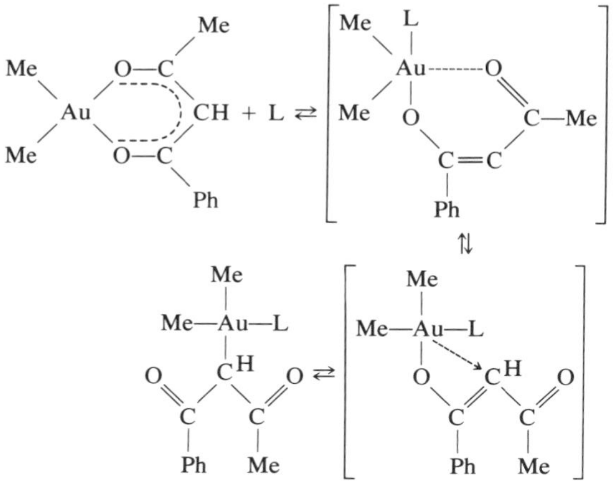

SCHEME 5

$$n\text{-Bu}_4\text{N}[\text{AuBr}(\text{C}_6\text{F}_5)] + \text{TlBr}(\text{C}_6\text{F}_5)_2 \rightarrow n\text{-Bu}_4[\text{AuBr}(\text{C}_6\text{F}_5)_3] + \text{TlBr} \qquad (79)$$

As a final example, the reaction of a binuclear gold(I) ylide complex with methyl iodide afforded (48, 56), a formal gold(II) center having three gold–carbon bonds [Eq. (80)].

$$\begin{array}{c} \text{CH}_2\text{—Au—CH}_2 \\ \diagup \qquad\qquad \diagdown \\ \text{Me}_2\text{P} \qquad\qquad\qquad \text{PMe}_2 \\ \diagdown \qquad\qquad \diagup \\ \text{CH}_2\text{—Au—CH}_2 \end{array} + \text{MeI} \rightarrow \begin{array}{c} \overset{\displaystyle \text{Me}}{\underset{\displaystyle |}{}} \\ \text{CH}_2\text{—Au—CH}_2 \\ \diagup \qquad | \qquad \diagdown \\ \text{Me}_2\text{P} \qquad\quad\; \text{PMe}_2 \\ \diagdown \qquad | \qquad \diagup \\ \text{CH}_2\text{—Au—CH}_2 \\ | \\ \text{I} \end{array} \qquad (80)$$

C. Diorganogold Complexes

The first gold(III) compound in which two organic groups were bonded to the metal center was prepared by the reaction of auric bromide with ethylmagnesium bromide (233–235). This was later shown from molecular weight measurements (236) to be the dimeric species [Au$_2$(μ-Br)$_2$Et$_4$]. The analogous chlorides were also prepared, as were complexes containing other alkyl groups (237, 238). Treatment with silver cyanide produced (237, 239) complexes of the composition ''AuR$_2$(CN),'' and a crystal structure of the n-propyl compound indicated (240) a tetrameric species (11) had been formed. Dipole moment measurements were performed (241) on [Au$_2$(μ-Br)$_2$Et$_4$] and [n-Pr$_2$Au(CN)]$_4$, and it was proposed (242) that the decomposition of the latter on standing gave n-hexane and a polymeric species containing alternate gold(I) and gold(III) units.

$$\begin{array}{ccc} n\text{-Pr} & & n\text{-Pr} \\ | & & | \\ n\text{-Pr—Au—C}\!\equiv\!\text{N—Au—}n\text{-Pr} \\ | & & | \\ \text{N} & & \text{C} \\ ||| & & ||| \\ \text{C} & & \text{N} \\ | & & | \\ n\text{-Pr—Au—N}\!\equiv\!\text{C—Au—}n\text{-Pr} \\ | & & | \\ n\text{-Pr} & & n\text{-Pr} \end{array}$$

(11)

The infrared spectrum of [Au$_2$(μ-I)$_2$Me$_4$] has been measured in solution and in the solid state (243, 244) and gives rise to identical band contours. This was taken to indicate that rotation about the gold–carbon bonds occurred in the solid state (which was frozen out at 90 K), and a barrier to rotation of 0.77 kcal mol^{-1} was deduced. It was suggested (245) that the lack of color in [Au$_2(\mu$-Br)$_2$Et$_4$] was due to the high energy of the gold(III) d$_{x^2-y^2}$ orbital, since the intense colors in AuX$_4^-$ salts were attributed to the band associated with electronic excitation from the halide p$_x$ to the gold(III) d$_{x^2-y^2}$ orbital.

Dialkylgold(III) dimers were produced (226, 227) by treatment of trial-kylgold(III) complexes with protic acids at low temperatures, the preparation being accompanied by elimination of alkane.

The reaction of [Au$_2(\mu$-Br)$_2$Et$_4$] with bromine produced a species (235, 237) of the constitution "AuBr$_2$Et," an analogous species also being isolated (246) from the thermal decomposition of [Au$_2(\mu$-Br)$_2$Me$_4$] in ether. These were shown to be dimeric (247). Reaction with ethylenediamine produced [Au(en)$_2$]Br$_3$ and [AuEt$_2$(en)]Br, which suggested that the complex had structure 12 rather than the more symmetrical one. This assignment was confirmed (248) by an X-ray diffraction study, which was carried out at 113 K in order to avoid decomposition. Thus a number of compounds of the type [AuX$_2$R]$_2$ do contain diorganogold(III) moieties.

$$\begin{array}{ccccc} \text{Et} & & \text{Br} & & \text{Br} \\ \diagdown & & \diagup\diagdown & & \diagup \\ & \text{Au} & & \text{Au} & \\ \diagup & & \diagdown\diagup & & \diagdown \\ \text{Et} & & \text{Br} & & \text{Br} \end{array}$$

(12)

Dialkylgold(III) dimers and tetramers containing a variety of bridging species have been prepared, often by metathetical anion exchange. The sulfato-bridged tetramer [Et$_4$Au$_2$(SO$_4$)]$_2$ was isolated (249), and a structure was postulated (250) in which close approach of gold atoms was circumvented. The complex reacted with ethylenediamine to give (249) a dimeric species containing a nine-membered ring, having bridging sulfato and ethylenediamine ligands.

Treatment of [Au$_2(\mu$-Br)$_2$Et$_4$] with the corresponding silver salt has led to complexes containing bridging thiocyanate (251), azide (252), acetate (253), trifluoromethanesulfonate (254), diphenylphosphate, and phenyl-phosphate and -arsonate (255) ligands. Cyanate and selenocyanate analogs were also prepared (256), the latter having a structure similar to the thiocyanate (13), whereas the cyanate had a N-bonded structure analogous to the azide complex (14). From their ^{1}H-NMR spectra, the thiocyanate and selenocyanate complexes were found to be stereochemically rigid at room temperature. The reaction of [Au$_2(\mu$-Br)$_2$Me$_4$] with sodium trimethylsiloxide also produced (257) a dimeric species containing bridging trimethylsiloxy groups, and an analogous structure was proposed

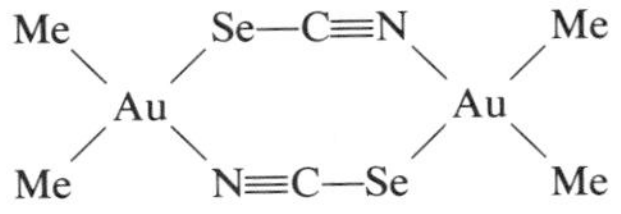

(13)

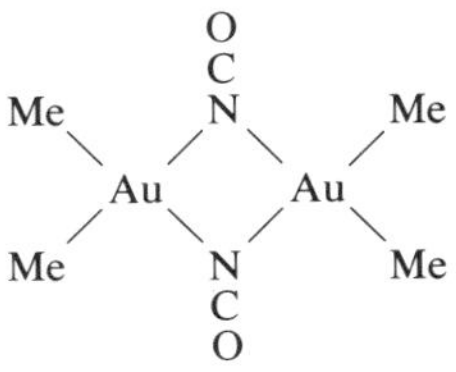

(14)

(258) for $[AuMe_2(OH)]_2$, obtained from the hydrolysis of $[AuMe_2(OH_2)_2]^+$ in aqueous sodium perchlorate. Dimethylgold(III) hydroxide, however, also formed a tetrameric species (15) whose crystal structure exhibited (259) short O—O distances due to hydrogen bonding. The eight-membered rings were found to stack together in columns.

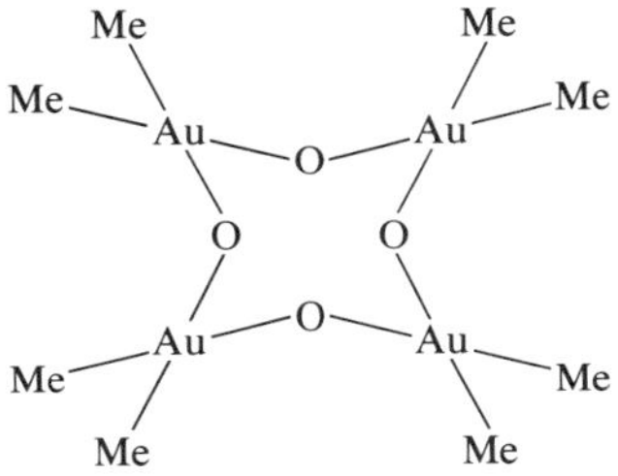

(15)

The dialkylgold(III) moiety has been stabilized by reaction with neutral ligands, either by cleavage of dimers or by various other routes. Since a large number of such species have been prepared, they are discussed here according to the nature of the donor atom.

The first complex of this type was prepared (235) by treating $[Au_2-(\mu\text{-}Br)_2Et_4]$ with ammonia to yield the adduct $[AuBrEt_2(NH_3)]$. The analogous pyridine complex was also synthesized (236), and in aqueous solutions of ammonia or pyridine the ionic species $[AuEt_2L_2]^+Br^-$ (L = NH_3 or C_5H_5N) were obtained, whereas the ethylenediamine analog $[AuEt_2(en)]^+Br^-$ was stable enough to be isolated (236, 238, 260). It was suggested (261) that an equilibrium existed between species of the type $[AuR_2(en)]^+X^-$ and dimeric compounds containing bridging ethylenediamine units, and the latter structures were proposed (239, 249) for the cyano and sulfato complexes. The interconversion of these was suggested (262) to proceed via five-coordinate species. It was later pointed out (263), however, that the complexes previously formulated as dimers, $[R_2AuX(\mu\text{-}en)XAuR_2]$ (X = halide or cyanide) were, in fact, ionic species of the type $[AuR_2(en)]^+ [AuX_2R_2]^-$.

Complexes containing unsymmetrical chelating ligands have also been prepared. Thus reaction of $[Au_2(\mu\text{-}Br)_2Et_4]$ with $H_2NCH_2CH_2SH$ in the presence of sodium ethoxide gave (264) a neutral compound $[Et_2\text{-}AuNH_2CH_2CH_2S]$, which when treated with ethyl bromide yielded the ionic complex $[Et_2AuNH_2CH_2CH_2SEt]^+Br^-$. Treatment of $[Au_2(\mu\text{-}I)_2Me_4]$ with 8-quinolinol (265) or salicylaldimine Schiff bases (266) produced monomeric compounds containing chelating N,O ligands, whereas bidentate Schiff bases gave bridged, dimeric products.

A ^{1}H-NMR study of the complex $[AuClMe_2(py)]$ in pyridine solution indicated (267) only one CH_3 resonance, which was suggested to be due to associative two-site exchange behavior [Eq. (81)]. Although the bipyridine complexes, formed by reaction of the ligand with $[Au_2(\mu\text{-}X)_2Me_4]$, were ionic species of the type $[AuMe_2(bipy)]^+X^-$, monomeric compounds were formed (268) with 2,7-dimethyl-1,8-naphthyridine. ^{1}H-NMR investigations revealed that below 200 K the gold center was associated with only one nitrogen donor atom, but that above 200 K there was rapid intramolecular two-site exchange [Eq. (82)]. Above ambient temperature, intermolecular ligand and/or anion exchange occurred, removing the nonequivalence of the gold–methyl groups.

$$
\begin{array}{ccc}
\underset{H_3C}{\overset{H_3C}{\diagdown}}\!\!Au\!\!\underset{Cl}{\overset{L}{\diagup}} & \underset{-L^*}{\overset{+L^*}{\rightleftharpoons}} & \underset{H_3C}{\overset{H_3C}{\diagdown}}\!\!Au\!\!\underset{L^*}{\overset{L}{\diagup}}\!\!-Cl & \underset{+L}{\overset{-L}{\rightleftharpoons}} & \underset{H_3C}{\overset{H_3C}{\diagdown}}\!\!Au\!\!\underset{L^*}{\overset{Cl}{\diagup}}
\end{array}
\qquad (81)
$$

$$
\text{Me—Au}\!\!\begin{array}{c}X\\|\\Me\end{array}\!\!\text{—N}\!\!\bigcirc\!\!\begin{array}{c}Me\\ \\ \end{array} \;\rightleftharpoons\; \text{Me—Au}\!\!\begin{array}{c}X\\|\\Me\end{array}\!\!\text{—N}\!\!\begin{array}{c}Me\\ \\N\end{array}\!\!\bigcirc\!\!\begin{array}{c} \\ \\Me\end{array}
\qquad (82)
$$

X = Cl, Br, I, OCN, SCN, SeCN, or CN

Complexes containing the dimethylgold(III) moiety with oxygen and sulfur donors have also been prepared. The reaction of $[Au_2(\mu\text{-}I)_2Me_4]$ with dibenzylsulfide produced (238) cis-$[AuIMe_2(SBz_2)]$, whereas treatment with Tl(acac) gave the O-bonded complex $[AuMe_2(acac)]$. With dithiocarbamates the halide-bridged species, $[Au_2(\mu\text{-}X)_2Me_4]$, yielded (253, 269) the $[AuMe_2(S_2CNR_2)]$ complexes. The dithiocarbonate ligand formed an analogous monomeric compound (253), whereas the reactions of $[Au_2(\mu\text{-}Br)_2Me_4]$ with $KSCO_2Et$ and trimethylgold(III) with MeCOSH gave dimeric and polymeric products, respectively.

Reaction of $[Au_2(\mu\text{-Cl})_2Me_4]$ with dimethyl sulfide or dimethyl selenide produced (270) cis-$[AuClMe_2L]$ (L = Me_2S or Me_2Se), but the latter underwent ligand exchange in solution, which was followed by ^{1}H-NMR spectroscopy. A higher activation energy for exchange was found for dimethyl selenide. With bidentate thioethers binuclear complexes were formed [Eq. (83)], although the reaction of dimethylgold(III) nitrate with $MeS(CH_2)_2SMe$ gave the ionic complex $[Me_2Au\overline{S(Me)CH_2CH_2SMe}]^+NO_3^-$.

$$[Au_2(\mu\text{-X})_2Me_4] + RS(CH_2)_nSR \rightarrow$$

$$\begin{array}{c} Me \quad Me \qquad\qquad Me \quad Me \\ \diagdown\diagup \qquad\qquad\qquad \diagdown\diagup \\ Au \qquad\qquad\qquad Au \\ \diagup\diagdown \qquad\qquad\qquad \diagup\diagdown \\ X \qquad S-(CH_2)_n-S \qquad X \\ \diagdown \qquad\qquad\qquad\qquad \diagdown \\ R \qquad\qquad\qquad\qquad R \end{array} \qquad (83)$$

X = Cl, Br, I, or SCN; n = 2 or 3

The decomposition of the gold(III) dithiocarbamates $[AuMe_2(S_2\text{-}CNMe_2)]$ and $[AuBr_2(S_2CNMe_2)]$ has been investigated (271) using photo-electron spectroscopy, and NMR studies of dimethylgold dithiocarbamates and aryl xanthates have given information (272) regarding rotation about the C—N and C—O bonds. By observing the coalescence of the ^{1}H signals for the Me_2Au moiety in complexes 16 and 17 in $CDCl_3$ solution were obtained barriers to rotation of 76 ± 4 and 50 ± 3 kJ mol^{-1}, respectively. Dimethylgold(III) complexes of α-amino acids and related ligands have been prepared (273), and their infrared spectra indicated that they were monomeric species containing bidentate ligands.

$$\begin{array}{cc} H_3C \quad S \quad Me & H_3C \quad S \quad Me \\ \diagdown\diagup\diagdown\diagup & \diagdown\diagup\diagdown \\ Au \quad C=N & Au \quad C=O \\ \diagup\diagdown\diagup\diagdown & \diagup\diagdown\diagup \\ H_3C \quad S \quad Ph & H_3C \quad S \quad Me \end{array}$$

(16)   (17)

Dimethylgold(III) species with coordinated hydroxide, nitrate and per-chlorate groups have been prepared (274), and the latter two found to exist in aqueous solution as cis-$[AuMe_2(OH_2)_2]^+X^-$. The ^{17}O-NMR spectrum of $[AuMe_2(ClO_4)]$ in aqueous solution exhibited (275) only one resonance down to 5°C, indicating that rapid exchange of water molecules occurred above this temperature.

Investigations of the vibrational spectra of cis-$[AuMe_2L_2]^{n+}$ species have revealed (276, 277) a trans influence series, by measuring the effectiveness of L in reducing $\nu(AuC)$, in the order Me_2S > Br> Cl > en > OH$^-$ > H_2O [which was the same as that found for the more widely studied platinum(II) systems]. It was also stated (276) that the most energetically favorable situation was one where two ligands were bound to the Me_2Au moiety by electrostatic interactions. The infrared and Raman spectra of $[Au_2(\mu\text{-}X)_2Me_4]$ and $[AuX_2Me_2]^-$ (X = Cl, Br, or I) were (278) very similar, and bridge stretching frequencies indicated that covalency in the Au—X bridges increased in the order Cl < Br < I.

Many diorganogold(III) compounds with tertiary phosphines and arsines are now known. It has been shown previously (Section III,B) that the reactions of [AuRL] complexes with alkyl halides usually produced [AuXL] and a triorganogold(III) species. In certain instances, however, the expected $[AuXR_2L]$ compounds were formed (46, 194, 195), notably in the reactions of [AuMeL] (L = PMe_3, PMe_2Ph, or $AsPh_3$) with methyl iodide.

Complexes of the type $[AuXMe_2L]$ (L = tertiary phosphine or arsine) have been prepared (228, 279) by addition of the ligand to $[Au_2(\mu\text{-}X)_2Me_4]$. With excess ligand the ionic complexes $[AuMe_2L_2]^+X^-$ are produced (279), and in both cases cis geometries were deduced from their ^{1}H-NMR and infrared spectra.

The complexes $[AuXMe_2L]$ (X = Cl or SCN; L = py, PPh_3, $AsPh_3$, or $SbPh_3$), $[AuMe_2L_2]^+ClO_4^-$ (L = PPh_3, $AsPh_3$, or $SbPh_3$), $[AuMe_2(dpe)]^+Cl^-$, and $[AuMe_2(dpe)]^+[AuCl_2Me_2]^-$ have been prepared (280), and their Raman and ^{1}H-NMR spectra show that each contains the cis-dimethylgold(III) moiety. The $[AuXMe_2(py)]$ (X = Cl or SCN) and $[AuMe_2(SCN)(AsPh_3)]$ complexes underwent rapid exchange of the neutral ligand at room temperature, as did the bis-ligand cations, but the bidentate ligand complexes showed no evidence for dissociation. The complexes $[AuMe_2L_2]^+X^-$ (X = Cl or I; L = PMe_3, PMe_2Ph, or $PMePh_2$) exhibited (281) phosphine exchange at 40°C, the rate of exchange increasing in the order PMe_3 < PMe_2Ph < $PMePh_2$. Exchange was faster with the iodide counterion, suggesting that anion attack was involved in the exchange process; indeed, a very slow exchange rate was observed for $[AuMe_2(PPh_3)_2]^+ClO_4^-$ where the noncoordinating perchlorate ion was involved.

Treatment of $[Au_2(\mu\text{-}I)_2Me_4]$ with a silver salt in THF yielded (282) the dimethylbis(solvent)gold(III) cation, which was converted by addition of the free ligand to $[AuMe_2L_2]^+X^-$ (L = tertiary phosphine, arsine, or stibine). Dimethylbis(phosphine)gold(III) cations were also produced by addition of excess ligand to $[AuMe_2X]$ [X = $CF_3COCHCOCH_3$ or $(CF_3CO)_2CH$] (232) or $[AuMe_2(OSO_2CF_3)(OH_2)]$ (254); it was previously noted (Section IV,B) that $[AuMe_2(acac)]$ with triphenylphosphine produced the C-bonded complex $[AuMe_2(acac)(PPh_3)]$ (232). With $CH_3C(CH_2PPh_2)_3$ (triphos), dimethylgold(III) chloride produced (283) the ionic material $[AuMe_2(triphos)]^+[AuCl_2Me_2]^-$, where it was assumed that only two of the phosphorus atoms were bound to the gold center in the cation.

The di-n-butylgold(III) complex [AuBr-n-Bu$_2$(PPh$_3$)] has been prepared (284) by addition of PPh$_3$ to the analogous ethylenediamine complex, and the reaction of [AuCl$_3$(PPh$_3$)] with ferrocenyllithium gave [AuCl-(C$_5$H$_4$FeC$_5$H$_5$)$_2$(PPh$_3$)]. Two ^{1}H-NMR signals were observed for the gold-substituted cyclopentadienyl rings.

Oxidative addition reactions of [AuCl(PPh$_3$)] with diarylthallium(III) reagents have led (187, 285, 286) to the complexes [AuClR$_2$(PPh$_3$)] (R = C$_6$F$_5$, C$_6$Cl$_5$, C$_6$F$_4$H, C$_6$F$_3$H$_2$, 4-C$_6$H$_4$F, or 3-C$_6$H$_4$CF$_3$), and a crystal structure determination was carried out for the pentafluorophenyl complex (287). Metathesis reactions of [AuCl(C$_6$F$_5$)$_2$(PPh$_3$)] allowed isolation (288) of the corresponding bromide and iodide complexes. The triphenylarsine analogs were also prepared (38), and the complexes [AuCl(C$_6$F$_5$)$_2$L] (L = PPh$_3$ or AsPh$_3$) reacted with silver perchlorate to yield [Au(ClO$_4$)(C$_6$F$_5$)$_2$L] (289, 290). These species readily undergo metathesis reactions to give the corresponding [AuX(C$_6$F$_5$)$_2$L] (X = CN, SCN, N$_3$, or HCO$_3$) complexes. With potassium sulfate they give binuclear species containing bridging sulfato ligands (289), and with neutral ligands [Au(C$_6$F$_5$)$_2$(PPh$_3$)L]$^+$ClO$_4^-$ (L = OPPh$_3$, OAsPh$_3$, ONC$_5$H$_5$, ONC$_9$H$_7$, NC$_9$H$_7$, PEt$_3$, P-n-Bu$_3$, or PMePh$_2$) (290). Metathesis reactions of the triphenylarsine complex gave (291) [AuX(C$_6$F$_5$)$_2$(AsPh$_3$)] (X = NO$_3$, MeCO$_2$, NO$_2$, CF$_3$CO$_2$, CN, SCN, or PhCO$_2$). Mixing [Au(ClO$_4$)(C$_6$F$_5$)$_2$(PPh$_3$)] with the analogous thiocyanate or azide complex produced (292) an ionic, binuclear species, [{Au-(C$_6$F$_5$)$_2$(PPh$_3$)}$_2$(μ-X)]$^+$ClO$_4^-$ (X = SCN or N$_3$). The dpe-bridged gold(I) complex underwent oxidative addition (183) with TlBr(C$_6$F$_5$)$_2$ to give [{AuCl(C$_6$F$_5$)$_2$}$_2$(μ-dpe)], which was converted to the bromide and iodide analogs by metathesis.

The reaction of [AuCl(CNPh)] with TlBr(C$_6$F$_5$)$_2$ produces (178) [AuCl-(C$_6$F$_5$)$_2$(CNPh)], whereas oxidative addition of halogens to [Au(C$_6$F$_5$)L] (L = CNR, C(NHR)OR′, or C(NHR)NHR′) gives the corresponding gold(III) compounds. The polymeric species [AuCl(C$_{12}$H$_8$)]$_n$, prepared from [AuCl$_3$(tht)] and Sn-n-Pr$_2$(C$_{12}$H$_8$), is cleaved (293) by an assortment of neutral and anionic ligands (Scheme 6). Anionic diarylgold(III) species have been prepared (12, 13) by oxidative addition to diarylaurates(I) of halogens, or by reaction with TlCl$_3$.

The analogous [AuBr$_2$Me$_2$]$^-$ anion is produced from the reaction of [Me$_2$Au(μ-Br)$_2$AuBr$_2$] with triphenylphosphine (248). It was expected that the products would be [AuBrMe$_2$(PPh$_3$)] and [AuBr$_3$(PPh$_3$)], but the ^{31}P-NMR spectrum of the reaction mixture indicated the presence of BrPPh$_3^+$ and [AuBr(PPh$_3$)]. Treatment of cis-[AuBrMe$_2$(PPh$_3$)] with bromine also caused formation of BrPPh$_3^+$, and in the ^{1}H-NMR spectrum, the methyl resonance associated with [Me$_2$Au(μ-Br)$_2$AuBr$_2$] was shifted on

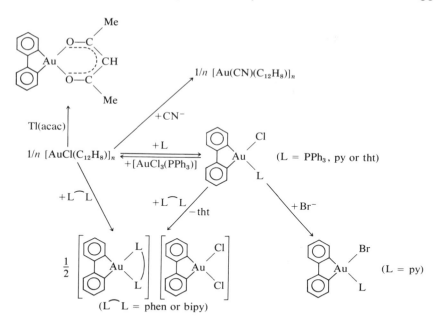

addition of BrPPh$_3^+$Br$^-$. Thus the same species was believed to be involved in each case (Scheme 7), and treatment of *cis*-[AuIMe$_2$(PPh$_3$)] with iodine gave the corresponding salt.

Other gold(III) compounds containing two gold–carbon bonds have

SCHEME 7

been isolated. These include bis(carbene)gold(III) complexes, obtained by oxidative addition of halogen to the corresponding bis(carbene)gold(I) compounds (*141, 144, 184*). The crystal structure of *trans*-[AuI₂{C-(NHC₆H₄Me-*p*)₂}₂]⁺ClO₄⁻ has been determined (*185*).

The reaction of gold(III) chloride with(LiCPh≡CPh—)₂ in aqueous medium produces (*294*) the 1-hydroxy-2,3,4,5-tetraphenylauracyclopentadiene dimer (**18**), the structure of which has been established by X-ray diffraction (*295*). The chloride-bridged analog was prepared (*296*) from [AuCl₃(tht)] and Et₂Sn(C₄Ph₄), and bridge cleavage reactions were performed with pyridine, triphenylphosphine, and phenanthroline. It was suggested that the last reaction produced the first five-coordinate organogold complex [AuCl(C₄Ph₄)(phen)].

$$\text{(18)}$$

D. *Monoorganogold Complexes*

It was noted in the previous section that the complexes [AuX₂R]₂ are unsymmetrical, containing a dialkylgold(III) moiety. In general, the dialkylgold(III) unit is of considerable stability, and by comparison gold(III) complexes containing only one gold–carbon bond are few in number.

Whereas the [AuX₂R]₂ complexes (R = alkyl) contain dialkylgold moieties, the analogous aryl complexes are of a symmetrical nature and have been prepared by the action of benzene (*297*) or substituted benzenes (*298, 299*) upon auric chloride. The complex [AuBr₂Ph]₂ has also been prepared by treating HAuCl₄ with phenyllithium (*284*). Auration of aromatic nitriles to give species of the form [AuCl₂R]₂ was claimed (*300*), but the products were later shown (*301*) to be the simple adducts [AuCl₃(NCR)].

Treatment of [AuCl₂Ph]₂ with diphenylzinc produced (*299*) species of composition Ph₃AuZn and Ph₂AuZnCl, as described in Section II,G. Cleavage of the dimers with neutral ligands led (*298*) to the complexes [AuCl₂RL] (R = Ph, *p*-MeC₆H₄ or *p*-ClC₆H₄; L = S-*n*-Pr₂, PPh₃, PMe₃, or py). For the *p*-tolyl complexes, thermal stability decreased in the order PPh₃ > py > S-*n*-Pr₂. Infrared measurements suggested cis configurations for the phosphine and sulfide complexes, as was indeed found from an X-ray diffraction determination of [AuCl$_2$Ph(S-n-Pr$_2$)] (*302*), but a trans geometry was indicated for the pyridine complex.

Oxidative addition reactions with halogens or thallium(III) chloride, for the complexes [AuRL], often produce [AuX$_2$RL] species. Thus [AuBr$_2$(C$_6$F$_5$)(PPh$_3$)] was one of the products of the reaction of [Au(C$_6$F$_5$)-(PPh$_3$)] with bromine (*33*). The addition of iodine produced [AuI$_2$(C$_6$F$_5$)-(PPh$_3$)] quantitatively (*180*), but with Cl$_2$ or Br$_2$ a mixture of products was obtained. The pentachloro- and pentabromophenyl analogs reacted cleanly (*36*) with X$_2$ (X = Cl, Br, or I) or TlCl$_3$ to give [AuX$_2$R(PPh$_3$)] (R = C$_6$Cl$_5$ or C$_6$Br$_5$).

The analogous triethylphosphine complexes [AuR(PEt$_3$)] (R = C$_6$F$_5$ or C$_6$Cl$_5$) also undergo oxidative addition (*182*) to give monoarylgold(III) species. With TlCl$_3$ both complexes react to give *cis*-[AuCl$_2$R(PEt$_3$)], and addition of iodine produces analogous compounds of unspecified geometry. With excess Cl$_2$, [Au(C$_6$F$_5$)(PEt$_3$)] yields *trans*-[AuCl$_2$(C$_6$F$_5$)-(PEt$_3$)], but with less than one equivalent of bromine the infrared spectrum indicated formation of the cis product. The pentachlorophenyl derivative, however, gave *trans*-[AuCl$_2$(C$_6$Cl$_5$)(PEt$_3$)] at $-10°$C, but afforded the cis isomer with chlorine addition at room temperature. With 0.5 molar equivalents of bromine the trans product was obtained.

When [Au(C$_6$Cl$_5$)(AsPh$_3$)] is treated with a halogen, it undergoes oxidative addition to yield [AuX$_2$(C$_6$Cl$_5$)(AsPh$_3$)] (X = Cl, Br, or I) (*181*), but the pentafluorophenyl analog suffers gold–carbon bond cleavage with bromine and iodine. In reaction with TlCl$_3$ the complex [AuCl$_2$(C$_6$X$_5$)-(AsPh$_3$)] (X = F or Cl) was formed.

The bridged complexes [{AuX$_2$(C$_6$F$_5$)}$_2$(μ-dpe)] (X = Cl or Br) are produced by oxidative addition (*183*), but with iodine gold–carbon bond cleavage takes place. The infrared spectrum of the chloro complex indicated trans geometry at both gold centers. Similar halogen oxidative addition reactions occur to give [AuX$_2$(C$_6$F$_5$)L] (X = Cl, Br, or I; L = tht, CNR or carbene) (*178, 179*). Treatment of the tetrahydrothiophene complexes with chloride or thiocyanate produces anionic monoarylgold(III) species, [AuX$_2$X'(C$_6$F$_5$)]$^-$ (X' = Cl or SCN). Such anions are also formed by oxidative addition of bromine to a monoarylaurate(I) to yield n-Bu$_4$N[AuBr$_3$(C$_6$F$_5$)] (*11, 12*).

Arylgold(III) complexes have also been obtained from the reactions of tetrachloroaurate(III) salts with substituted phenylhydrazines (*303, 304*). Thus were prepared the compounds R$_4$N[AuCl$_3$Ar] (R = Et or n-Bu; Ar = C$_6$H$_5$, p-ClC$_6$H$_4$, p-BrC$_6$H$_4$, or p-NO$_2$C$_6$H$_4$), which were characterized by their ^{1}H-NMR and vibrational spectra. Arylgold(III) species have been postulated as intermediates (*305*) in the thermal decomposition of [AuCl$_3$(azobenzene)].

Monoorganogold(III) compounds have been synthesized where the

(19)   (20)

coordinated ligands contain a vinyl group. Thus complexes **19** and **20** are obtained by addition of bromine to the corresponding [AuBrL] species (*306, 307*). The analogous alkoxy-substituted compounds are prepared by refluxing [AuBr$_3$L] (L = o-CH$_2$═CHC$_6$H$_4$PPh$_2$ or o-CH$_2$═CHCH$_2$C$_6$H$_4$-PPh$_2$) in methanol or ethanol, though two isomers were found in the latter case (*307*). Reaction of auric bromide with allylmethyl sulfide produces (*308*) an alkylgold(III) complex containing a five-membered ring [Eq. (84)], formed essentially by insertion of the olefinic moiety into the gold–bromine bond.

$$\text{Au}_2\text{Br}_6 + \text{MeSCH}_2\text{CH}═\text{CH}_2 \rightarrow \qquad\qquad (84)$$

Other gold(III) complexes containing one gold–carbon bond include the isonitrile complexes [AuX$_3$(CNR)] (*131*) and the trimeric imino compounds [Au$_3${C(OR)═NR'}$_3$X$_6$] (X = Br or I), which have been prepared (*155*) by oxidative addition of halogen to the gold(I) species.

Several mixed gold(I)/gold(III) olefin and acetylene complexes have been mentioned (Section II,E), so only species containing gold(III) centers alone are considered here. The species [AuCl$_3$(1,3-cod)] has been prepared (*309*), but was found to be unstable, and it was suggested that the diolefin functioned as a monodentate ligand. The reaction of gold(III) chloride with dimethylacetylene initially produces (*310*) red π-complexes [Au$_2$Cl$_6$(MeC≡CMe)$_n$] (n = 1 or 2), but these undergo rearrangement to give the products of insertion of the acetylene into a gold–chlorine bond. With excess dimethylacetylene, reduction to gold(I) occurs. The reactions of auric chloride or HAuCl$_4$ with hexamethyl Dewarbenzene gave delocalized carbonium ions (*77, 311*) which were stabilized by the tetrachloroaurate anion. Finally, gold(III) carborane complexes have been obtained (*312*), but it is doubtful whether there are any gold–carbon interactions involved.

V
REACTIONS OF ORGANOGOLD(III) COMPLEXES

A. Reductive Elimination Reactions

The earliest example of the reductive elimination of an alkane from a gold(III) compound concerned (239) the tetramers $[AuR_2(CN)]_4$ (R = Et or n-Pr). On heating these complexes, stepwise loss of butane or hexane, respectively, occurs to yield $[AuR_2(CN)Au(CN)]_2$ and finally $[Au(CN)]$. It was at first thought (261) that reductive elimination occurs from $[n\text{-}Pr_2BrAu(\mu\text{-}en)AuBr\text{-}n\text{-}Pr_2]$ to give a gold(I)/gold(III) dimer, but the reformulation (263) of the starting complex as $[AuR_2(en)][AuX_2R_2]$ allows the reaction to be interpreted as producing the AuX_2^- anion. Ethane is produced when $[Au_2(\mu\text{-}Br)_2Me_4]$ is refluxed in diethyl ether (246) or when trimethylgold(III) is allowed to warm above $-40°C$ (227). Attempted isolation of tris(pentafluorophenyl)gold(III) produced the biaryl and gold metal (37).

The mechanism of reductive elimination of alkanes from trialkyl(tertiary phosphine)gold(III) complexes has been thoroughly investigated by Kochi and co-workers (231). It was found that decomposition of cis-$[AuMe_2R(PPh_3)]$ (R = Et, n-Pr, or i-Pr) produced ethane and RCH_3, whereas the trans isomer produced RCH_3 only. These findings were consistent with reductive coupling of two cis ligands. The decomposition was retarded by excess triphenylphosphine, suggesting that phosphine dissociation was involved since build-up of an unreactive $[AuMe_2R(PPh_3)_2]$ species was not found. The role of trialkylgold(III) species as reactive intermediates has been treated theoretically (313), and reductive elimination from a trigonal intermediate that maintained its stereochemical integrity was proposed [Eq. (85)].

$$AuMe_2R \begin{array}{c} \longrightarrow R—Me + Au—Me \\ \longrightarrow Me—Me + Au—R \end{array} \qquad (85)$$

Symmetry requirements for concerted reductive elimination of dialkyls have been considered (314), and for trialkylgold(III) species reductive elimination from a trigonal, three-coordinate intermediate was found to be symmetry forbidden. Solvent participation or the involvement of T-shaped species, however, was suggested as possible. Charge transfer to the high-oxidation state gold(III) center and reductive elimination from such a charge transfer state was proposed as an alternative reaction pathway.

In the presence of free triphenylphosphine, only propane was obtained from the decomposition of $[AuMe_2Et(PPh_3)]$, and it was suggested (231) that isomerization to the trans isomer was faster than reductive elimination under these conditions. Indeed, isomerization and reductive elimination are inextricably linked for these species (231, 313, 315), and the former is discussed more fully in the next section.

Isotopic labeling in the reaction of [AuMe$_3$(PR$_3$)] with HX (X = ClO$_4$, SO$_3$CF$_3$, NO$_3$, Cl, Br, I, CO$_2$CF$_3$, or CO$_2$Me) indicated that only the alkyl group cis to PR$_3$ was removed (316), a result consistent with the high trans influence of alkyl groups. The elimination of methane was suggested to occur by an electrophilic mechanism involving proton transfer to carbon with concomitant Au—C cleavage, as an oxidative addition–reductive elimination sequence is unlikely for square planar gold(III) species.

Reductive elimination of alkanes has been observed from a variety of complexes. The thermal decomposition of [AuMe$_3$(PPh$_3$)] produces (171) ethane and [AuMe(PPh$_3$)], and the same process has been postulated (194) (Section III,B) for the reaction of [AuMe(PPh$_3$)] with methyl iodide. Reductive elimination of ethane occurs (281) from [AuMe$_2$L$_2$]$^+$ (L = PMe$_3$, PMe$_2$Ph, or PMePh$_2$) cations; in the more general case of the [AuR$_2$L$_2$]$^+$ cations, the ease of reductive elimination depends (282) on the alkyl group (Bu ∼ Et > Me), the nature of the donor atom (SbPh$_3$ > AsPh$_3$ > PPh$_3$), and the bulk of the neutral ligand (PPh$_3$ > PMePh$_2$ > PMe$_2$Ph > PMe$_3$). The last series, however, also reflects the nucleophilicities of the ligands and may simply parallel the findings for the variation of the donor atom.

The dimethylbis(ylide)gold(III) complex [AuMe$_2$(CH$_2$PMe$_3$)$_2$]$^+$Br$^-$ on heating undergoes reductive elimination (221) of ethane to produce the bis(ylide)gold(I) cation. Binuclear ylide complexes also eliminate ethane on heating (48, 56) to yield the analogous gold(I) species [Eq. (86)].

$$\text{(86)}$$

Finally, hydrazine reduction of [AuCl(C$_6$F$_5$)$_2$(PPh$_3$)] yields (285) [Au(C$_6$F$_5$)(PPh$_3$)], and it was found in general that elimination of chloride occurs more readily upon reduction than loss of the pentafluorophenyl group does.

B. Isomerization Reactions

Two types of isomerization reaction have been identified for trialkyl(triphenylphosphine)gold(III) complexes, namely, isomerization at one (or more) of the alkyl groups and isomerization at the square planar gold center.

The complex trans-[AuMe$_2$-t-Bu(PPh$_3$)], prepared by addition of methyl iodide to the mixed dialkylaurate(I), was found (230) to convert spontaneously to the corresponding isobutyl complex in solution. First-order kinetic behavior was observed, and the rate was diminished by the addition of free triphenylphosphine, and so a dissociative mechanism was proposed (Scheme 8). The isopropyl analog did not react similarly at room temperature, and heating caused reductive elimination.

The complexes cis- and trans-[AuMe$_2$R(PPh$_3$)] (R = Et, n-Pr, or i-Pr) were separately prepared (Section V,A), and their reductive elimination reactions in the presence of added PPh$_3$ indicated (231) that isomerization to the trans isomer occurred prior to reductive elimination. Some isobutane was detected in the reductive elimination from cis- and trans-[AuMe$_2$-n-Pr(PPh$_3$)], suggesting that some n-propyl to isopropyl group conversion occurred in a fashion analogous to that shown in Scheme 8.

Although the rates of alkyl isomerization and reductive elimination were adversely affected by addition of free PPh$_3$, the rate of cis–trans rearrangement of [AuMe$_2$Et(PPh$_3$)] was unchanged, and it was suggested at one stage (315) that a unimolecular process involving spontaneous inversion at the gold nucleus was operative. It was later proposed (313) that the coordinatively unsaturated trialkylgold(III) species was a common intermediate in the isomerization and reductive elimination processes. Based on kinetic results, it was calculated that isomerization between these T-shaped configurations occurred 100 times faster than reductive elimination. Molecular orbital calculations indicated that T-shaped species were of lowest energy, whereas Y-shaped moieties represented energy troughs in the interconversion of the T-shaped species. The mechanism depicted in Scheme 9 was proposed for the isomerization process.

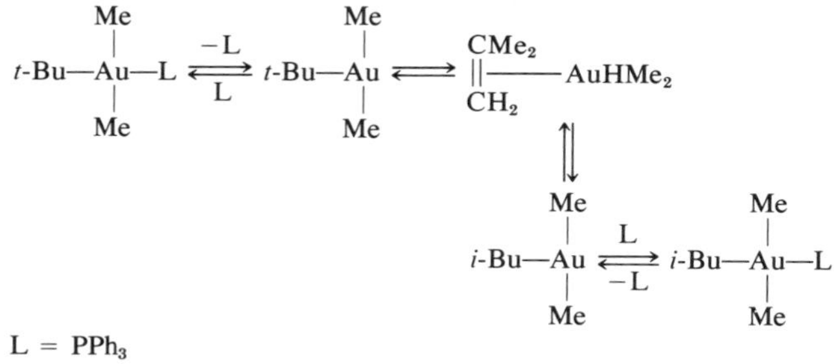

SCHEME 8

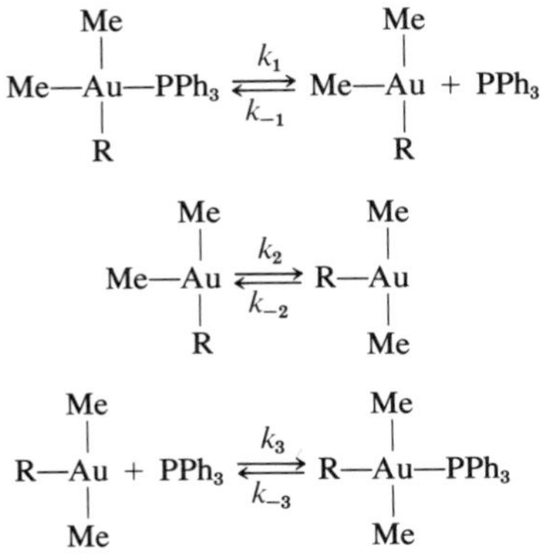

$$\begin{array}{ccc}
\underset{\underset{R}{|}}{\overset{\overset{Me}{|}}{Me-Au-PPh_3}} & \underset{k_{-1}}{\overset{k_1}{\rightleftharpoons}} & \underset{\underset{R}{|}}{\overset{\overset{Me}{|}}{Me-Au}} + PPh_3
\end{array}$$

$$\begin{array}{ccc}
\underset{\underset{R}{|}}{\overset{\overset{Me}{|}}{Me-Au}} & \underset{k_{-2}}{\overset{k_2}{\rightleftharpoons}} & \underset{\underset{Me}{|}}{\overset{\overset{Me}{|}}{R-Au}}
\end{array}$$

$$\begin{array}{ccc}
\underset{\underset{Me}{|}}{\overset{\overset{Me}{|}}{R-Au}} + PPh_3 & \underset{k_{-3}}{\overset{k_3}{\rightleftharpoons}} & \underset{\underset{Me}{|}}{\overset{\overset{Me}{|}}{R-Au-PPh_3}}
\end{array}$$

SCHEME 9

C. Insertion Reactions of Unsaturated Molecules

Organogold(III) complexes, unlike their lower valence counterparts, do not react with olefins or nonterminal acetylenes (197, 198), and with terminal acetylenes, such as $CF_3C\equiv CH$, cleavage of a gold–carbon bond [Eq. (87)], rather than insertion takes place. Insertion of sulfur dioxide does occur, however, to give the complex cis-[AuMe$_2$(SO$_2$Me)(PPh$_3$)]. Although it has been suggested (198) that reaction of [AuMe(PPh$_3$)] with SO$_2$ proceeds via a sulfur dioxide adduct, a direct attack of SO$_2$ on the methyl group was proposed for the gold(III) case.

$$[AuMe_3(PPh_3)] + CF_3C\equiv CH \rightarrow cis\text{-}[AuMe_2(C\equiv CCF_3)(PPh_3)] + CH_4 \qquad (87)$$

With cis-[AuMe$_2$Ph(PMe$_3$)], where the possibility exists for insertion of sulfur dioxide into the gold–methyl or gold–phenyl bond, quantitative reaction occurs (317) to give [AuMePh(SO$_2$Me)(PMe$_3$)] [Eq. (88).]. Similar competition reactions achieve the same result with cis-[PtMePh(PMePh$_2$)$_2$], but these observations are in contrast to those for certain main group metal compounds (318). With cis-[Au(CH$_3$)$_2$(CD$_3$)-(PMe$_3$)] a 50 : 50 mixture of products was obtained [Eq. (89)] (317). In each case, the high trans influence of the organic group is doubtless instrumental in directing insertion toward the gold–carbon bond cis to the tertiary phosphine.

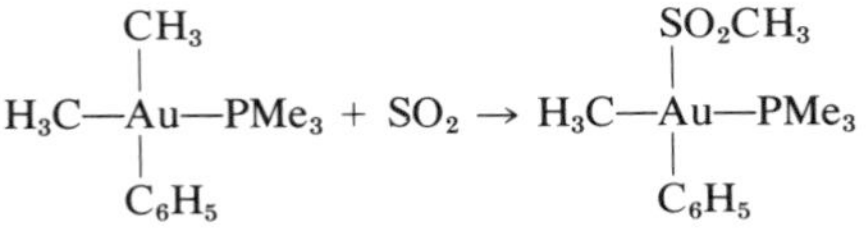

$$\underset{\underset{C_6H_5}{|}}{\overset{\overset{CH_3}{|}}{H_3C-Au-PMe_3}} + SO_2 \rightarrow \underset{\underset{C_6H_5}{|}}{\overset{\overset{SO_2CH_3}{|}}{H_3C-Au-PMe_3}} \qquad (88)$$

$$2 \; H_3C-\overset{\underset{\displaystyle |}{CD_3}}{\underset{\underset{\displaystyle |}{CH_3}}{Au}}-PMe_3 + 2 \; SO_2 \rightarrow H_3C-\overset{\underset{\displaystyle |}{SO_2CD_3}}{\underset{\underset{\displaystyle |}{CH_3}}{Au}}-PMe_3 + H_3C-\overset{\underset{\displaystyle |}{CD_3}}{\underset{\underset{\displaystyle |}{SO_2CH_3}}{Au}}-PMe_3 \quad (89)$$

D. Ligand-Substitution Reactions

It has already been mentioned (Section IV,C) that [AuClMe$_2$(py)] (267), and complexes of the type (281) [AuMe$_2$L$_2$]$^+$X$^-$ undergo ligand exchange in solution at ambient temperature. Tertiary phosphine exchange in [AuMe$_3$(PR$_3$)] complexes was found to occur (17); indeed, this point has been alluded to previously in discussing the mechanisms of isomerization and reductive elimination (Section V,A and B).

Ligand-substitution reactions have been employed to prepare a number of organogold(III) complexes. Thus displacement of the poorly nucleophilic tetrahydrothiophene ligand allowed (179) the complexes [AuX$_2$-(C$_6$F$_5$)L] (X = Cl, Br, or I; L = PPh$_3$, AsPh$_3$, CNC$_6$H$_4$Me-p, or phenanthroline) and [{AuX$_2$(C$_6$F$_5$)}$_2$(μ-dpe)] to be isolated. Similarly, substitution of the weakly bound ligand in [AuCl(C$_{12}$H$_8$)(tht)] by phenanthroline or bipyridine gave (293) the ionic complexes, [Au(C$_{12}$H$_8$)-(L$\frown$L)]$^+$[AuCl$_2$(C$_{12}$H$_8$)]$^-$.

E. Reactions Involving Cleavage of the Gold–Carbon Bond(s)

Treatment of dimeric dialkylgold(III) bromides with bromine causes cleavage of the gold–carbon bonds at one gold center to give [R$_2$Au(μ-Br)$_2$AuBr$_2$] (235, 237, 238). Protic acids (226) convert trialkylgold(III) species to the corresponding [Au$_2$(μ-X)$_2$R$_4$] complexes and cleave the gold–carbon bonds (319) in the tetrakis(tetrazolato)gold(III) complexes Ph$_4$As[Au(CN$_4$R)$_4$].

A bimolecular electrophilic substitution reaction has been proposed (211) for the interaction of [AuMe$_3$(PPh$_3$)] with mercuric bromide, which results in cleavage of one of the gold–methyl bonds [Eq. (90)]. This reaction was shown (206) to give the cis product and an oxidative addition–reductive elimination sequence was proposed as the reaction mechanism, although oxidative addition to a gold(III) center must be considered unlikely. Indeed, a competitive reaction of cis-[AuMe$_2$(C$_6$H$_4$Me-p)(PPh$_3$)] with HgCl$_2$ and a variety of other electrophiles produced (320) cis-[AuClMe$_2$(PPh$_3$)] only, suggesting that an S$_E$2 mechanism is operative.

An oxidative addition–reductive elimination sequence is expected to result in gold–methyl bond cleavage, so the latter was dismissed as a possible mechanism.

$$[AuMe_3(PPh_3)] + HgBr_2 \rightarrow [AuBrMe_2(PPh_3)] + MeHgBr \qquad (90)$$

The reactions of organogold(I) complexes with benzenethiol and -selenol were shown to proceed by a radical chain mechanism (Section III,E). Radical scavengers had no effect on the slower reactions of $[AuMe_3L]$ (L = PMe_3, PMe_2Ph, $PMePh_2$, or PPh_3), which proceeded by a nonradical mechanism to yield (207–209) cis-$[AuMe_2(EPh)L]$ (E = S or Se). On standing in CH_2Cl_2 solution for several weeks the complex cis-$[AuMe_2(SPh)(PMePh_2)]$ underwent further reaction to give (208) $[Me_2Au-(\mu\text{-}SPh)_2AuMe_2]$ and Ph_2MePS, whereas cis-$[AuMe_2(SePh)(PMe_2Ph)]$ after several days had produced (209) a detectable amount of $[Me_2Au(\mu\text{-}SePh)_2AuMe_2]$. With diphenylphosphine, trimethyl(dimethylphenylphosphine)gold(III) produced the dimeric complex $[Au_2(\mu\text{-}PPh_2)_2Me_4]$ [Eq. (91)], but no reaction took place with diphenylarsine.

$$2\,[AuMe_3(PMe_2Ph)] + 2\,Ph_2PH \rightarrow [Au_2(\mu\text{-}PPh_2)_2Me_4] + 2\,CH_4 + 2\,PMe_2Ph \qquad (91)$$

Finally, $[AuMe_3(PMe_2Ph)]$ reacted with NO_2 to give (321) cis-$[AuMe_2-(NO_2)(PMe_2Ph)]$, although the fate of the displaced alkyl group was uncertain, with methane and nitromethane both being detected. Reaction with NO also occurred, but it was much slower.

VI

COMPARISONS OF BONDING, SPECTRA, AND REACTIVITY OF ORGANOGOLD(I) AND ORGANOGOLD(III) COMPLEXES

The nature of the bonding in organogold compounds has been investigated by X-ray diffraction, as well as by a variety of spectroscopic techniques. Mössbauer and ESCA studies have proved particularly informative in determining the oxidation state of the gold center, and vibrational and NMR spectra have been utilized in assigning molecular geometries.

A number of organogold complexes have been subjected to X-ray crystallographic studies, and sufficient data have been accumulated for meaningful comparisons of bond lengths to be made. A series of complexes of the type $[AuR(PPh_3)]$ have been studied, and the bond length data for

TABLE I

BOND LENGTHS IN COMPLEXES OF THE TYPE [AuR(PPh₃)]

Complex	l(Au—C) (Å)	l(Au—P) (Å)	Reference
[AuMe(PPh₃)]	2.124(28)	2.279(8)	*322*
[Au(C₆F₅)(PPh₃)]	2.07(2)	2.27(1)	*323*
[Au{C₆H₃(OMe)₂-2,6}(PPh₃)]	2.050(4)	2.284(1)	*324*
[Au(CN)(PPh₃)]	1.85(4)	2.27(1)	*325*
[AuCl(PPh₃)]		2.243(3)	*326*
[(Ph₃PAu)₂C₄F₆]	2.05(6)	2.28(1)	*72*

these linear species are given in Table I. It can be seen that the length of the gold–carbon bond is dependent on the nature of the organic ligand. Thus the bonds in the arylgold(I) complexes are shorter by about 0.05 Å than that in [AuMe(PPh₃)], indicating a small but significant degree of π-overlap in the former. A more pronounced effect is observed in [Au(CN)(PPh₃)]. The gold–phosphorus bond lengths are very similar, and slightly greater than in [AuCl(PPh₃)].

Although studies of trans influence involving spectroscopic and crystallographic techniques have been extensive for square planar platinum(II) complexes (*327–329*), such investigations have so far been very limited for the isoelectronic gold(III) systems, and for the linear gold(I) compounds. Indeed, it might be expected that the latter would provide the best measure of trans influence, since complications due to cis ligands would not arise. For the complexes in Table I, the gold–phosphorus bond in [AuCl(PPh₃)] is significantly shorter than when PPh₃ lies trans to an organic group, a finding in keeping with the high trans influence of organic ligands versus chloride anion found for square planar complexes. As stated above, the Au—P distances are similar for all the organogold(I) species, and this bond length does not exhibit sufficient sensitivity to the nature of the trans ligand to allow a trans influence series for the respective organic ligands to be compiled.

The complex [Au(C≡CPh)(NH₂-*i*-Pr)] had a relatively short gold–carbon bond (1.935 Å), (*64*) whereas the three gold–carbon bonds in the trimeric compound [Au{C(OEt)=NC₆H₄Me-*p*}]₃ were of magnitudes (*151*) 1.935(28), 1.953(25), and 1.975(26) Å. In each case the bonds were significantly shorter than in the alkylgold(I) complex [AuMe(PPh₃)], indicating the existence of some double bond character. In [Au(CN)(CNMe)] (*136*) the cyano–gold bond was of length 2.01 Å, whereas a value of 1.98 Å was found for the gold–isonitrile interaction, suggesting that greater π-overlap occurs with the isonitrile ligand.

In the 3-center 2-electron cation [C₅H₅FeC₅H₄(AuPPh₃)₂]⁺ the two gold–carbon distances (104) are 2.13(4) and 2.27(4) Å, indicating that the two gold centers are unsymmetrically bonded to the cyclopentadienyl ring. The Au—C bond lengths are slightly greater than in "simple" alkyl-gold(I) species. The gold–gold distance of 2.768(3) Å was considered to be sufficiently short to imply some interaction between the metal centers. This contrasts with the complexes $[(Ph_3PAu)_2C_4F_6]$ (72) and $[Au\{C(OEt)=NC_6H_4Me-p\}]_3$ (151) in which all the gold–gold separations are greater than 3.2 Å. In $[Au\{C_6H_3(OMe)_2-2,6\}(PPh_3)]$ a C—Au—P angle of 172.7° was found (324), and gold–oxygen distances of 3.231(4) and 2.961(4) Å suggest that the deviation from linearity is due to a weak gold–oxygen interaction.

Several crystallographic studies of organogold(III) complexes have been made, and a compilation of data is made in Table II. The gold–carbon bond in cis-$[AuCl_2Ph(S-n-Pr_2)]$ is only fractionally shorter than in the alkylgold(III) complexes, whereas in the bis(pentafluorophenyl)gold(III) compound the Au—C bonds are significantly longer. The two C_6F_5 rings make angles of 76 and 77° with the plane of the molecule, and some steric

TABLE II

BOND LENGTHS IN ORGANOGOLD(III) COMPLEXES

Complex	1(Au—C) (Å)	Reference
cis-$[AuCl_2Ph(S-n-Pr_2)]$	2.00	302
cis-$[AuCl(C_6F_5)_2(PPh_3)]$	2.18 (cis to Cl)	287
	2.12 (trans to Cl)	
cis-$[AuMe_2(SO_3CF_3)(OH_2)]$	2.017 (trans to H_2O)	254
	2.019 (cis to H_2O)	
$[AuMe_2(OH)]_4$	2.05 ± 0.08	259
$trans$-$[AuI_2\{C(NHC_6H_4Me-p)_2\}_2]^+ClO_4^-$	2.07(2), 2.09(2)	185
	2.10	306
	2.13	306

crowding may result in the lengthening of the Au—C bonds. In the cationic bis(carbene)gold(III) species $trans$-$[AuI_2\{C(NHC_6H_4Me)_2\}_2]^+$ the gold–carbon bond lengths are slightly greater than in the methylgold(III) complexes; this is likely to be due to the trans arrangement of the two high trans-influence carbene ligands.

In the mixed gold(I)–gold(III) complex $[LAuC_4F_6AuMe_2L]$ (L = PMe_3) (10) the C—Au(I) bond length (199) is 2.10 Å and the corresponding C—Au(III) bond is 0.07 Å shorter, whereas the mutually trans methyl groups give rise to relatively long gold–carbon bonds (2.11 and 2.14 Å). The gold–methyl bonds lie nearly perpendicular to the plane containing the gold and phosphorus atoms, and it was suggested (199) that this configuration should facilitate π-overlap between the gold(III) and vinyl carbon atoms, thus accounting for the shorter C—Au(III) distance. The gold–gold separation is 3.31 Å, indicating that no such interaction is involved.

Bond lengths for several ylide complexes are given in Table III. The gold(I) complexes have Au—C distances slightly shorter than in $[AuMe(PPh_3)]$, which may indicate some charge delocalization and π-overlap within the ring systems. The gold–gold distances indicate no significant interaction, whereas for the formal gold(III) compound $[Au_2\{\mu$-$(CH_2)_2PEt_2\}_2Cl_2]$ a short Au—Au separation (2.597 Å) indicates the existence of some metal–metal bonding. In this complex fairly short gold–carbon bond lengths (1.942 Å) are also found.

The mononuclear dimethylgold(III) ylide complexes exhibit longer Au—CH_2 bonds, due to the high trans influence of the methyl group, whereas for the methyl groups the gold–carbon distances are slightly shorter than in $[AuMe(PPh_3)]$.

The use of Mössbauer and ESCA techniques has been of great value in the study of ylide complexes. Thus in the complexes $[Au_2\{\mu$-$(CH_2)_2PEt_2\}_2Br_n]$ (n = 0, 2, or 4), ESCA studies gave (189, 190) Au($4f_{7/2}$) binding energies of 84.2, 85.9, and 86.7 eV, respectively, which allowed identification of gold in the uni-, di-, and trivalent states. A formal oxidation state, namely gold(II), was therefore able to be assigned to the metal centers in the complex $[Au_2\{\mu$-$(CH_2)_2PEt_2\}_2(SCSNEt_2)_2]$, which had a binding energy of 85.2 eV.

Similarly, ^{197}Au Mössbauer spectra, where the isomer shift decreases with increasing oxidation state, yielded (189) isomer shift values of 3.76, 2.67, and 2.02 mm sec^{-1}, respectively, for the complexes $[Au_2\{\mu$-$CH_2\}_2$-$PEt_2\}_2Br_n]$ (n = 0, 2, or 4). For each of the complexes $[LAuC(CF_3)$=$C(CF_3)AuMe_2L]$ (L = PMe_3 or PMe_2Ph) two doublets were observed (199) in the ^{197}Au Mössbauer spectrum (Table IV), and the magnitudes of the isomer shifts and quadrupole splittings were consistent with gold(I)

TABLE III
BOND LENGTHS IN GOLD YLIDE COMPLEXES

Complex	$l(Au—C)$ (Å)	Other distances (Å)	Reference
Et$_2$P⟨CH$_2$AuCH$_2$⟩$_2$PEt$_2$	2.09(3), 2.10(3)	Au--Au 3.023	57
(CH$_2$CH$_2$)$_2$P⟨CH$_2$AuCH$_2$⟩$_2$P(CH$_2$CH$_2$)$_2$		Au--Au 3.01–3.07	59
Cl—Et$_2$P⟨CH$_2$AuCH$_2$⟩$_2$PEt$_2$—Cl	1.942	Au--Au 2.597	188
Me$_2$Au(CH$_2$—PMe$_2$)$_2$CH	2.147, 2.113	Au-CH$_3$ 2.081, 2.098	225
Me$_2$Au(CH$_2$—PMe$_2$)$_2$N	2.154, 2.117	Au-CH$_3$ 2.104, 2.087	225

TABLE IV
MÖSSBAUER SPECTRA FOR THE COMPLEXES [Au$_2$Me$_2$L$_2$(C$_4$F$_6$)]

Complex	I.S. (mm sec^{-1})	Q.S. (mm sec^{-1})	Oxidation state
trans-[(PhMe$_2$P)Me$_2$AuC$_4$F$_6$Au(PMe$_2$Ph)]	5.11 ± 0.01	9.13 ± 0.04	III
	4.13 ± 0.04	9.44 ± 0.03	I
trans-[(Me$_3$P)Me$_2$AuC$_4$F$_6$Au(PMe$_3$)]	5.16 ± 0.06	9.27 ± 0.05	III
	4.37 ± 0.04	9.18 ± 0.02	I
cis-[(Me$_3$P)Me$_2$AuC$_4$F$_6$Au(PMe$_3$)]	5.28 ± 0.26	9.07 ± 0.13	III
	4.14 ± 0.26	9.17 ± 0.10	I

and gold(III) centers. Since a large amount of data has been accumulated from ^{197}Au Mössbauer spectra, approximate relationships between isomer shifts (I.S.) and quadrupole splittings (Q.S.) for gold(I) [Eq. (92)] and gold(III), [Eq. (93)] compounds have been observed (330). In this manner

$$Q.S. = I.S. + 5.6 \text{ mm sec}^{-1} \tag{92}$$

$$Q.S. = 2.0 \text{ I.S.} - 0.5 \text{ mm sec}^{-1} \tag{93}$$

the assignments in Table IV were made, and it was these values which first suggested the nature of this mixed gold(I)–gold(III) complex (199). For the complex [Au{C(CF$_3$)=C(CF$_3$)Me}(PMe$_2$Ph)], formed by reductive elimination from the mixed oxidation state species, only one doublet was observed (200) and the values were consistent with a gold(I) center. The compounds [AuMe(PMe$_3$)] and [AuMe$_3$(PMe$_2$Ph)] also produced I.S. and Q.S. values that fitted well in Eqs. (92) and (93), respectively.

The ^{197}Au Mössbauer spectra have been recorded for a series of complexes of the type [AuX(PPh$_3$)] (331), and the order of increasing isomer shift is I < Br < Cl < O$_2$CMe ~ N$_3$ < Me. This was found to parallel the spectrochemical series for these anionic ligands.

The photoelectron spectra of the complexes [AuMe(PMe$_3$)] and [AuMe$_3$L] (L = PMe$_3$, PMe$_2$Ph, PMePh$_2$, or PPh$_3$) have been recorded (332), and the vertical ionization potentials of the phosphorus lone pairs suggest that the donor abilities of the neutral ligands lie in the order PPh$_3$ ~ PMePh$_2$ > PMe$_2$Ph > PMe$_3$. This order is supported by the 5d and σ(MeAu) ionization potentials for [AuMe$_3$L] and by ultraviolet and NMR spectroscopic data. However, it is opposite to that deduced from the reactivities of [AuMeL] toward oxidative addition and the usually assumed order of σ-donor ability, which suggests that the phenomenon is of a steric nature. The lowest ionization energy of [AuMe(PMe$_3$)] is higher than for [AuMe$_3$(PMe$_3$)], possibly due to the unfavorable configuration of mutually trans methyl groups in the latter.

The operation of trans influence in linear gold(I) complexes has been investigated (135) using ^{35}Cl nuclear quadrupole resonance spectroscopy. It has already been noted (Section II,H) that for an extensive series of compounds of the type [AuClL], the ^{35}Cl NQR frequencies are sensitive to the nature of the trans ligand, and σ-bonding effects were concluded to be dominant. This behavior contrasts with the observations for square planar platinum(II) complexes, where π-bonding and cis influence were also important factors.

It has been shown (Section V,A and B) that phosphine dissociation is involved in the decomposition of organogold(III) complexes. Addition of

free PPh$_3$ slowed the thermal decomposition of [AuR$_3$(PPh$_3$)] complexes
(*231*), but greater retardation occurred for the analogous alkylgold(I)
species. Thermal decompositions of the AuMe$_2^-$ and AuMe$_4^-$ anions were
found (*10*) to be slower than for the mono- and trimethyl(triphenylphosphine)gold complexes, and molecular orbital calculations suggest that
fragmentation of tetramethylaurate(III), although it should be a symmetry
allowed process, is less favored than the analogous reaction of trimethyl-
gold(III) due to ligand–ligand interactions (*10*).

A considerable difference in reactivity toward gold–carbon bond cleavage is found for methylgold(I) and -gold(III) compounds. Thus, for the
reactions with benzenethiol, which proceed by elimination of methane,
the following order of reactivity is obtained (*207, 208*):

[AuMe(PMe$_3$)] > [AuMe(PMePh$_2$)] > [AuMe(PPh$_3$)] ≫ [AuMe$_3$(PMe$_3$)]
> [AuMe$_3$(PMePh$_2$)] ≫ *cis*-[AuMe$_2$(SPh)(PMe$_3$)]

The gold(I) complexes were shown to react via a radical chain mechanism, and it was suggested that such a process would not occur for
gold(III) because initial addition of a SPh · radical would lead to an unstable Au(IV) intermediate. A similar reactivity series was found for the
reactions with phenylselenol (*209*). The reactions of [AuMe(PMe$_2$Ph)]
with diphenylphosphine and diphenylarsine are also faster than those of
the trimethylgold(III) analog, but it was suggested that initial phosphine
displacement might be involved and that the order of reactivity simply
reflects the tendency to undergo ligand-substitution reactions. The products of the reactions of [AuMe(PMe$_2$Ph)] with NO or NO$_2$ are unstable
(*321*), but again their formation is more rapid than with the analogous
trimethylgold(III) complex.

The greater reactivity of organogold(I) complexes is also evident in
their reactions with unsaturated molecules. Thus methyl(triphenylphosphine)gold(I) readily reacts with fluorinated olefins and acetylenes (*197,
198, 200*), whereas no reaction occurs with [AuMe$_3$(PMe$_2$Ph)]. It was suggested that reaction proceeds via a π-complex, and the filled d orbitals of
the gold(III) compound are too stable to interact with olefins or acetylenes
in this manner. On the other hand, the dimethylaurate(I) anion did not
undergo any reaction with olefins either (*8*).

Both methylgold(I) and trimethylgold(III) complexes react with terminal acetylenes to yield methane and gold acetylide complexes (*198*). Reactions with sulfur dioxide proceed for each type of complex (*198, 317*), but
the reactions of the gold(I) compounds are again faster. The products of
the reactions with SO$_2$ are the result of insertion, and it was suggested that
the greater reactivity of the gold(I) species was due to their ability to form

an initial SO_2 adduct, whereas direct attack at the methyl group might occur for [AuMe₃(PMe₂Ph)].

VII

COMPARISONS OF ORGANOGOLD COMPLEXES WITH THOSE OF OTHER METALS

A number of comparative studies have been made of organogold compounds and related complexes of neighboring metals, notably platinum and mercury as well as the other group IB elements. These comparisons have included such aspects of their chemistry as thermal stability and reactivity, bonding modes, and spectroscopic data.

An early study concerned the ease of isolation of fluorophenyl complexes (34) of gold(I) and platinum(II). The complexes cis-[PtR₂L₂] (R = p-FC₆H₄ or m-FC₆H₄; L = PR′₃, AsR′₃, or SbR′₃) were readily prepared, whereas [Au(C₆H₄F-m)(PPh₃)] was less stable and the analogous p-fluorophenyl complex was not isolable. The last compound had been successfully synthesized previously, however (33).

Dimethylgold(III) complexes of the types cis-[AuMe₂XL] (X = Cl or SCN; L = py, PPh₃, AsPh₃, or SbPh₃) and cis-[AuMe₂L₂]⁺ClO₄⁻ (L = PPh₃, AsPh₃, or SbPh₃) were found (280) to undergo rapid neutral ligand exchange at room temperature; such ligand dissociation occurred more readily than for isoelectronic platinum(II) complexes. The behavior of dimethylgold(III) species has been likened (267, 280) to that of allylpalladium(II) complexes.

The interactions of methylmercuric iodide, [Au₂(μ-I)₂Me₄], and [PtIMe₃]₄ with ammonia and pyridine have been investigated (267). The metal–iodide stretching frequencies in the complexes formed in pyridine solution decreased in the order MeHgI > Me₂AuI(py) > PtMe₃I(py)₂, suggesting that the degree of covalency decreased in this order. The [AuMe₂(NH₃)₂]⁺ cation was of considerably lower stability than the mercury and platinum species formed in liquid ammonia. The dimethylgold(III) moiety (278) was similar to methylmercury(II) (333), however, in that it was stabilized by a variety of donor ligands, whereas the dimethylplatinum(II) moiety was more satisfactorily stabilized by heavier donor atoms.

Interactions of [AuMe₂(ClO₄)] and [PtMe₃(ClO₄)] with water were studied by ¹⁷O-NMR spectroscopy (275), and exchange of H₂O molecules was found to be more rapid for the [AuMe₂(OH₂)₂]⁺ cation than for

$[PtMe_3(OH_2)_3]^+$. Two water resonances were observed up to 85°C for the $[Pt(NH_3)_2(OH_2)_2]^{2+}$ cation, indicating that exchange was extremely slow.

The trans-influence series $Me_2S > Br > Cl > en > OH^- > H_2O$ was obtained (276) from the vibrational spectra of complexes of the type $[AuMe_2X_2]^{n+}$, and this order paralleled that found for platinum(II). In contrast to the findings for platinum(II), however, the role of π-bonding was suggested to be minimal.

The Raman spectrum of the dimethylaurate(I) anion indicated a linear structure (219), and the bands were found to correlate with those of Me_2Hg, Me_2Tl^+, and Me_2Pb^{2+}. The metal–carbon bond strength decreased in the order $Au > Hg > Tl > Pb$, opposite the expected trend in terms of nuclear charge.

Electrophilic attack leading to transition metal–alkyl bond cleavage occurs via interaction with the highest occupied molecular orbital. Photoelectron spectroscopic studies of [AuMeL], [AuMe$_3$L], and cis-[PtMe$_2$L$_2$] (L = tertiary phosphine) complexes (332) have shown that the HOMO in the platinum(II) compounds is a 5d orbital, so cleavage should proceed by an oxidative addition–reductive elimination sequence. The σ(AuMe) orbital is the HOMO for the gold complexes, so an S_E2 mechanism should be operative. In mixed alkyl–aryl complexes of gold(III) and platinum(II), aryl–gold bond cleavage occurred (320) by an S_E2 mechanism, whereas an oxidative addition–reductive elimination sequence took place to cause cleavage of the alkyl–platinum bond.

Methyl for halogen exchange reactions occurred readily with palladium(II), platinum(II), gold(I), and gold(III) compounds (213). An S_E2 mechanism was proposed, and the reactivities of selected complexes toward methyl–platinum bond cleavage in cis-[PtMe$_2$(PMe$_2$Ph)$_2$] were found to lie in the following order:

$$[AuCl(PMe_2Ph)] > \textit{cis-}[PdCl_2(PMe_2Ph)_2] \gg \textit{cis-}[PtCl_2(PMe_2Ph)_2]$$

The methylating power of various metal complexes followed the order:

$$\textit{cis-}[PtMe_2(PMe_2Ph)_2], [AuMe(PMe_2Ph)] > [AuMe_3(PMe_2Ph)]$$
$$\gg \textit{trans-}[PdClMe(PMe_2Ph)_2], \textit{trans-}[PtClMe(PMe_2Ph)_2], \textit{cis-}[AuClMe_2(PMe_2Ph)]$$

Methyl–metal bond cleavage also occurred in the reactions of gold(I), gold(III), and platinum(II) compounds with thiols and selenols (207–209). The reactions of gold(I) and platinum(II) complexes were much faster than those of gold(III), due to the operation of a radical chain mechanism in the former cases. For the reactions with diphenylphosphine, however, the following order of reactivity was found (209):

$$[AuMe(PMe_2Ph)] \gg [AuMe_3(PMe_2Ph)] > cis\text{-}[PtMe_2(PMe_2Ph)_2]$$

A different mechanism from the above was suspected, and the order found paralleled the reactivities toward ligand-substitution reactions, so initial displacement of dimethylphenylphosphine was proposed.

Transfer of a methyl group from methyl vitamin B_{12} to Hg(II), Tl(III), Pt(II), and Au(I) was found to occur (334). An acid–base reaction path was proposed for mercury and thallium, whereas an oxidation–reduction sequence involving Pt(IV) and Au(III) intermediates was suggested.

Competitive reactions with mixtures of the nucleotides GMP, CMP, AMP, and UMP provided information regarding the reactivities and selectivities of several transition metal moieties (335). Of the groups studied, the most extensive reaction occurred with $trans\text{-}(NH_3)_2Pd(II)$, whereas the $cis\text{-}Me_2Au(III)$ moiety exhibited the lowest reactivity. MeHg(II) showed high selectivity for attack at N—H bonds, whereas the $cis\text{-}(NH_3)_2Pt(II)$ group showed a total preference for the purines, GMP and AMP. For the gold(III) and palladium(II) species intermediate behavior in terms of selectivity was observed.

The types of compounds formed by gold(I) and gold(III) often differ from those of other metals due to the constraints imposed by coordination number and electron count at the metal. Thus, for example, whereas π-bonded cyclopentadienyl complexes of palladium and platinum are numerous (336), and a copper(I) species of this type is known (337), cyclopentadienyl complexes of univalent (94, 96, 97) and trivalent (228) gold have invariably been found to be σ-bonded. ^{1}H-NMR studies have shown that fluxional behavior, similar to that in dicyclopentadienylmercury, was involved (228).

Copper(I), gold(I), and platinum(II) were found to form complexes with cyclooctyne (88), whereas with n-4-octyne only platinum species proved to be isolable. With cis,trans-1,3-cyclooctadiene copper(I) formed dimeric compounds (309), the latter involving a bridging diolefin moiety, whereas mononuclear complexes were formed with gold(III) and silver(I). The gold complex, [AuCl₃(1,3-cod)], was thermally unstable, and it was suggested that the diolefin acted as a monodentate ligand, whereas in [Ag(NO₃)(1,3-cod)] both olefinic groups were coordinated to the metal. Bridging diolefin moieties were proposed (84) for the polymeric complexes [Au₂X₂(1,5-cod)]ₙ (X = Cl or Br), a bonding mode not found for binuclear 1,5-cyclooctadiene complexes of such metals as rhodium and platinum.

In the mode of bonding of bifunctional ylides, gold was again found to differ from other metals. The complexes [M{CH(PPh₂)₂}₂] (M = Ni, Pd, or Pt) reacted (62, 338) with Et₂MeP=CH₂ to yield mononuclear products

[Eq. (94)], whereas the analogous polymeric gold complex $[Au\{CH(PPh_2)_2\}]_n$, which had been prepared similarly, gave a binuclear species [Eq. (95)] identified by its NMR, infrared, and mass spectra (62). With t-$Bu_2MeP{=}CH_2$, the chloride-bridged iridium(I) complex $[Ir_2(\mu\text{-}Cl)_2(cod)_2]$ yielded $[Ir\{(CH_2)_2P\text{-}t\text{-}Bu_2\}(cod)]$ (339), whereas the bisylide complex $[PtMe_2(CH_2PMe\text{-}t\text{-}Bu_2)_2]$ was obtained by reaction of $[PtMe_2\text{-}(cod)]$. Thus, although anomalous behavior of other metals does occur, the strong preference of gold(I) for digonal geometry often produces unusual compounds.

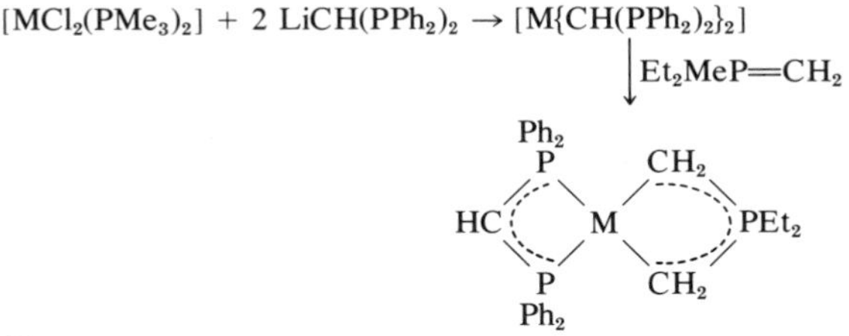

$$[MCl_2(PMe_3)_2] + 2\ LiCH(PPh_2)_2 \rightarrow [M\{CH(PPh_2)_2\}_2]$$

(94)

M = Ni, Pd, or Pt

$$[Au\{CH(PPh_2)_2\}]_n + Et_2MeP{=}CH_2 \rightarrow$$

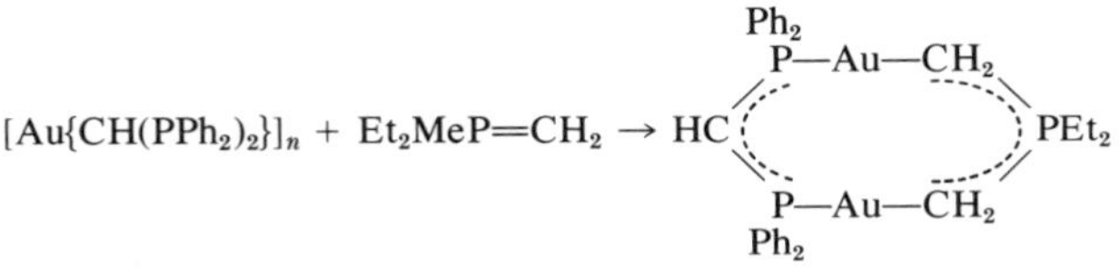

(95)

With the ligand bis(trimethylphosphoranylidene)methane $(Me_3P{=}C{=}PMe_3)$ reaction (61) of $[AuMe(PMe_3)]$ yielded $[(MeAu)_2C(PMe_3)_2]$ (21), whereas elimination of methane occurred with trimethylgallium to give $[GaMe_2\{CH(PMe_2CH_2)_2\}]$ (22) and dialkylzinc and -cadmium species yielded the corresponding alkane and the bis-ligand complexes (23).

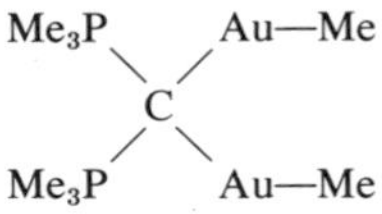

(21) 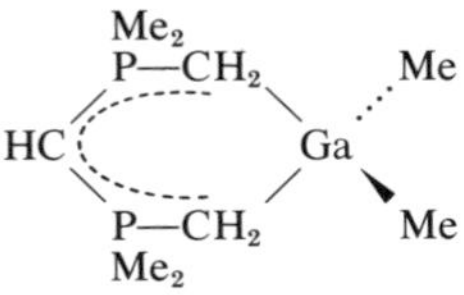   (22) 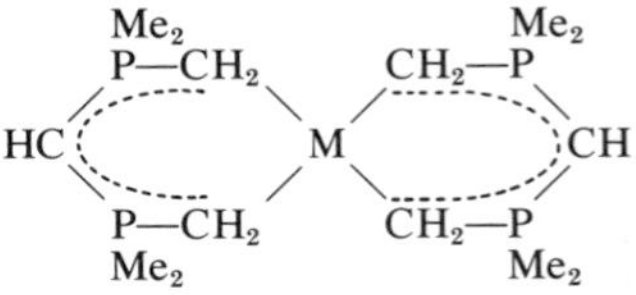   (23)   M = Zn or Cd

Lithium tetramethylaluminate reacted with $[Me_3PCHPMe_3]^+F^-$ to yield (223) the aluminium analog of 22, and the analogous dimethyl(ligand)gold-(III) complexes were obtained (223, 224) by the reactions of $[Au_2(\mu\text{-}Cl)_2\text{-}Me_4]$ with $Me_3P{=}C{=}PMe_3$ and $[Me_3P{\cdots}N{\cdots}PMe_3]^+Cl^-$ [Eqs. (73) and

(74)]. With both of these ligands, dialkylzinc and -cadmium compounds gave tetrahedral bis-ligand complexes with structures analogous to 23.

The similarity between the dimethylgallium(III) and -gold(III) complexes noted above is not an isolated example. Despite the tetrahedral geometry adopted by dimethylgallium(III) complexes (340), it has been pointed out that in many ways the chemistries of the $Me_2Ga(III)$ and $Me_2Au(III)$ moieties exhibit a striking resemblance (278, 340, 341).

A final comparison relates to the mode of attachment of ambidentate ligands. The complexes [Au(acac)(PR_3)] (PR_3 = PPh_3, PEtPh_2, or PEt_3), prepared by treatment of the corresponding chloride with acetylacetonatethallium, contained C-bonded ligands (90); carbon-bonded β-diketonate complexes had previously been found for platinum(II) (342). Analogous copper(I) and silver(I) β-diketonates were found to be O-bonded (89), while [PtClH(acac)_2] was found to contain an O-bonded acetylacetonate ligand (91). It was noted (Section IV,B) that [AuMe_2-(acac)] contained an oxygen-bonded ligand (232), which would facilitate achievement of the usual four-coordination, but addition of tertiary phosphine readily produced the C-bonded complex [AuMe_2(acac)(PR_3)]. Thus gold(I), being the softest acid, forms only carbon-bonded β-diketonate complexes, whereas gold(III) and platinum(II) tend to form species of this type, although under certain circumstances they will adopt O-bonded structures.

The sulfinate complexes obtained by insertion of sulfur dioxide into the gold–methyl bonds (198) of [AuMe(PMe_3)] and [AuMe_3(PMe_2Ph)] contained S-bonded ligands. The complexes [M(SO_2Ph)(PPh_3)_2] (M = Cu or Ag), [Au(SO_2Ph)(PPh_3)], and [Au(SO_2R)(PCy_3)] (R = Me, Et, or Ph) were prepared (202), and the copper and silver complexes contained O-bonded sulfinate ligands, whereas in each of the gold compounds the ligand was S-bonded.

Thus it may be appreciated that a number of factors can affect the types of organometallic compounds formed by uni- and trivalent gold. Among the most important appear to be (1) the strong tendencies to form two-coordinate, linear, and four-coordinate, square planar complexes, respectively, and (2) the soft acid behavior of these metal centers, particularly gold(I). Although these have acted as restraints upon the number of complexes which have been isolated, an extensive organic chemistry of gold now exists.

REFERENCES

1. D. H. Brown and W. E. Smith, *Chem. Soc. Rev.* 9, 217 (1980).
2. A. Feldt, *Klin. Wochenschr.* 5, 299 (1926).

3. R. J. Puddephatt, "The Chemistry of Gold" (Topics in Inorganic and General Chemistry (R. J. H. Clark, ed.), Vol. 16). Elsevier, Amsterdam, 1978.
4. B. Armer and H. Schmidbaur, *Angew. Chem. Int. Ed. Engl.* **9**, 101 (1970).
5. H. Schmidbaur, *Angew. Chem. Int. Ed. Engl.* **15**, 728 (1976).
6. R. Uson, A. Laguna, and J. Vicente, *Synth. React. Inorg. Met. Org. Chem.* **7**, 463 (1977).
7. A. Tamaki and J. K. Kochi, *J. Organomet. Chem.* **51**, C39 (1973).
8. A. Tamaki and J. K. Kochi, *J. Chem. Soc. Dalton Trans.* p. 2620 (1973).
9. G. W. Rice and R. S. Tobias, *Inorg. Chem.* **15**, 489 (1976).
10. S. Komiya, T. A. Albright, R. Hoffmann, and J. K. Kochi, *J. Am. Chem. Soc.* **99**, 8440 (1977).
11. R. Uson, A. Laguna, and J. Vicente, *J. Chem. Soc. Chem. Commun.* p. 353 (1976).
12. R. Uson, A. Laguna, and J. Vicente, *J. Organomet. Chem.* **131**, 471 (1977).
13. R. Uson, A. Laguna, J. Garcia, and M. Laguna, *Inorg. Chim. Acta* **37**, 201 (1979).
14. R. G. Pearson and P. E. Fidgore, *J. Am. Chem. Soc.* **102**, 1541 (1980).
15. G. Calvin, G. E. Coates, and P. S. Dixon, *Chem. Ind. (London)* p. 1628 (1959).
16. H. Schmidbaur, A. Shiotani, and H. F. Klein, *J. Am. Chem. Soc.* **93**, 1555 (1971).
17. H. Schmidbaur and A. Shiotani, *Chem. Ber.* **104**, 2821 (1971).
18. H. Schmidbaur, A. Shiotani, and H. F. Klein, *Chem. Ber.* **104**, 2831 (1971).
19. H. Schmidbaur and R. Franke, *Chem. Ber.* **105**, 2985 (1972).
20. H. Schmidbaur and A. Shiotani, *J. Am. Chem. Soc.* **92**, 7003 (1970).
21. B. Wozniak, J. D. Ruddick, and G. Wilkinson, *J. Chem. Soc. A* p. 3116 (1971).
22. H. Schmidbaur, A. Wohlleben, F. Wagner, O. Orama, and G. Hüttner, *Chem. Ber.* **110**, 1748 (1977).
23. H. Schmidbaur and Y. Inoguchi, *Chem. Ber.* **113**, 1646 (1980).
24. L. G. Vaughan, *J. Am. Chem. Soc.* **92**, 730 (1970).
25. L. G. Vaughan, *J. Organomet. Chem.* **190**, C56 (1980).
26. M. Aresta and G. Vasapollo, *J. Organomet. Chem.* **50**, C51 (1973).
27. G. Van Koten and J. G. Noltes, *J. Organomet. Chem.* **80**, C56 (1974).
28. H. P. Abicht and K. Issleib, *J. Organomet. Chem.* **149**, 209 (1978).
29. R. Uson, A. Laguna, and P. Brun, *J. Organomet. Chem.* **182**, 449 (1979).
30. A. Tamaki and J. K. Kochi, *J. Organomet. Chem.* **61**, 441 (1973).
31. A. N. Nesmeyanov, E. G. Perevalova, O. B. Afanasova, and K. I. Grandberg, *Bull. Acad. Sci. USSR, Div. Chem. Sci. (Engl. Transl.)* **27**, 1689 (1978).
32. A. N. Nesmeyanov, E. G. Perevalova, V. V. Krivykh, A. N. Kosina, K. I. Grandberg, and E. I. Smyslova, *Bull. Acad. Sci. USSR, Div. Chem. Sci. (Engl. Transl.)* **21**, 618 (1972).
33. L. G. Vaughan and W. A. Sheppard, *J. Am. Chem. Soc.* **91**, 6151 (1969).
34. A. D. Westland and M. Northcott, *Can. J. Chem.* **48**, 2907 (1970).
35. D. I. Nichols, *J. Chem. Soc. A* p. 1216 (1970).
36. R. Uson and A. Laguna, *Synth. React. Inorg. Met. Org. Chem.* **5**, 17 (1975).
37. L. G. Vaughan and W. A. Sheppard, *J. Organomet. Chem.* **22**, 739 (1970).
38. R. Uson, P. Royo, and A. Laguna, *J. Organomet. Chem.* **69**, 361 (1974).
39. S. Numata, H. Kurosawa, and R. Okawara, *J. Organomet. Chem.* **102**, 259 (1975).
40. A. N. Nesmeyanov, E. G. Perevalova, D. A. Lemenovskii, V. P. Dyadchenko, and K. I. Grandberg, *Bull. Acad. Sci. USSR, Div. Chem. Sci. (Engl. Transl.)* **23**, 1587 (1974).
41. A. N. Nesmeyanov, E. G. Perevalova, V. P. Dyadchenko, and K. I. Grandberg, *Bull. Acad. Sci. USSR, Div. Chem. Sci. (Engl. Transl.)* **23**, 2779 (1974).
42. V. P. Dyadchenko, *Vestn. Mosk. Univ., Khim.* **17**, 358 (1976); *C.A.* **86**, 5567 (1977).
43. A. N. Nesmeyanov, K. I. Grandberg, E. I. Smyslova, and E. G. Perevalova, *Bull. Acad. Sci. USSR, Div. Chem. Sci. (Engl. Transl.)* **21**, 2324 (1972).

44. A. N. Nesmeyanov, E. G. Perevalova, E. I. Smyslova, V. P. Dyadchenko, and K. I. Grandberg, *Bull. Acad. Sci. USSR, Div. Chem. Sci. (Engl. Transl.)* **26**, 2417 (1977).
45. V. I. Rozenberg, R. I. Gorbacheva, E. I. Smyslova, K. I. Grandberg, V. A. Nikanorov, Y. G. Bundel, and O. A. Reutov, *Dokl. Akad. Nauk SSSR* **225**, 1082 (1975).
46. A. Johnson and R. J. Puddephatt, *J. Chem. Soc. Dalton Trans.* p. 1360 (1976).
47. J. D. Kennedy, W. McFarlane, and R. J. Puddephatt, *J. Chem. Soc. Dalton Trans.* p. 745 (1976).
48. H. Schmidbaur, *Acc. Chem. Res.* **8**, 62 (1975).
49. H. Schmidbaur and R. Franke, *Angew. Chem., Int. Ed. Engl.* **12**, 416 (1973).
50. Y. Yamamoto and Z. Kanda, *Bull. Chem. Soc. Jpn.* **52**, 2560 (1979).
51. P. A. Arnup and M. C. Baird, *Inorg. Nucl. Chem. Lett.* **5**, 65 (1969).
52. H. Schmidbaur and R. Franke, *Chem. Ber.* **108**, 1321 (1975).
53. H. Schmidbaur and K. H. Raethlein, *Inorg. Synth.* **18**, 140 (1978).
54. Y. Yamamoto, *Chem. Lett.* p. 311 (1980).
55. H. Schmidbaur, J. R. Mandl, and A. Wohlleben-Hammer, *Z. Naturforsch., Teil B* **33**, 1325 (1978).
56. H. Schmidbaur and R.Franke, *Inorg. Chim. Acta* **13**, 85 (1975).
57. H. Schmidbaur, J. R. Mandl, W. Richter, V. Bejenke, A. Frank, and G. Hüttner, *Chem. Ber.* **110**, 2236 (1977).
58. H. Schmidbaur and H. P. Scherm, *Chem. Ber.* **110**, 1576 (1977).
59. H. Schmidbaur, H. P. Scherm, and U. Schubert, *Chem. Ber.* **111**, 764 (1978).
60. H. Schmidbaur and H. P. Scherm, *Z. Naturforsch., Teil B* **34**, 1347 (1979).
61. H. Schmidbaur and O. Gasser, *Angew. Chem., Int. Ed. Engl.* **15**, 502 (1976).
62. H. Schmidbaur and J. R. Mandl, *Angew. Chem. Int. Ed. Engl.* **16**, 640 (1977).
63. G. E. Coates and C. Parkin, *J. Chem. Soc.* p. 3220 (1962).
64. P. W. R. Corfield and H. M. M. Shearer, *Acta Crystallogr.* **23**, 156 (1967).
65. R. Nast and U. Kirner, *Z. Anorg. Allg. Chem.* **330**, 311 (1964).
66. M. C. Barral, R. Jimenez, E. Royer, V. Moreno, and A. Santos, *An. Quim.* **74**, 585 (1978); *C. A.* **89**, 163696 (1978).
67. E. T. Blues, D. Bryce-Smith, I. W. Lawston, and G. D. Wall, *J. Chem. Soc. Chem. Commun.* p. 513 (1974).
68. V. I. Sokolov, K.I. Grandberg, V. V. Bashilov, D. A. Lemenovskii, P. V. Petrovskii, and O. A. Reutov, *Dokl. Akad. Nauk SSSR* **214**, 393 (1974).
69. A. N. Nesmeyanov, E. G. Perevalova, M. V. Ovchinnikov, and K. I. Grandberg, *Bull. Acad. Sci. USSR, Div. Chem. Sci. (Engl. Transl.)* **24**, 2165 (1975).
70. A. N. Nesmeyanov, E. G. Perevalova, Y. Y. Snakin, and K. I. Grandberg, *Bull. Acad. Sci. USSR, Div. Chem. Sci. (Engl. Transl.)* **27**, 1698 (1978).
71. C. M. Mitchell and F. G. A. Stone, *J. Chem. Soc. D* p. 1263 (1970).
72. C. J. Gilmore and P. Woodward, *J. Chem. Soc. D* p. 1233 (1971).
73. A. J. Chalk, *J. Am. Chem. Soc.* **86**, 4733 (1964).
74. R. Hüttel, H. Reinheimer, and H. Dietl, *Chem. Ber.* **99**, 462 (1966).
75. R. Hüttel and H. Dietl, *Angew. Chem., Int. Ed. Engl.* **4**, 438 (1965).
76. R. Hüttel and H. Reinheimer, *Chem. Ber.* **99**, 2778 (1966).
77. R. Hüttel, P. Tauchner, and H. Forkl, *Chem. Ber.* **105**, 1 (1972).
78. R. Hüttel, H. Reinheimer, and K. Nowak, *Tetrahedron Lett.* p. 1019 (1967).
79. R. Hüttel, H. Reinheimer, and K. Nowak, *Chem. Ber.* **101**, 3761 (1968).
80. P. Tauchner and R. Hüttel, *Chem. Ber.* **107**, 3761 (1974).
81. H. Coutelle and R. Hüttel, *J. Organomet. Chem.* **153**, 359 (1978).
82. S. Komiya and J. K. Kochi, *J. Organomet. Chem.* **135**, 65 (1977).
83. R. G. Salomon and J. K. Kochi, *J. Am. Chem. Soc.* **95**, 1889 (1973).
84. T. J. Leedham, D. B. Powell, and J. G. V. Scott, *Spectrochim. Acta, Part A* **29**, 559 (1973).

85. D. McIntosh and G. A. Ozin, *J. Organomet. Chem.* **121**, 127 (1976).
86. T. Ziegler and A. Rauk, *Inorg. Chem.* **18**, 1558 (1979).
87. R. Hüttel and H. Forkl, *Chem. Ber.* **105**, 1664 (1972).
88. G. Wittig and S. Fischer, *Chem. Ber.* **105**, 3542 (1972).
89. D. Gibson, B. F. G. Johnson, and J. Lewis, *J. Chem. Soc. A* p. 367 (1970).
90. D. Gibson, B. F. G. Johnson, J. Lewis, and C. Oldham, *Chem. Ind. (London)* p. 342 (1966).
91. D. Gibson, *Coord. Chem. Rev.* **4**, 225 (1969).
92. J. Bailey, *J. Inorg. Nucl. Chem.* **35**, 1921 (1973).
93. A. N. Nesmeyanov, K. I. Grandberg, V. P. Dyadchenko, D. A. Lemenovskii, and E. G. Perevalova, *Bull. Acad. Sci. USSR, Div. Chem. Sci. (Engl. Transl.)* **23**, 1138 (1974).
94. R. Hüttel, U. Raffay, and H. Reinheimer, *Angew. Chem. Int. Ed. Engl.* **6**, 862 (1967).
95. G. Wilkinson and T. S. Piper, *J. Inorg. Nucl. Chem.* **2**, 32 (1956).
96. C. H. Campbell and M. L. H. Green, *J. Chem. Soc. A* p. 3282 (1971).
97. G. Ortaggi, *J. Organomet. Chem.* **80**, 275 (1974).
98. A. N. Nesmeyanov, E. G. Perevalova, D. A. Lemenovskii, A. N. Kosina, and K. I. Grandberg, *Bull. Acad. Sci. USSR, Div. Chem. Sci. (Engl. Transl.)* **18**, 1876 (1969).
99. E. G. Perevalova, D. A. Lemenovskii, O. B. Afanasova, V. P. Dyadchenko, K. I. Grandberg, and A. N. Nesmeyanov, *Bull. Acad. Sci. USSR, Div. Chem. Sci. (Engl. Transl.)* **21**, 2522 (1972).
100. A. N. Nesmeyanov, K. I. Grandberg, T. V. Baukova, A. N. Kosina, and E. G. Perevalova, *Bull. Acad. Sci. USSR, Div. Chem. Sci. (Engl. Transl.)* **18**, 1879 (1969).
101. E. G. Perevalova, D. A. Lemenovskii, K. I. Grandberg, and A. N. Nesmeyanov, *Dokl. Akad. Nauk SSSR* **199**, 832 (1971).
102. A. N. Nesmeyanov, K. I. Grandberg, D. A. Lemenovskii, O. B. Afanasova, and E. G. Perevalova, *Bull. Acad. Sci. USSR, Div. Chem. Sci. (Engl. Transl.)* **22**, 856 (1973).
103. E. G. Perevalova, D. A.Lemenovskii, K. I. Grandberg, and A. N. Nesmeyanov, *Dokl. Akad. Nauk SSSR* **202**, 93 (1972).
104. V. G. Andrianov, Y. T. Struchkov, and E. R. Rossinskaja, *J. Chem. Soc. Chem. Commun.* p. 338 (1973).
105. E. G. Perevalova, D. A. Lemenovskii, T. V. Baukova, E. I. Smyslova, K. I. Grandberg, and A. N. Nesmeyanov, *Dokl. Akad. Nauk SSSR* **206**, 883 (1972).
106. K. I. Grandberg, A.N. Nesmeyanov, V. P. Dyadchenko, A. N. Red'kin, and E. G. Perevalova, *Bull. Acad. Sci. USSR, Div. Chem. Sci. (Engl. Transl.)* **28**, 1891 (1979).
107. A. N. Nesmeyanov, E. G. Perevalova, O. B. Afanasova, and K. I. Grandberg, *Bull. Acad. Sci. USSR, Div. Chem. Sci. (Engl. Transl.)* **23**, 456 (1974).
108. A. N. Nesmeyanov, E. G. Perevalova, O. B. Afanasova, M. N. Elinson, and K. I. Grandberg, *Bull. Acad. Sci. USSR, Div. Chem. Sci. (Engl. Transl.)* **24**, 408 (1975).
109. E. G. Perevalova, D. A. Lemenovskii, K. I. Grandberg, and A. N. Nesmeyanov, *Dokl. Akad. Nauk SSSR* **203**, 1320 (1972).
110. A. N. Nesmeyanov, E. G. Perevalova, K. I. Grandberg, D. A. Lemenovskii, T. V. Baukova, and O. B. Afanasova, *Vestn. Mosk. Univ., Khim.* **14**, 387 (1973); *C.A.* **79**, 146613 (1973).
111. A. N. Nesmeyanov, E. G. Perevalova, K. I. Grandberg, D. A. Lemenovskii, T. V. Baukova, and O. B. Afanasova, *J. Organomet. Chem.* **65**, 131 (1974).
112. A. N. Nesmeyanov, E. G. Perevalova, O. B. Afanasova, M. V. Tolstaya, and K. I. Grandberg, *Bull. Acad. Sci. USSR, Div. Chem. Sci. (Engl. Transl.)* **27**, 969 (1978).
113. K. I. Grandberg, T. V. Baukova, E. G. Perevalova, and A. N. Nesmeyanov, *Dokl. Akad. Nauk SSSR* **206**, 1355 (1972).
114. A. N. Nesmeyanov, E. G. Perevalova, T. V. Baukova, and K. I. Grandberg, *Bull. Acad. Sci. USSR, Div. Chem. Sci. (Engl. Transl.)* **23**, 830 (1974).

115. A. N. Nesmeyanov, E. G. Perevalova, O. B. Afanasova, and K. I. Grandberg, *Bull. Acad. Sci. USSR, Div. Chem. Sci. (Engl. Transl.)* **27**, 973 (1978).
116. A. N. Nesmeyanov, E. G. Perevalova, K. I. Grandberg, and D. A. Lemenovskii, *Bull. Acad. Sci. USSR, Div. Chem. Sci. (Engl. Transl.)* **23**, 1068 (1974).
117. P. W. J. De Graaf, J. Boersma, and G. J. M. Van der Kerk, *J. Organomet. Chem.* **78**, C19 (1974).
118. P. W. J. De Graaf, J. Boersma, and G. J. M. Van der Kerk, *J. Organomet. Chem.* **127**, 391 (1977).
119. P. W. J. De Graaf, A. J. De Koning, J. Boersma, and G. J. M. Van der Kerk, *J. Organomet. Chem.* **141**, 345 (1977).
120. G. van Koten and J. G. Noltes, *J. Organomet. Chem.* **82**, C53 (1974).
121. G. van Koten and J. G. Noltes, *J. Chem. Soc. Chem. Commun.* p. 940 (1972).
122. A. J. Leusink, G. van Koten, J. W. Marsman, and J. G. Noltes, *J. Organomet. Chem.* **55**, 419 (1973).
123. G. van Koten and J. G. Noltes, *J. Am. Chem. Soc.* **101**, 6593 (1979).
124. G. van Koten and J. G. Noltes, *J. Organomet. Chem.* **171**, C39 (1979).
125. G. van Koten, J. T. B. H. Jastrzebski, T. B. H. Johann, and J. G. Noltes, *J. Chem. Soc. Chem. Commun.* p. 203 (1977).
126. G. van Koten, J. T. B. H. Jastrzebski, and J. G. Noltes, *Inorg. Chem.* **16**, 1782 (1977).
127. G. van Koten, J. T. B. H. Jastrzebski, and J. G. Noltes, *J. Organomet. Chem.* **148**, 317 (1978).
128. G. van Koten, C. A. Schaap, J. T. B. H. Jastrzebski, and J. G. Noltes, *J. Organomet. Chem.* **186**, 427 (1980).
129. R. Uson, A. Laguna, and P. Brun, *J. Organomet. Chem.* **197**, 369 (1980).
130. J. G. Noltes, *J. Organomet. Chem.* **100**, 177 (1975).
131. A. Sacco and M. Freni, *Gazz. Chim. Ital.* **86**, 195 (1956).
132. J. A. McCleverty and M. M. M. Da Mota, *J. Chem. Soc. Dalton Trans.* p. 2571 (1973).
133. J. Browning, P. L. Goggin, and R. J. Goodfellow, *J. Chem. Res. Miniprint* 4201 (1978).
134. F. Bonati and G. Minghetti, *Gazz. Chim. Ital.* **103**, 373 (1973).
135. P. G. Jones and A. F. Williams, *J. Chem. Soc. Dalton Trans.* p. 1430 (1977).
136. S. Esperas, *Acta Chem. Scand., Ser. A* **30**, 527 (1976).
137. F. Bonati and G. Minghetti, *Synth. React. Inorg. Met. Org. Chem.* **1**, 299 (1971).
138. G. Banditelli, F. Bonati, and G. Minghetti, *Gazz. Chim. Ital.* **107**, 267 (1977).
139. F. Bonati and G. Minghetti, *J. Organomet. Chem., 59*, 403 (1973).
140. R. Uson, A. Laguna, J. Vicente, J. Garcia, and B. Bergareche, *J. Organomet. Chem.* **173**, 349 (1979).
141. G. Minghetti and F. Bonati, *J. Organomet. Chem.* **54**, C62 (1973).
142. G. Minghetti, L. Baratto, and F. Bonati, *J. Organomet. Chem.* **102**, 397 (1975).
143. J. E. Parks and A. L. Balch, *J. Organomet. Chem.* **57**, C103 (1973).
144. J. E. Parks and A. L. Balch, *J. Organomet. Chem.* **71**, 453 (1974).
145. B. Cetinkaya, P. Dixneuf, and M. F. Lappert, *J. Chem. Soc. Chem. Commun.* p. 206 (1973).
146. B. Cetinkaya, P. Dixneuf, and M. F. Lappert, *J. Chem. Soc. Dalton Trans.* p. 1827 (1974).
147. K. Bartel and W. P. Fehlhammer, *Angew. Chem. Int. Ed. Engl.* **13**, 599 (1974).
148. G. Minghetti and F. Bonati, *Angew. Chem. Int. Ed. Engl.* **11**, 429 (1972).
149. G. Minghetti and F. Bonati, *Gazz. Chim. Ital.* **102**, 205 (1972).
150. G. Minghetti and F. Bonati, *Inorg. Chem.* **13**, 1600 (1974).
151. A. Tiripicchio, M. Tiripicchio Camellini, and G. Minghetti, *J. Organomet. Chem.* **171**, 399 (1979).
152. G. Minghetti, F. Bonati, and M. Massobrio, *J. Chem. Soc. Chem. Commun.* p. 260 (1973).

153. G. Minghetti, F. Bonati, and M. Massobrio, *Inorg. Chem.* **14**, 1974 (1975).
154. G. Banditelli, F. Bonati, and G. Minghetti, *Gazz. Chim. Ital.* **110**, 317 (1980).
155. A. L. Balch and D. J. Doonan, *J. Organomet. Chem.* **131**, 137 (1977).
156. F. Bonati and G. Minghetti, *J. Organomet. Chem.* **60**, C43 (1973).
157. F. Bonati, G. Minghetti, and G. Banditelli, *Synth. React. Inorg. Met. Org. Chem.* **6**, 383 (1976).
158. W. Manchot and H. Gall, *Ber. Dtsch. Chem. Ges. B* **58**, 2175 (1925).
159. M. S. Kharasch and H. S. Isbell, *J. Am. Chem. Soc.* **52**, 2919 (1930).
160. D. B. Dell'Amico and F. Calderazzo, *Gazz. Chim. Ital.* **103**, 1099 (1973).
161. D. B. Dell'Amico, F. Calderazzo, and F. Marchetti, *J. Chem. Soc. Dalton Trans.* p. 1829 (1976).
162. D. B. Dell'Amico, F. Calderazzo, and G. Dell'Amico, *Gazz. Chim. Ital.* **107**, 101 (1977).
163. M. I. Bruce, *J. Organomet. Chem.* **44**, 209 (1972).
164. J. Browning, P. L. Goggin, R. J. Goodfellow, M. G. Norton, A. J. M. Rattray, B. F. Taylor, and J. Mink, *J. Chem. Soc. Dalton Trans.* p. 2061 (1977).
165. D. McIntosh and G. A. Ozin, *Inorg. Chem.,* **16**, 51 (1977).
166. H. Huber, D. McIntosh, and G. A. Ozin, *Inorg. Chem.* **16**, 975 (1977).
167. J. W. A. van der Velden, J. J. Bour, F. A. Vollenbroek, P. T. Beurskens, and J. M. M. Smits, *J. Chem. Soc. Chem. Commun.* p. 1162 (1979).
168. L. Malatesta, L. Naldini, G. Simonetta, and F. Cariati, *Chem. Commun.* p. 212 (1965).
169. W. Beck, P. Swoboda, K. Feldl, and E. Schuierer, *Chem. Ber.* **103**, 3591 (1970).
170. U. Nagel, K. Peters, H. G. v. Schnering, and W. Beck, *J. Organomet. Chem.* **185**, 427 (1980).
171. G. E. Coates and C. Parkin, *J. Chem. Soc.* p. 421 (1963).
172. P. Jutzi and H. Heusler, *J. Organomet. Chem.* **114**, 265 (1976).
173. F. Glockling and V. B. Mahale, *J. Chem. Res. Miniprint* p. 2169 (1978).
174. P. W. N. M. Van Leeuwen, R. Kaptein, R. Huis, and C. F. Roobeek, *J. Organomet. Chem.* **104**, C44 (1976).
175. F. G. Mann and D. Purdie, *J. Chem. Soc.* p. 1235 (1940).
176. J. L. Burmeister and N. J. De Stefano, *Inorg. Chem.* **10**, 998 (1971).
177. B. T. Heaton and R. J. Kelsey, *Inorg. Nucl. Chem. Lett.* **11**, 363 (1975).
178. R. Uson, A. Laguna, J. Vicente, J. Garcia, B. Bergareche, and P. Brun, *Inorg. Chim. Acta* **28**, 237 (1978).
179. R. Uson, A. Laguna, and B. Bergareche, *J. Organomet. Chem.* **184**, 411 (1980).
180. R. Uson, A. Laguna, and J. Pardo, *Synth. React. Inorg. Met. Org. Chem.* **4**, 499 (1974).
181. R. Uson, A. Laguna, and J. Vicente, *J. Organomet. Chem.* **86**, 415 (1975).
182. R. Uson, A. Laguna, and J. Vicente, *Synth. React. Inorg. Met. Org. Chem.* **6**, 293 (1976).
183. R. Uson, A. Laguna, J. Vicente, and J. Garcia, *J. Organomet. Chem.* **104**, 401 (1976).
184. G. Minghetti, F. Bonati, and G. Banditelli, *Inorg. Chem.* **15**, 1718 (1976).
185. L. Manojlović-Muir, *J. Organomet. Chem.* **73**, C45 (1974).
186. G. Minghetti and F. Bonati, *J. Organomet. Chem.* **73**, C43 (1974).
187. R. Uson, A. Laguna, and T. Cuenca, *J. Organomet. Chem.* **194**, 271 (1980).
188. H. Schmidbaur, J. R. Mandl, A. Frank, and G. Hüttner, *Chem. Ber.* **109**, 466 (1976).
189. H. Schmidbaur, J. R. Mandl, F. E. Wagner, D. F. Van de Vondel, and G. P. Van der Kelen, *J. Chem. Soc. Chem. Commun.* p. 170 (1976).
190. D. F. Van de Vondel, G. P. Van der Kelen, H. Schmidbaur, A. Wolleben, and F. E. Wagner, *Phys. Scr.* **16**, 364 (1977).
191. H. Schmidbaur and J. R. Mandl, *Naturwissenschaften* **63**, 585 (1976).
192. A. Johnson and R. J. Puddephatt, *Inorg. Nucl. Chem. Lett.* **9**, 1175 (1973).

193. A. Tamaki and J. K. Kochi, *J. Organomet. Chem.* **40**, C81 (1972).
194. A. Tamaki and J. K. Kochi, *J. Organomet. Chem.* **64**, 411 (1974).
195. A. Johnson and R. J. Puddephatt, *J. Organomet. Chem.* **85**, 115 (1975).
196. C. M. Mitchell and F. G. A. Stone, *J. Chem. Soc. Dalton Trans.* p. 102 (1972).
197. A. Johnson, R. J. Puddephatt, and J. L. Quirk, *J. Chem. Soc. Chem. Commun.* p. 938 (1972).
198. A. Johnson and R. J. Puddephatt, *J. Chem. Soc. Dalton Trans.*, p. 1384 (1977).
199. J. A. J. Jarvis, A. Johnson, and R. J. Puddephatt, *J. Chem. Soc. Chem. Commun.* p. 373 (1973).
200. A. Johnson and R. J. Puddephatt, *J. Chem. Soc. Dalton Trans.* p. 980 (1978).
201. J. Bailey and M. J. Mays, *J. Organomet. Chem.* **63**, C24 (1973).
202. M. J. Mays and J. Bailey, *J. Chem. Soc. Dalton Trans.* p. 578 (1977).
203. K. I. Grandberg, E. I. Smyslova, and A. N. Kosina, *Bull. Acad. Sci. USSR, Div. Chem. Sci. (Engl. Transl.)* **22**, 2721 (1973).
204. E. G. Perevalova, T. V. Baukova, E. I. Goryunov, and K. I. Grandberg, *Bull. Acad. Sci. USSR, Div. Chem. Sci. (Engl. Transl.)* **19**, 2031 (1970).
205. H. Schmidbaur and A. A. M. Aly, *Angew. Chem. Int. Ed. Engl.* **19**, 71 (1980).
206. A. Shiotani and H. Schmidbaur, *J. Organomet. Chem.* **37**, C24 (1972).
207. N. G. Hargreaves, A. Johnson, R. J. Puddephatt, and L. H. Sutcliffe, *J. Organomet. Chem.* **69**, C21 (1974).
208. A. Johnson and R. J. Puddephatt, *J. Chem. Soc. Dalton Trans.*, p. 115 (1975).
209. R. J. Puddephatt and P. J. Thompson, *J. Organomet. Chem.* **117**, 395 (1976).
210. R. Kaptein, P. W. N. M. Van Leeuween, and R. Huis, *J. Chem.Soc. Chem. Commun.* p. 568 (1975).
211. B. J. Gregory and C. K. Ingold, *J. Chem. Soc. B* p. 276 (1969).
212. A. N. Nesmeyanov, E. G. Perevalova, M. V. Ovchinnikov, Y. Y. Snakin, and K. I. Grandberg, *Bull. Acad. Sci. USSR, Div. Chem. Sci. (Engl. Transl.)* **27**, 1695 (1978).
213. R. J. Puddephatt and P. J. Thompson, *J. Chem. Soc. Dalton Trans.* p. 1810 (1975).
214. G. W. Rice and R. S. Tobias, *J. Organomet. Chem.* **86**, C37 (1975).
215. L. J. Farrugia, J. A. K. Howard, P. Mitrprachachon, J. L. Spencer, F. G. A. Stone, and P. Woodward, *J. Chem. Soc. Chem. Commun.* p. 260 (1978).
216. R. J. Puddephatt and P. J. Thompson, *J. Organomet. Chem.* **166**, 251 (1979).
217. C. S. Gibson, *Nature (London)* **131**, 130 (1933).
218. A. Burawoy and C. S. Gibson, *J. Chem. Soc.* p. 324 (1936).
219. G. W. Rice and R. S. Tobias, *Inorg. Chem.* **14**, 2402 (1975).
220. G. W. Rice and R. S. Tobias, *J. Am. Chem. Soc.* **99**, 2141 (1977).
221. H. Schmidbaur and R. Franke, *Inorg. Chim. Acta* **13**, 79 (1975).
222. J. P. Fackler, Jr. and C. Paparizos, *J. Am. Chem. Soc.* **99**, 2363 (1977).
223. H. Schmidbaur, O. Gasser, C. Krueger, and J. C. Sekutowski, *Chem. Ber.* **110**, 3517 (1977).
224. H. Schmidbaur, H. J. Fueller, V. Bejenke, A. Frank, and G. Hüttner, *Chem. Ber.* **110**, 3536 (1977).
225. C. Krueger, J. C. Sekutowski, R. Goddard, H. J. Fueller, O. Gasser, and H. Schmidbaur, *Isr. J. Chem.* **15**, 149 (1976–1977).
226. L. A. Woods and H. Gilman, *Proc. Iowa Acad.Sci.* **49**, 286 (1942).
227. H. Gilman and L. A. Woods, *J. Am. Chem.Soc.* **70**, 550 (1948).
228. S. W. Krauhs, G. C. Stocco, and R. S. Tobias, *Inorg. Chem.* **10**, 1365 (1971).
229. C. F. Shaw and R. S. Tobias, *Inorg. Chem.* **12**, 965 (1973).
230. A. Tamaki and J. K. Kochi, *J. Chem. Soc. Chem. Commun.* p. 423 (1973).
231. A. Tamaki, S. A. Mageninis, and J. K. Kochi, *J. Am. Chem. Soc.* **95**, 6487 (1973).
232. S. Komiya and J. K. Kochi, *J. Am. Chem. Soc.* **99**, 3695 (1977).

233. W. J. Pope and C. S. Gibson, *Proc. Chem. Soc., London* **23**, 245 (1907).
234. W. J. Pope and C. S. Gibson, *Proc. Chem. Soc., London* **23**, 295 (1907).
235. W. J. Pope and C. S. Gibson, *J. Chem. Soc.* **91–92**, 2061 (1907).
236. C. S. Gibson and J. L. Simonsen, *J. Chem. Soc.* p. 2531 (1930).
237. M. S. Kharasch and H. S. Isbell, *J. Am. Chem. Soc.* **53**, 2701 (1931).
238. F. H. Brain and C. S. Gibson, *J. Chem. Soc.* p. 762 (1939).
239. A. Burawoy, C. S. Gibson, and S. Holt, *J. Chem. Soc.* p. 1024 (1935).
240. R. F. Phillips and H. M. Powell, *Proc. R. Soc. London, Ser.* A **173**, 147 (1939).
241. A. Burawoy, C. S. Gibson, G. C. Hampson, and H. M. Powell, *J. Chem. Soc.* p. 1690 (1937).
242. C. S. Gibson, *Proc. R. Soc. London, Ser.* A **173**, 160 (1939).
243. R. C. Leech, D. B. Powell, and N. Sheppard, *Spectrochim. Acta* **21**, 559 (1965).
244. R. C. Leech, D. B. Powell, and N. Sheppard, *Spectrochim. Acta* **22**, 1931 (1966).
245. A. Chakravorty, *Naturwissenschaften* **48**, 643 (1961); *C.A.* **56**, 5539 (1962).
246. L. A. Woods, *Iowa State Coll. J. Sci.* **19**, 61 (1944).
247. A. Burawoy and C. S. Gibson, *J. Chem. Soc.* p. 860 (1934).
248. S. Komiya, J. C. Huffman, and J. K. Kochi, *Inorg. Chem.* **16**, 1253 (1977).
249. C. S. Gibson and W. T. Weller, *J. Chem. Soc.* p. 102 (1941).
250. R. V. G. Ewens and C. S. Gibson, *J. Chem. Soc.* p. 109 (1941).
251. W. L. G. Gent and C. S. Gibson, *J. Chem. Soc.* p. 1835 (1949).
252. W. Beck, W. P. Fehlhammer, P. Poellmann, and R. S. Tobias, *Inorg. Chim. Acta* **2**, 467 (1968).
253. M. Bergfeld and H. Schmidbaur, *Chem. Ber.* **102**, 2408 (1969).
254. S. Komiya, J. C. Huffman, and J. K. Kochi, *Inorg. Chem.* **16**, 2138 (1977).
255. M. E. Foss and C. S. Gibson, *J. Chem. Soc.* p. 3075 (1949).
256. F. Stocco, G. C. Stocco, W. M. Scovell, and R. S. Tobias, *Inorg. Chem.* **10**, 2639 (1971).
257. H. Schmidbaur and M. Bergfeld, *Inorg. Chem.* **5**, 2069 (1966).
258. S. J. Harris and R. S. Tobias, *Inorg. Chem.* **8**, 2259 (1969).
259. G. E. Glass, J. H. Konnert, M. G. Miles, D. Britton, and R. S. Tobias, *J. Am. Chem. Soc.* **90**, 1131 (1968).
260. C. S. Gibson and W. M. Colles, *J. Chem. Soc.* p. 2407 (1931).
261. A. Burawoy and C. S. Gibson, *J. Chem. Soc.* p. 219 (1935).
262. C. S. Gibson, *Br. Assoc. Adv. Sci., Rep.* p. 35 (1938); *C.A.* **33**, 2838 (1939).
263. M. E. Foss and C. S. Gibson, *J. Chem. Soc.* p. 3063 (1949).
264. R. V. G. Ewens and C. S. Gibson, *J. Chem. Soc.* p. 431 (1949).
265. E. Rivarola, G. C. Stocco, P. B. Lassandro, and R. Barbieri, *J. Organomet. Chem.* **14**, 467 (1968).
266. K. S. Murray, B. E. Reichert, and B. O. West, *J. Organomet. Chem.* **61**, 451 (1973).
267. H. Hagnauer, G. C. Stocco, and R. S. Tobias, *J. Organomet. Chem.* **46**, 179 (1972).
268. H. Schmidbaur and K. C. Dash, *J. Am. Chem. Soc.* **95**, 4855 (1973).
269. H. J. A. Blaauw, R. J. F. Nivard, and G. J. M van der Kerk, *J. Organomet. Chem.* **2**, 236 (1964).
270. H. Schmidbaur and K. C. Dash, *Chem. Ber.* **105**, 3662 (1972).
271. P. M. T. M. van Attekum and J. M. Trooster, *J. Chem. Soc. Dalton Trans.* p. 201 (1980).
272. C. Paparizos and J. P. Fackler, Jr, *Inorg. Chem.* **19**, 2886 (1980).
273. R. S. Tobias, C. E. Rice, W. Beck, B. Purucker, and K. Bartel, *Inorg. Chim. Acta* **35**, 11 (1979).
274. M. G. Miles, G. E. Glass, and R. S. Tobias, *J. Am. Chem. Soc.* **88**, 5738 (1966).
275. G. E. Glass, W. B. Schwabacher, and R. S. Tobias, *Inorg. Chem.* **7**, 2471 (1968).

276. W. M. Scovell and R. S. Tobias, *Inorg. Chem.* **9**, 945 (1970).
277. R. S. Tobias, *Inorg. Chem.* **9**, 1296 (1970).
278. W. M. Scovell, G. C. Stocco, and R. S. Tobias, *Inorg. Chem.* **9**, 2682 (1970).
279. A. Shiotani and H. Schmidbaur, *Chem. Ber.* **104**, 2838 (1971).
280. G. C. Stocco and R. S. Tobias, *J. Am. Chem. Soc.* **93**, 5057 (1971).
281. C. F. Shaw, J. W. Lundeen, and R. S. Tobias, *J. Organomet. Chem.* **51**, 365 (1973).
282. P. L. Kuch and R. S. Tobias, *J. Organomet. Chem.* **122**, 429 (1976).
283. G. C. Stocco, L. Pellerito and N. Bertazzi, *Inorg. Chim. Acta* **12**, 67 (1975).
284. E. G. Perevalova, K. I. Grandberg, D. A. Lemenovskii, and T. V. Baukova, *Bull. Acad. Sci. USSR, Div. Chem. Sci. (Engl. Transl.)* **20**, 1967 (1971).
285. R. S. Nyholm and P. Royo, *J. Chem. Soc. D* p. 421 (1969).
286. P. Royo and R. Serrano, *J. Organomet. Chem.* **144**, 33 (1978).
287. R. W. Baker and P. Pauling, *J. Chem. Soc. D* p. 745 (1969).
288. R. Uson, P. Royo, and A. Laguna, *Inorg. Nucl. Chem. Lett.* **7**, 1037 (1971).
289. R. Uson, P. Royo, and A. Laguna, *Synth. React. Inorg. Met. Org. Chem.* **3**, 237 (1973).
290. R. Uson, A. Laguna, and J. L. Sanjoaquin, *J. Organomet. Chem.* **80**, 147 (1974).
291. R. Uson, A. Laguna, and J. Buil, *J. Organomet. Chem.* **85**, 403 (1975).
292. R. Uson, A. Laguna, and M. V. Castrillo, *Synth. React. Inorg. Met. Org. Chem.* **9**, 317 (1979).
293. R. Uson, J. Vicente, J. A. Cirac, and M.T. Chicote, *J. Organomet. Chem.* **198**, 105 (1980).
294. E. H. Braye, W. Hübel, and I. Caplier, *J. Am. Chem. Soc.* **83**, 4406 (1961).
295. M. Peteau-Boisdenghien, J. Meunier-Piret, and M. Van Meerssche, *Cryst. Struct. Commun.* **4**, 375 (1975).
296. R. Uson, J. Vicente, and M. T. Chicote, *Inorg. Chim. Acta* **35**, L305 (1979).
297. M. S. Kharasch and H. S. Isbell, *J. Am. Chem. Soc.* **53**, 3053 (1931).
298. K. S. Liddle and C. Parkin, *J. Chem. Soc. Chem. Commun.* p. 26 (1972).
299. P. W. J. De Graaf, J. Boersma, and G. J. M. van der Kerk, *J. Organomet. Chem.* **105**, 399 (1976).
300. M. S. Kharasch and T. M. Beck, *J. Am. Chem. Soc.* **56**, 2057 (1934).
301. F. Calderazzo and D. B. Dell'Amico, *J. Organomet. Chem.* **76**, C59 (1974).
302. M. McPartlin and A. J. Markwell, *J. Organomet. Chem.* **57**, C25 (1973).
303. P. Braunstein, *J. Chem. Soc. Chem. Commun.* p. 851 (1973).
304. P. Braunstein and R. J. H. Clark, *Inorg. Chem.* **13**, 2224 (1974).
305. R. Hüttel and A. Konietzny, *Chem. Ber.* **106**, 2098 (1973).
306. M. A. Bennett, K. Hoskins, W. R. Kneen, R. S. Nyholm, P. B. Hitchcock, R. Mason, G. B. Robertson, and A. D. C. Towl, *J. Am. Chem. Soc.* **93**, 4591, 4592 (1971).
307. D. I. Hall, J. H. Ling, and R. S. Nyholm, *Struct. Bonding (Berlin)* **15**, 3 (1973).
308. P. K. Monaghan and R. J. Puddephatt, *Inorg. Chim. Acta* **15**, 231 (1975).
309. H. A. Tayim and A. Vassilian, *Inorg. Nucl. Chem. Lett.* **8**, 215 (1972).
310. R. Hüttel and H. Forkl, *Chem. Ber.* **105**, 2913 (1972).
311. P. Tauchner and R. Hüttel, *Tetrahedron Lett.* **46**, 4733 (1972).
312. L. F. Warren and M. F. Hawthorne, *J. Am. Chem. Soc.* **90**, 4823 (1968).
313. S. Komiya, T. A. Albright, R. Hoffmann, and J. K. Kochi, *J. Am. Chem. Soc.* **98**, 7255 (1976).
314. B. Akermark and A. Ljungqvist, *J. Organomet. Chem.* **182**, 59 (1979).
315. A. Tamaki, S. A. Magennis, and J. K. Kochi, *J. Am. Chem. Soc.* **96**, 6140 (1974).
316. S. Komiya and J. K. Kochi, *J. Am. Chem. Soc.* **98**, 7599 (1976).
317. R. J. Puddephatt and M. A. Stalteri, *J. Organomet. Chem.* **193**, C27 (1980).
318. U. Kunze and J. D. Koola, *J. Organomet. Chem.* **80**, 281 (1974).
319. W. Beck, K. Burger, and W. P. Fehlhammer, *Chem. Ber.* **104**, 1816 (1971).

320. J. K. Jawad and R. J. Puddephatt, *J. Chem. Soc. Chem. Commun.* p. 892 (1977).
321. R. J. Puddephatt and P. J. Thompson, *J. Chem. Soc. Dalton Trans.* p. 2091 (1976).
322. P. D. Gavens, J. J. Guy, M. J. Mays, and G. M. Sheldrick, *Acta Crystallogr., Sect. B* **33**, 137 (1977).
323. R. W. Baker and P. J. Pauling, *J. Chem. Soc. Dalton Trans.* p. 2264 (1972).
324. P. E. Riley and R. E. Davis, *J. Organomet. Chem.* **192**, 283 (1980).
325. P. L. Bellon, M. Manassero, and M. Sansoni, *Ric. Sci.* **39**, 173 (1969).
326. N. C. Baenziger, W. E. Bennett, and D. M. Soboroff, *Acta Crystallogr., Sect. B* **32**, 962 (1976).
327. T. G. Appleton, H. C. Clark, and L. E. Manzer, *Coord. Chem. Rev.* **10**, 335 (1973).
328. F. R. Hartley, *Chem. Soc. Rev.* **2**, 163 (1973).
329. L. Manojlović-Muir and K. W. Muir, *Inorg. Chim. Acta* **10**, 47 (1974).
330. R. J. Puddephatt, "The Chemistry of Gold" (Topics in Inorganic and General Chemistry (R. J. H. Clark, ed.) Vol. 16), p. 231. Elsevier, Amsterdam, 1978.
331. J. S. Charlton and D. I. Nichols, *J. Chem. Soc. A* p. 1484 (1970).
332. J. Behan, R. A. W. Johnstone, and R. J. Puddephatt, *J. Chem. Soc. Chem. Commun.* p. 444 (1978).
333. P. L. Goggin and L. A. Woodward, *Trans. Faraday Soc.* **62**, 1423 (1966).
334. R. J. Williams, G. Agnes, S. Bendle, H. A. O. Hill, and F. R. Williams, *J. Chem. Soc. D* p. 850 (1971).
335. M. R. Moller, M. A. Bruck, T. O'Connor, F. J. Armatis, Jr., E. A. Knolinski, N. Kottmair, and R. S. Tobias, *J. Am. Chem. Soc.* **102**, 4589 (1980).
336. R. J. Cross and R. Wardle, *J. Chem. Soc. A* p. 2000 (1971).
337. F. A. Cotton and T. J. Marks, *J. Am. Chem. Soc.* **91**, 7281 (1969).
338. J.-M. Bassett, J. R. Mandl, and H. Schmidbaur, *Chem. Ber.* **113**, 1145 (1980).
339. G. Blaschke, H. Schmidbaur, and W. C. Kaska, *J. Organomet. Chem.* **182**, 251 (1979).
340. G. E. Glass and R. S. Tobias, *J. Organomet. Chem.* **15**, 481 (1968).
341. L. Pellerito and R. S. Tobias, *Inorg. Chem.* **9**, 953 (1970).
342. J. Lewis, R. F. Long, and C. Oldham, *J. Chem. Soc.* p. 6740 (1965).

Arsonium Ylides

HUANG YAOZENG (Y. Z. HUANG)* and SHEN
YANCHANG (Y. C. SHEN)

Shanghai Institute of Organic Chemistry
Academia Sinica
Shanghai, People's Republic of China

I

INTRODUCTION

In 1953 Wittig and Geissler (*100*) reported that methylene triphenylphosphorane reacted with benzophenone to form 1,1-diphenylethene and triphenylphosphine oxide. This experiment marked the birth of the Wittig reaction, a novel method for the conversion of carbonyl groups to olefins, and the entry of ylides into the arsenal of important synthetic tools. Since

* Formerly spelled Huang Yao-Tseng.

then, a great number of reports concerning the preparation of phosphonium ylides, the synthetic applications thereof and the elucidation of their reaction mechanism have appeared in the literature.

However, certain phosphonium ylides, such as those with an electron-withdrawing substituent in the alkylidene moiety, are relatively unreactive toward certain substrates such as ketones (22, 77, 95). This led us to consider whether arsonium ylides might be preferable to phosphonium ylides in certain reactions (48, 94). The overlap of the p orbitals of carbon with d orbitals of arsenic is less effective than with d orbitals of phosphorus. Therefore the "covalent" canonical form (1a) should make a smaller contribution to the overall structure of arsonium ylides than to that of the corresponding phosphonium ylides.

$$Ph_3As{=}CRR' \longleftrightarrow Ph_3\overset{+}{As}{-}\overset{-}{C}RR'$$

$$(1a) \qquad\qquad (1b)$$

This has been supported by X-ray crystallography (102) which showed that methylene triphenylphosphorane possesses planar geometry, whereas methylene triphenylarsorane has pseudotetrahedral geometry. Furthermore, because the As—C bond is weaker than the P—C bond, when a reaction starting from a heteronium ylide finally undergoes elimination of the tertiary arsine or phosphine, it may be anticipated that the arsonium ylide would react more rapidly. The difference in behavior between phosphonium and arsonium ylides may also be correlated with the difference in strength of the P—O and As—O bonds (18). The strength of the former provides a considerable driving force in many reactions, including those of the Wittig type.

The chemistry of arsonium ylides was first reviewed by Johnson (52) in 1966, by Samaan (78) in 1978, and by Bansal and Sharma (7) recently. In the present chapter, emphasis is placed on studies of the chemistry of arsonium ylides at our Institute and at the University of Science and Technology of Shanghai. However, our work was interrupted and some of our findings were delayed in publication for about 15 years.

II

METHODS OF PREPARATION

A. From Arsonium Salts ("Salt Method")

Most of the arsonium ylides reported have been prepared by the so-called salt method, which involves the quaternization of a suitable arsine with an appropriate halogen compound followed by the elimination of HX from the resulting arsonium salt with a suitable base. The choice of base and reaction medium depends upon the nature of the ylide to be formed. In general, a stronger base is necessary here than for phosphonium ylides (*1, 52, 54, 73*).

$$R_3As + R'CH_2X \rightarrow [R_3\overset{+}{As}-CH_2R']X^- \xrightarrow{\text{base}} R_3As=CHR' \longleftrightarrow R_3\overset{+}{As}-\overset{-}{C}HR' \quad (1)$$
$$\text{(2a)} \qquad\qquad \text{(2b)}$$

R = alkyl, aryl

R' = H, alkyl, aryl, acyl, carboalkoxy, CN, etc.

Methylene triphenylarsorane (*11, 31, 34, 39*) has been prepared *in situ* from the salt in ether (*34, 39*), benzene (*34, 39, 60*), tetrahydrofuran (*31*), or dimethyl sulfoxide (*11*) by reaction with butyl- or phenyllithium (*34, 39*), sodium hydride (*31*), methylsulfinylcarbanion (*11*), or sodamide (*102*). Ethylidene triphenylarsorane has been prepared by reaction with methylsulfinylcarbanion (*11*) whereas fluorenylidene-(9)-trimethyl- and fluorenylidene-(9)-dimethylbenzylarsorane (*101*) have been made using phenyllithium in ether. Sodium methoxide in methanol has been used successfully (*41, 47, 48, 94*) to generate ylides from their corresponding salts when R' is an electron-withdrawing group (e.g., COOR, COR, or CN).

Treatment of an aqueous alcoholic solution of 9-fluorenyltriphenylarsonium bromide with dilute ammonia or dilute sodium hydroxide afforded fluorenylidene-(9)-triphenylarsorane (*51*). Using the same method, 2,7-dibromofluorenylidene-(9)-triphenylarsorane was also prepared (*93*).

Benzylidene triphenylarsoranes have been prepared *in situ* from their corresponding salts, either with butyl- or phenyllithium in ether (*55, 76*) or with sodium alkoxide in alcohol (*97*). Carbomethoxybenzylidene triphenylarsorane (**3**) has been prepared from the corresponding salt in chloroform with ammonia (*74*). With the same method or with sodium alkoxide in alcohol, acetylmethyl-, phenacyl-, and carbomethoxymethyltriphenylarsonium salts afforded the corresponding ylides (*28, 48, 70, 73, 74, 94*).

$$[Ph_3As^+-CH(C_6H_5)-COOCH_3]Br^- \xrightarrow[-NH_4Br]{NH_3/CHCl_3} Ph_3As=C(C_6H_5)COOCH_3 \quad (2)$$
$$\text{(3)}$$

Although, in principle, any arsonium ylide could be prepared via the "slat method" as demonstrated above, attempts to prepare a ylide from dimethyldibenzylarsonium salt (**4**) with ethereal phenyllithium failed. The product of the reaction after quenching with water was stilbene (**5**). Apparently, a Stevens rearrangement occurred during the reaction (*101*).

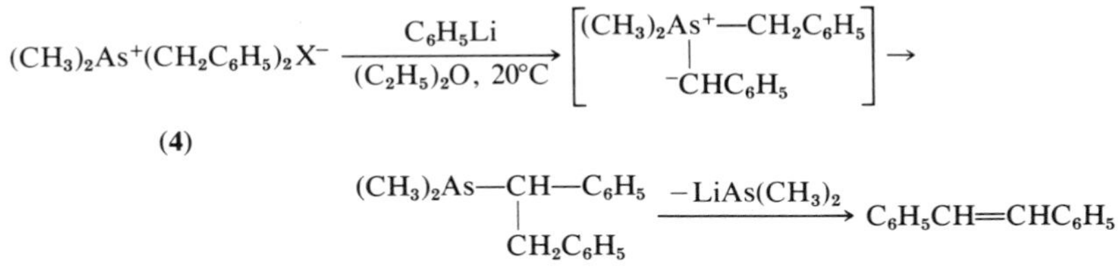

$$(CH_3)_2As^+(CH_2C_6H_5)_2X^- \xrightarrow[\ (C_2H_5)_2O,\ 20°C\ ]{C_6H_5Li} \left[\begin{matrix} (CH_3)_2As^+\!-\!CH_2C_6H_5 \\ | \\ ^-CHC_6H_5 \end{matrix} \right] \rightarrow$$

(4)

(3)

$$\begin{matrix} (CH_3)_2As\!-\!CH\!-\!C_6H_5 \\ | \\ CH_2C_6H_5 \end{matrix} \xrightarrow{-LiAs(CH_3)_2} C_6H_5CH\!=\!CHC_6H_5$$

(5)

B. *From Tertiary Arsine Dihalide and Tertiary Arsine Oxides ("Arsine Oxide Method")*

A facile method for the preparation of a variety of stabilized arsonium ylides in good yield has been developed by the action of active methylene compounds with tertiary arsine oxide or tertiary arsine dihalide. Thus triphenyl-arsine dihalides react with a number of active methylene compounds in the presence of a tertiary amine to afford arsonium ylides (6) (40). The reaction of triphenylarsine oxide with active methylene compounds in the presence of either acetic anhydride or triethylamine–phosphorus pentoxide gave rise to arsonium ylides (6) (32, 36, 65, 67).

$$Ph_3AsCl_2 + H_2CXY \xrightarrow{(C_2H_5)_3N/C_6H_6} Ph_3As\!=\!CXY$$
(6)

(4)

$$X = CN,\ SO_2\!-\!C_6H_5, COOCH_3, NO_2$$

$$Y = CN,\ SO_2\!-\!C_6H_5,\ C_6H_5$$

$$Ph_3AsO + H_2CXY \xrightarrow[40-48\%]{(CH_3CO)_2O} Ph_3As\!=\!CXY$$
(6)

(5)

$$Y = X = COCH_3,\ COPh,\ C_6H_5,\ CN$$

$$XY = -CH\!=\!CH\!-\!CH\!=\!CH-$$

The reaction of triphenylarsine oxide with strongly acidic methylene compounds in acetic anhydride affords acetylated ylides (7) (32, 67), e.g.,

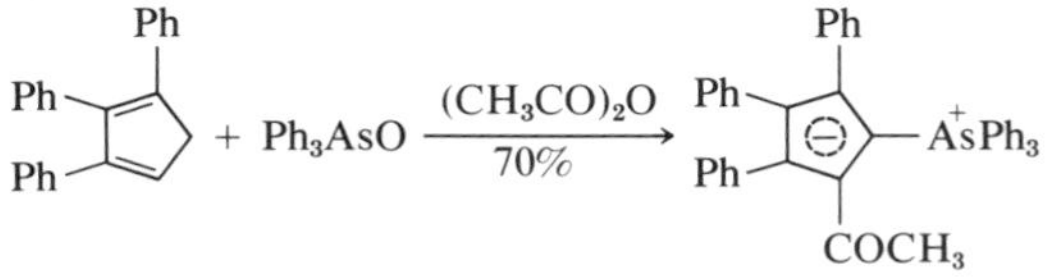

(6)

In triethylamine–phosphorus pentoxide the same reactants afforded triphenylcyclopentadiene triphenylarsorane (8) (67). Similarly, triphenylarsine oxide reacts with nitromethane in the presence of triethylamine–phosphorus pentoxide to give nitromethylene triphenylarsorane (9), which gives the acetylated ylide (10) with acetic anhydride (32).

$$\text{Ph}_3\text{AsO} + \text{CH}_3\text{NO}_2 \longrightarrow \begin{array}{l} \xrightarrow{\text{Ac}_2\text{O}} \text{Ph}_3\text{As}=\text{C(NO}_2)\text{Ac} \\ \qquad\qquad\qquad (\mathbf{10}) \\ \qquad\qquad\qquad\uparrow \text{Ac}_2\text{O} \\ \xrightarrow[\text{P}_2\text{O}_5]{(\text{C}_2\text{H}_5)_3\text{N}} \text{Ph}_2\text{As}=\text{CHNO}_2 \\ \qquad\qquad\qquad (\mathbf{9}) \end{array} \qquad (8)$$

It is necessary that the methylene reactant be a fairly strong carbon acid; less acidic compounds such as phenylacetonitrile, acetophenone, or fluorene do not react under these conditions (32). Sulfoxides also condense readily with reactive methylene compounds in the presence of dicyclohexylcarbodiimide (16) but triphenylarsine oxide did not form ylides under identical conditions. Triphenylarsine sulfide did not react with methylene compounds under any of the above conditions (32).

However, the reaction of tri-n-butylarsine oxide with dibenzoylmethane or with dimethyl malonate in triethylamine and in the presence of phosphorus pentoxide gave the corresponding ylides (11), respectively, (20, 40).

$$(\text{C}_4\text{H}_9)_3\text{AsO} + \text{H}_2\text{CR}_2 \xrightarrow{(\text{C}_2\text{H}_5)_3\text{N}/\text{P}_2\text{O}_5} (\text{C}_4\text{H}_9)_3\text{As}=\text{CR}_2 \qquad (9)$$
$$(\mathbf{11})$$
$$\text{R} = \text{COC}_6\text{H}_5, \text{COOCH}_3$$

In the reversible Wittig reaction, triphenylarsine oxide reacted with electron-deficient acetylene derivatives to form stable ylides. Thus triphenylarsine oxide reacted readily with methyl propiolate, ethyl phenylpropiolate, dimethyl acetylenedicarboxylate, and hexafluoro-2-butyne as well as dicyanoacetylene to give arsonium ylides (12). The reaction temperatures required ranged from −70°C in the case of dicyanoacetylene to 130°C in the case of ethyl phenylpropiolate (15).

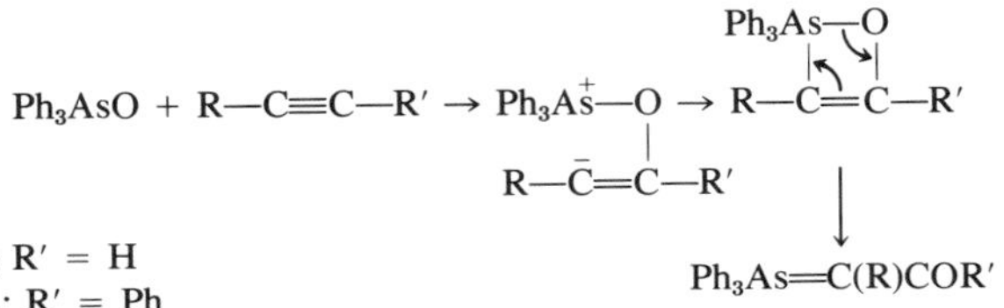

$$Ph_3AsO + R-C\equiv C-R' \rightarrow Ph_3\overset{+}{A}s-O \rightarrow R-\overset{Ph_3As-O}{\underset{}{C=C}}-R'$$

$$R-\overset{-}{C}=C-R'$$

R = COOCH₃; R' = H
R = COOC₂H₅; R' = Ph
R = R' = COOCH₃
R = R' = CF₃
R = R' = CN

$$Ph_3As=C(R)COR'$$
$$(12)$$

$$(10)$$

C. From Tertiary Arsines

Arsonium ylides have also been prepared from the decomposition of diazonium compounds in the presence of a tertiary arsine. Thus tetraphenylcyclopentadiene triphenylarsorane (13) was obtained by heating diazotetraphenylcyclopentadiene at its melting point in the presence of triphenylarsine (66).

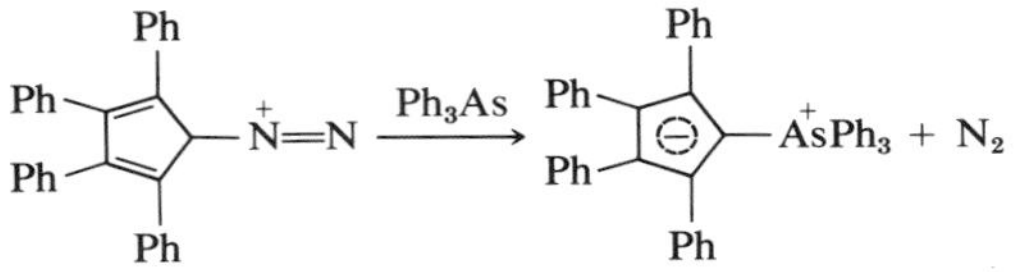

$$(11)$$

$$(13)$$

This method was extended to different diazonium salts and several arsonium ylides (14) were prepared (23, 32). The reaction is greatly facilitated by the presence of copper, copper–bronze, or copper salts. For example, attempts to prepare the bis(carbethoxy)methylene ylide by thermolysis of diethyl diazomalonate in the presence of triphenylarsine without the presence of a catalyst proved abortive, whereas this ylide was obtained in 61% yield if the reactants were heated at 150°C with copper–bronze (32).

$$Ph_3As + N_2=C(COOC_2H_5)_2 \xrightarrow[-N_2]{Cu/\Delta} Ph_3As=C(COOC_2H_5)_2$$
$$(14)$$

$$(12)$$

The reaction is assumed to involve initial formation of a carbene, by decomposition of the diazo compound with loss of nitrogen, followed by reaction of the electron-deficient carbene with the lone pair of electrons of the arsenic atom. Thermolysis of diazo compounds in copper-catalyzed reactions is known to provide singlet carbenes or carbenoid species (17).

Reaction of triphenylarsine and 2 mol of dimethyl acetylenedicarboxylate in water at 100°C afforded (2-oxo-1,2-dicarbomethoxyethylene)triphenylarsorane (15) (38). The reaction mechanism is not yet clear.

$$Ph_3As + 2\ H_3COOCC{\equiv}CCOOCH_3 \xrightarrow[\ CH_3OOCCH{=}CHCOOCH_3\ ]{H_2O}$$

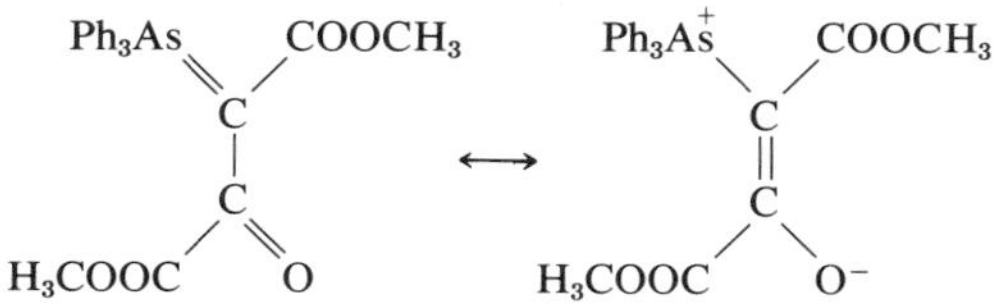

(13)

(15a)   (15b)

D. *From Other Ylides ("Transylidation")*

The methylene trimethylarsorane (17), which has up until now been difficult to make, was easily obtained on treating trimethylsilylmethylene trimethylarsorane (16) with trimethylsilanol (84).

$$Me_3As{=}CHSiMe_3 + Me_3SiOH \rightarrow Me_3SiOSiMe_3 + Me_3As{=}CH_2 \quad (14)$$
$$(16) \qquad\qquad\qquad\qquad\qquad\qquad (17)$$

In the presence of methanol, the ylide gave tetramethylarsonium methoxide (81).

$$Me_3As{=}CHSiMe_3 + MeOH \rightarrow Me_3SiOMe + Me_3As{=}CH_2 \quad (15)$$
$$(16) \qquad\qquad\qquad\qquad\qquad (17)$$

$$Me_3As{=}CH_2 + MeOH \rightarrow Me_4AsOMe$$
$$(17)$$

Alkylidene triphenylarsoranes (18) which still carry one hydrogen atom on the alkylidene moiety reacted with anhydrides or acyl halides to form stable ylides (19) (20, 32, 35, 36, 56, 58, 68).

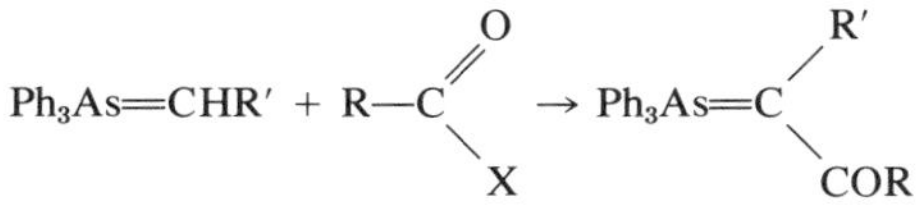

(16)

(18)   (19)

R = alkyl, aryl, H
X = halogen, —O—CO—R, —O—alkyl

Benzoylation of benzoylmethylene triphenylarsorane (20) with benzoyl bromide gave a kinetically controlled acylated product, which on treatment with sodium acetate in chloroform afforded thermodynamically controlled dibenzoylmethylene triphenylarsorane (21) (56). Acylation with carbonic acid anhydride (32, 56), phenylisocyanate (32), or chloroformic ester (32) gave in no case O-acylated product. Similarly, reaction with acetic anhydride afforded 1,3-dioxo-1-phenyl-butylidene-(2)-triphenylarsorane (56).

$$
\underset{(20)}{Ph_3\overset{+}{As}-CH=C\underset{Ph}{\overset{O^-}{\diagup}}}
\xrightarrow[\text{(PhCO)}_2\text{O}]{\text{PhCOBr}}
\left[Ph_3\overset{+}{As}-CH=C\underset{OCOPh}{\overset{Ph}{\diagup}} \right] Br^-
\tag{17}
$$

$$
\xrightarrow{\text{NaOAc}} \underset{(21)}{Ph_3As=C(COPh)_2}
$$

Cyclopentadiene triphenylarsorane (22) reacted with aromatic diazonium salts to give coupling products (24).

$$
\underset{(22)}{Ph_3\overset{+}{As}-\overset{\ominus}{\langle \bigcirc \rangle}} + [Ph-\overset{+}{N}_2]\, X^- \rightarrow \underset{\underset{Ph-N}{\overset{\|}{N}}}{Ph_3As=\langle \bigcirc \rangle}
\tag{18}
$$

Under the conditions of the Vilsmeyer reaction, 2,3,4-triphenyl-cyclopentadiene triphenylarsorane (8) was converted to its formylated product (23) (68).

$$
\underset{(8)}{Ph_3As=} \quad \xrightarrow{\text{POCl}_3/\text{DMF}} \quad \underset{(23)}{Ph_3As=}
\tag{19}
$$

p-Bromobenzoylmethylene triphenylarsorane (24) reacted with phenyl-sulfine to give sulfinylated product (97), whereas carbethoxymethylene triphenylarsorane reacted with phenylsulfene under the same conditions to form its benzyl sulfonyl derivative accompanied with ethyl cinnamate and triphenylarsine (97).

Ph_3As=$CHCOC_6H_4Br(p)$
     (24)

$+ C_6H_5$—CH=SO → Ph_3As=$C(COC_6H_4Br)SOCH_2C_6H_5$ (20)

Ph_3As=$CHCOOC_2H_5 + C_6H_5CH$=SO_2 → Ph_3As=$C(COOC_2H_5)SO_2CH_2C_6H_5$
                              $+ C_6H_5CH$=$CHCOOC_2H_5 + Ph_3As$ (21)

Thermolysis of 1-(4-nitrobenzoyl)aziridine with benzoylmethylene triphenylarsorane in toluene afforded [4-(4-nitrobenzoylamino)-1-oxo-1-phenylbutylidene-(2)]triphenylarsorane in 43% yield (37).

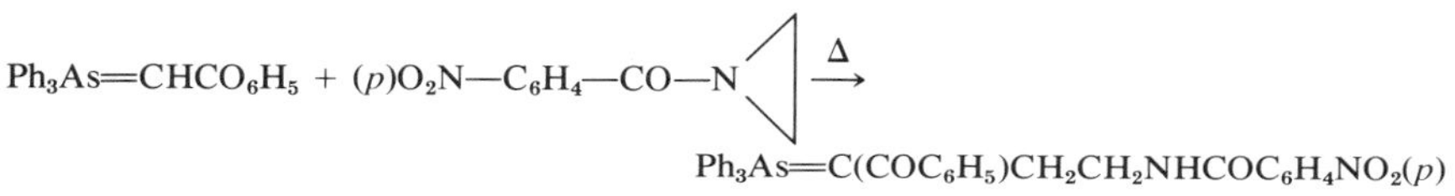

Ph_3As=$CHCO_6H_5 + (p)O_2N$—C_6H_4—CO—N$\overset{\Delta}{\longrightarrow}$

Ph_3As=$C(COC_6H_5)CH_2CH_2NHCOC_6H_4NO_2(p)$

Thermal treatment of carbethoxymethylene- or p-bromobenzoylmethylene triphenylarsorane (25) with dimethyl acetylenedicarboxylate gave rise to stable ylides (97).

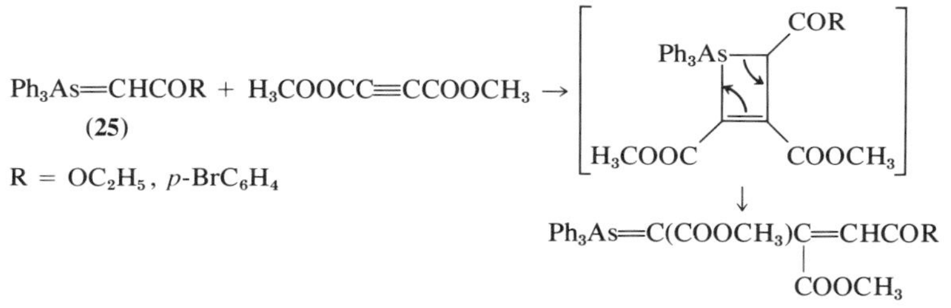

Ph_3As=$CHCOR + H_3COOCC$≡$CCOOCH_3$ →
     (25)

R = OC_2H_5, p-BrC_6H_4

$$\begin{bmatrix} & & COR \\ & Ph_3As & \\ H_3COOC & & COOCH_3 \end{bmatrix}$$ (22)

↓

Ph_3As=$C(COOCH_3)C$=$CHCOR$
                    |
                  $COOCH_3$

Arsoranes that contain fluoroacyl groups have been obtained by the methods described in the following paragraphs.

Reactions of carbomethoxymethylene triphenylarsorane with perfluoroacyl chlorides gave perfluoroacylcarbomethoxymethylene triphenylarsoranes (26) (42).

$R_FCOCl + Ph_3As$=$CHCO_2CH_3$ → $(Ph_3\overset{+}{A}s$—$CHCO_2CH_3)Cl^- \xrightarrow{Ph_3As=CHCO_2CH_3}$
                              |
                          O=C—R_F

$R_F = CF_3$

$R_F = F_3CCF_2CF_2$ $(Ph_3\overset{+}{A}s$—$CH_2CO_2CH_3) Cl^- +$

$Ph_3\overset{+}{A}s$—C—CO_2CH_3 (23)
                  ‖
          $\overset{-}{O}$—C—R_F
              (26)

Carbomethoxymethylene or benzoylmethylene triphenylarsorane reaction with perfluoropropylene oxide to give the corresponding fluorinated ylides (27). The following mechanism was suggested (44):

$$\text{Ph}_3\text{As}=\text{CH}-\overset{\displaystyle O}{\overset{\|}{\text{C}}}\diagdown_R + \text{F}_3\text{C}-\text{CF}-\text{CF}_2 \rightarrow \left[\begin{array}{c} \text{Ph}_3\overset{+}{\text{As}}-\overset{H}{\underset{}{\text{C}}}-\overset{\displaystyle O}{\overset{\|}{\text{C}}}\diagdown_R \\[4pt] \overset{\bar{}}{\text{O}} \\[2pt] \text{F}_3\text{C}-\text{CF}-\text{C}-\text{F} \\ :\text{F} \end{array} \right] \rightarrow$$

R = OMe
R = Ph   (24)

$$\text{Ph}_3\text{As}=\text{CH}-\overset{\displaystyle O}{\overset{\|}{\text{C}}}\diagdown_R + \text{F}_3\text{CCF}_2\text{CFO} \rightarrow \left[\begin{array}{c} \text{Ph}_3\overset{+}{\text{As}}-\text{CH}-\overset{\displaystyle O}{\overset{\|}{\text{C}}}\diagdown_R \\ \text{O}=\text{C}-\text{CF}_2\text{CF}_3 \end{array} \right]^{F^-} \quad \text{Ph}_3\text{As}=\text{CH}-\overset{\displaystyle O}{\overset{\|}{\text{C}}}\diagdown_R \longrightarrow$$

$$\text{Ph}_3\overset{+}{\text{As}}-\text{C}-\overset{\displaystyle O}{\overset{\|}{\text{C}}}\diagdown_R \quad + \quad \left[\text{Ph}_3\overset{+}{\text{As}}-\text{CH}_2-\overset{\displaystyle O}{\overset{\|}{\text{C}}}\diagdown_R \right] F^-$$
$$\overset{\|}{\text{O}^-\text{C}-\text{CF}_2\text{CF}_3}$$
$$(\mathbf{27})$$

Carbomethoxymethylene triphenylarsorane reacted with fluoroolefins at room temperature to afford **28** in moderate yields (49). With 4,4-dichlorohexafluorobutene it gave carbomethoxy(4,4-dichloropentafluorobuten-1-yl)-methylene triphenylarsorane, which was easily converted to carbomethoxy(4,4-dichloro-2,3,3,4-tetrafluorobutanoyl)-methylene triphenylarsorane either on passing through an alumina column or heating with moistened benzene. The reaction products of carbomethoxymethylene triphenylarsorane with tetrafluoroethylene and with perfluoropropylene on passing through an alumina column gave rise to the corresponding fluoroacyl derivatives (**28**). The following mechanism was suggested:

$$\text{Ph}_3\text{As}=\text{CHCO}_2\text{CH}_3 + \text{CF}_2=\text{CFR}_F \rightarrow \left[\begin{array}{c} \text{Ph}_3\overset{+}{\text{As}}-\text{CHCO}_2\text{CH}_3 \\ \text{F}-\text{CF}-\text{CFR}_F \end{array} \right] \xrightarrow{-\text{HF}}$$

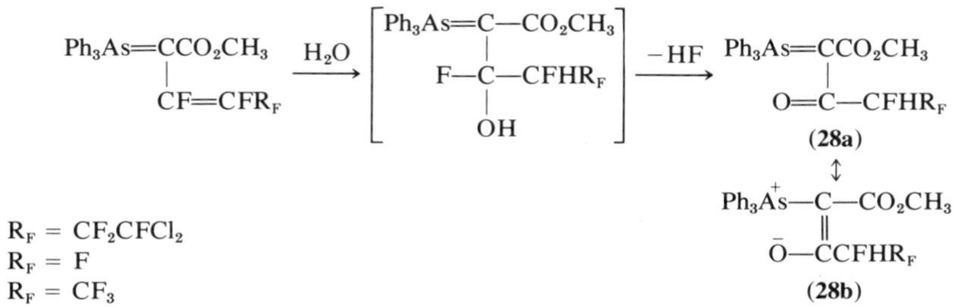

$$Ph_3As\!=\!CCO_2CH_3 \atop CF\!=\!CFR_F \quad \xrightarrow{H_2O} \quad \left[\begin{array}{c} Ph_3As\!=\!C\!-\!CO_2CH_3 \\ | \\ F\!-\!C\!-\!CFHR_F \\ | \\ OH \end{array} \right] \quad \xrightarrow{-HF} \quad Ph_3As\!=\!CCO_2CH_3 \atop O\!=\!C\!-\!CFHR_F$$

$$\text{(28a)}$$

$$\updownarrow$$

$$Ph_3\overset{+}{As}\!-\!C\!-\!CO_2CH_3 \atop \overset{-}{O}\!-\!\overset{\|}{C}CFHR_F$$

$$\text{(28b)}$$

(25)

$R_F = CF_2CFCl_2$
$R_F = F$
$R_F = CF_3$

The formation of *trans*-alkene was interpreted by conformational analysis:

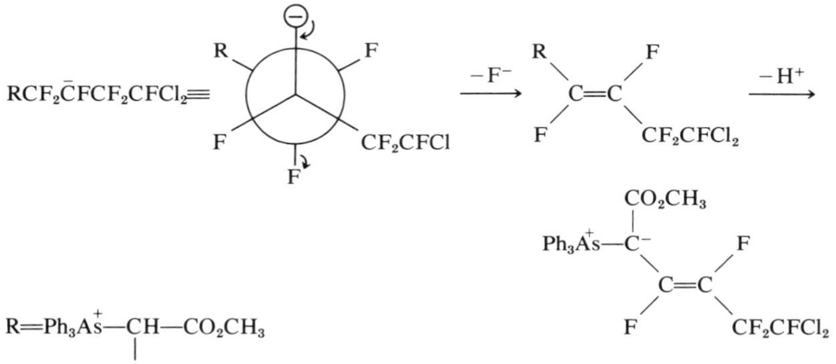

$$RCF_2\overset{-}{C}FCF_2CFCl_2 \equiv$$

$$\xrightarrow{-F^-} \quad \begin{array}{c} R \\ \diagdown \\ F \end{array} C\!=\!C \begin{array}{c} F \\ \diagup \\ CF_2CFCl_2 \end{array} \quad \xrightarrow{-H^+}$$

(26)

$$Ph_3\overset{+}{As}\!-\!\overset{\overset{\displaystyle CO_2CH_3}{|}}{\underset{}{C^-}} \begin{array}{c} F \\ \diagdown \\ \end{array} \atop C\!=\!C \begin{array}{c} F \\ \diagup \end{array} \begin{array}{c} \\ CF_2CFCl_2 \end{array}$$

$R\!=\!Ph_3\overset{+}{As}\!-\!CH\!-\!CO_2CH_3$

The benzoylmethylene triphenylarsorane reacted with sulfur trioxide to give **29** and with bromine to give **30** (74).

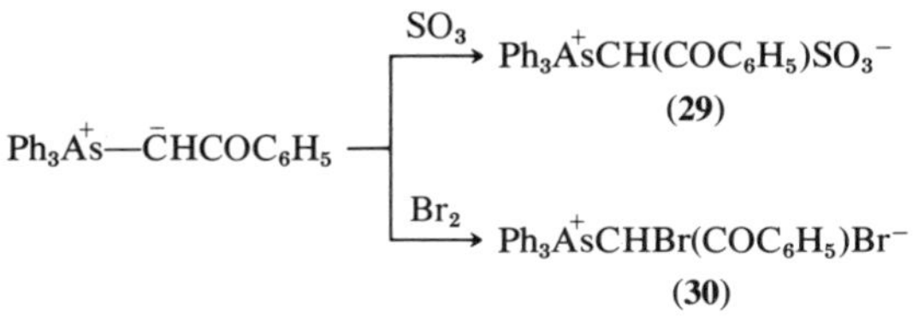

$$Ph_3\overset{+}{As}\!-\!\overset{-}{C}HCOC_6H_5 \quad \begin{array}{c} \xrightarrow{SO_3} \quad Ph_3\overset{+}{As}CH(COC_6H_5)SO_3{}^- \\ \text{(29)} \\ \\ \xrightarrow{Br_2} \quad Ph_3\overset{+}{As}CHBr(COC_6H_5)Br^- \\ \text{(30)} \end{array}$$

(27)

III

STRUCTURE AND PHYSICAL PROPERTIES

The first arsonium ylide was prepared in 1902 (70), but its structure was not ascertained until 1950 (63). In the past decade, UV, IR, NMR, X-ray diffraction, and ESCA techniques have been developed and applied to verify the arsonium ylide structure. Nevertheless, more information is needed in order to clarify the relationship of the structure and physical properties of arsonium ylides to the phosphonium ylides. Furthermore, most of the structural studies are now only applicable to stabilized arsonium ylides.

A. *Ylide Structure*

Arsonium ylides can be represented generally by the canonical forms **31a** and **31b**. By the application of X-ray diffraction to 2-acetyl-3,4,5-triphenylcyclopentadiene triphenylarsorane, Ferguson and Rendle (*21*) established that the canonical form **7c** makes a significant contribution to the ground state structure.

Figure 1 shows the principal bond lengths and angles.

The bond lengths of C-1—C-5 and C-3—C-4 (double bond in **7c**) are shorter than that of C-2—C-3 and C-4—C-5 (double bond in **7a**) indicating that the ground state population of **7c** exceeds that of **7a**. Furthermore, the C-1—C-2 distance (1.445 Å) is significantly less than the single bond length of cyclooctatetraene (1.462(1) Å) vindicating the inclusion of **7b**, whereas the C-1—As bond length, (1.881 Å), being shorter than a normal phenyl arsonium distance (1.897 Å), is in accordance with the participation of **7a**. By comparing these bond lengths with standard ones, the population densities of **7a** (30–35%), **7b** (20–30%), and **7c** (40–45%) were deduced. The oxygen atom interacts appreciably with the arsenic, resulting in the distortion of the tetrahedral-type geometry implied in **7a** and **7b** toward a trigonal bipyramidal configuration.

Yamamoto and Schmidbaur (*102*) determined the ^{13}C-NMR spectra of

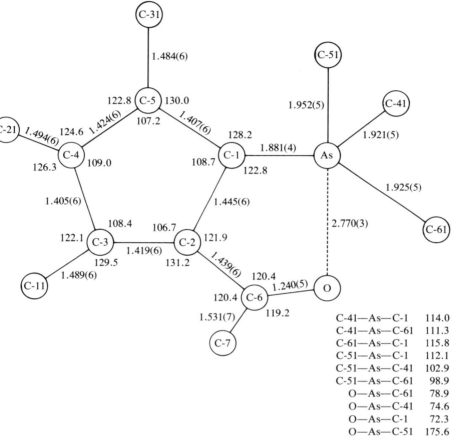

FIG. 1. Principal distances (with e.s.d.s. in parenthesis) and angles. From Ferguson and Rendle (21).

methylene triphenylarsorane (Ph$_3$AsCH$_2$) and found that the bonding of the ylidic carbon of arsenic was most likely unchanged (sp^3), whereas that for phosphorus was changed (sp^3 → sp^2). Therefore, the arsenic ylidic carbon has pseudotetrahedral geometry, whereas the phosphorus one has planar geometry.

This result can be explained on the basis of a markedly reduced π-bonding contribution in the arsenic series. X-ray data on arsenic ylides further verified this assumption by showing increased As—C distances and a nonplanar geometry at the ylidic carbon.

Ylide conformations have been determined by ^{1}H-NMR studies. Dale and Froyen (20) studied the ^{1}H-NMR spectrum of carbomethoxymethylene triphenylarsorane (Ph_3As=$CRCOOCH_3$) and found that the cis–trans equilibrium varied with temperature.

$$Ph_3As=C\underset{\underset{CH_3O}{\diagdown}}{\overset{\overset{R}{\diagup}}{}}C=O \quad \rightleftharpoons \quad Ph_3As=C\underset{\underset{O}{\diagdown}}{\overset{\overset{R}{\diagup}}{}}C-OCH_3$$

$$(E) \qquad\qquad\qquad (Z)$$

The free energy of activation of rotation in these ylides was calculated and was found to increase in the following order: H > Ph > CN. Among these compounds, the cisoid structure (Z) was shown to be the major contributing form.

Gosney and Lloyd (32) studied the ^{1}H-NMR spectra of diketo ylides and suggested that these ylides existed as Z,Z isomers. The Z,Z conformation is favored because the two negatively charged O atoms are not only disposed to interact with the arsonium atom but also to be as far apart as possible. The ^{1}H-NMR spectrum of ylide 32 showed a broad singlet at δ 7.6 ppm indicating hydrogen bonding between the carbonyl group and amide proton. Thus there is Z,E conformation in this arsonium ylide.

(32a)   (32b)

On the basis of IR spectra, we suggested that the resonance structures of perfluoropropionyl carbomethoxy ylides **33** and perfluoropropionyl benzoylylides **34** are different (*44*).

(33)   M = P, As   (34)

B. *Physical Properties*

1. *Basicity*

The pK_a values of a series of conjugated acids of the arsonium ylides have been determined by potentiometric titration (*73*). A decrease in the basicity of the ylides with an increase in substituent electronegativity was observed. The effect of the substituents agrees well with the Hammett equation. Arsonium ylides are more basic (200–230 times) than the corresponding phosphonium compounds.

$$Ph_3As\overset{+}{=}CH_2-C(O)C_6H_4-X(p)$$

Substituent X	OCH_3	CH_3	H	Cl	Br	NO_2
pK_a	9.16	8.97	8.52	8.02	7.95	6.67

Johnson and Amel (*53*) studied the pK_a values of a series of phenacylonium salts and found that the acidifying effects of the onium group increased in the following order: arsonium < phosphonium < sulfonium. The acidifying effects of the onium atom substituents increase in the order: methyl < *n*-butyl < phenyl.

2. *Ultraviolet Spectra*

The ultraviolet spectra of some stable arsonium ylides have been studied (*32*). The absorption maxima of the ylides occurred in the region

240–350 nm. As the conjugation is enhanced by means of electron-withdrawing substituents in the carbanionic moieties, a bathochromic shift of this absorption maxima is observed. Freeman *et al.* (*25*) also studied the ultraviolet spectra of phosphonium, arsonium, stibonium, and bismuthonium tetraphenylcyclopentadienylides.

$$Ph_3\overset{+}{X} - \underset{Ph}{\overset{Ph}{\bigcirc}}^{Ph}_{Ph}$$

X	P	As	Sb	Bi	
γ_{max}	288	291	349	525	596
Solvent	C_2H_5OH	C_2H_5OH	$CHCl_3$	C_2H_5OH	C_6H_6

On going down the periodic table (from P to As, Sb, and Bi), the absorption maxima are greatly shifted toward longer wavelengths. The shift is consistent with a decreasing contribution from the covalent form or the increasing contribution by the dipolar form of the ylides with increasing atomic number of the heteroatom.

3. *Infrared Spectra*

Johnson. and Schubert (*56*) reported the carbonyl absorption of phenacylidene triphenylarsorane ($Ph_3As=CHCOC_6H_5$) at 1570 cm^{-1}, whereas that for the ylide salt ($Ph_3\overset{+}{As}-CH_2COC_6H_5Br^-$) was 1660 cm^{-1}. These results indicate that the enolate structure (**20b**) makes a significant contribution.

$$Ph_3\overset{+}{As}-\overset{-}{C}HCOC_6H_5 \longleftrightarrow Ph_3\overset{+}{As}-CH=\overset{\overset{\bar{O}}{/}}{C}-C_6H_5$$

$$\text{(20a)} \qquad\qquad\qquad \text{(20b)}$$

Gosney and Lloyd (*32*) studied the infrared spectra of some stable arsonium ylides containing electron-withdrawing substituents in the alkylidene moiety and found that the stretching frequencies of the carbonyl group of these ylides were uniformly low due to the delocalization of negative charge onto these groups. All these stretching frequencies were lower than those in the corresponding phosphonium ylides.

4. Nuclear Magnetic Resonance Spectra

Johnson and Schubert (56) reported that the ¹H-NMR spectrum of phenacylidene triphenylarsorane showed 20 aryl protons at δ 7.2–8.1 and a single methine proton at δ 4.75. The latter signal was quite broad unless the sample and solvent were perfectly dried, probably because of proton exchange. The ¹³C-NMR spectra of some arsoranes were also studied (26, 102). The data are listed in Table I.

The ¹³C-NMR spectra of methylene triphenylarsorane and its salts showed that J_{H-C} of the methyl group is only slightly reduced upon deprotonation of the arsonium cation, suggesting that the bonding of the ylidic carbon is unchanged (sp³).

Fronza et al. (26) found an increase in the H—C-1 coupling constant in going from phosphorus (164.5 Hz in Ph₃P=CHCOCH₃) to arsenic (173.5 Hz in Ph₃As=CH—COCH₃) indicating increased s-character of the H—C-1 bond in the arsenic ylide in the series. The corresponding salts follow the same trend (130.0 for phosphorus, 135.0 for arsenic). The H—C-1 coupling constant in acetylmethylene triphenylarsorane is 173.5 Hz, 36.8 Hz larger than in methylene triphenylarsorane. This large increase is due partly to the carbonyl stabilization, but even more likely due to the change in C-1 hybridization from sp³ to sp². Thus the ylidic carbon in nonstabilized arsonium ylides is pseudotetrahedral (sp³) (83, 102), whereas that in stabilized arsonium ylides is planar (sp²).

TABLE I

¹³C CHEMICAL SHIFTS AND $'J_{H-C}$ FOR ARSONIUM SALTS AND YLIDES

Compound	C-1 (ppm)	C-2 (ppm)	$'J_{H-C-1}$ (Hz)	Reference
(1) {Cl⁻	7.6(q)		142.0	
Ph₃A⁺sCH₃ {Br⁻	8.8(q)		142.0	102
{I⁻	9.9(q)		142.0	
(1) Ph₃As=CH₂	1.6(t)		136.7	
(1) (2) Ph₃A⁺s—CH₂—COPhBr⁻	42.6	192.8		
				26
(1) (2) Ph₃As=CH—COPh	57.1	181.8		
(1) (2) Ph₃A⁺s—CH₂—COCH₃Br⁻	44.2	201.9	135.0	
				26
(1) (2) Ph₃As=CH—COCH₃	56.9	187.0	173.5	

5. Photoelectron Spectra

The photoelectron spectra of the arsonium ylides $(CH_3)_3As{=}CH_2$ and its derivatives $(CH_3)_3As{=}CHSi(CH_3)_3$ and $(CH_3)_3As{=}C[Si(CH_3)_3]_2$ have been studied (86). Vertical ionization potentials of arsonium and phosphonium ylides are listed in the following tabulation.

Ylide	As=C (eV)	As—C (eV)
$(CH_3)_3As{=}CH_2$	6.72	11.4
$(CH_3)_3As{=}CHSi(CH_3)_3$	6.56	11.4
$(CH_3)_3As{=}C[Si(CH_3)_3]_2$	6.66	10.9/→
	P=C (eV)	P—C (eV)
$(CH_3)_3P{=}CH_2$	6.81	11.8
$(CH_3)_3P{=}CHSi(CH_3)_3$	6.81	11.7
$(CH_3)_3P{=}C[Si(CH_3)_3]_2$	6.92	11.1/→

The π-ionization energies of arsonium ylides are very low, as is typical for ylides. The arsonium ylides are slightly more easily ionized than the phosphorus ylides, the difference amounting to about 0.1 eV: $\Delta IE_\pi(As{=}CH_2/P{=}CH_2) = 0.1$ eV. The differences in σ-ionization energies between As—C and P—C are also significant: $\Delta IE_\sigma(As{-}C/P{-}C) \approx 0.2\text{–}0.4$ eV.

The nature of the bonding in some stable arsonium and phosphonium ylides ($Ph_3X{=}CHCOR$: R = alkoxy, Ph; X = P, As) was studied by photoelectron spectroscopy (19). The data showed a partial positive charge on phosphorus and arsenic and little change in the charge on the heteroatoms by changing R from alkoxy to phenyl.

6. Mass Spectra

Only a few reports on studies of the mass spectra of arsonium ylides have appeared in the literature. Gosney and Lloyd (32) studied the mass spectra of bis(carbethoxy)methylene triphenylarsorane and nitromethylene triphenylarsorane and found that preliminary fragmentation resulted in loss of the carbanionic moiety. We studied the mass spectra of 11 arsonium ylides and of triphenylarsine difluoride (30).

$$(C_6H_5)_3\overset{+}{As}{-}\overset{-}{C}\overset{\displaystyle COR^1}{\underset{\displaystyle R^2}{<}}$$

(35)

No.	R^1	R_2
1	OCH_3	H
2	C_6H_5	H
3	C_6H_5	$CF{=}CF{-}CF_3$
4	OCH_3	$CF{=}CF{-}CF_2CFCl_2$

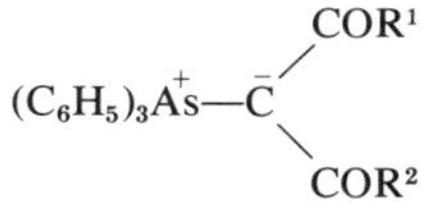

$$(C_6H_5)_3\overset{+}{As}{-}\overset{-}{C}\overset{\displaystyle COR^1}{\underset{\displaystyle COR^2}{\big\langle}}$$

(36)

No.	R^1	R^2
5	OCH_3	CHF_2
6	OCH_3	$CHFCl$
7	OCH_3	$CHFCF_3$
8	OCH_3	CF_2CF_3
9	OCH_3	$CHFCF_2CFCl_2$
10	C_6H_5	$CHFCF_3$
11	C_6H_5	CF_2CF_3

The following mechanism was suggested for the fragmentation of the parent ylides 1 and 2:

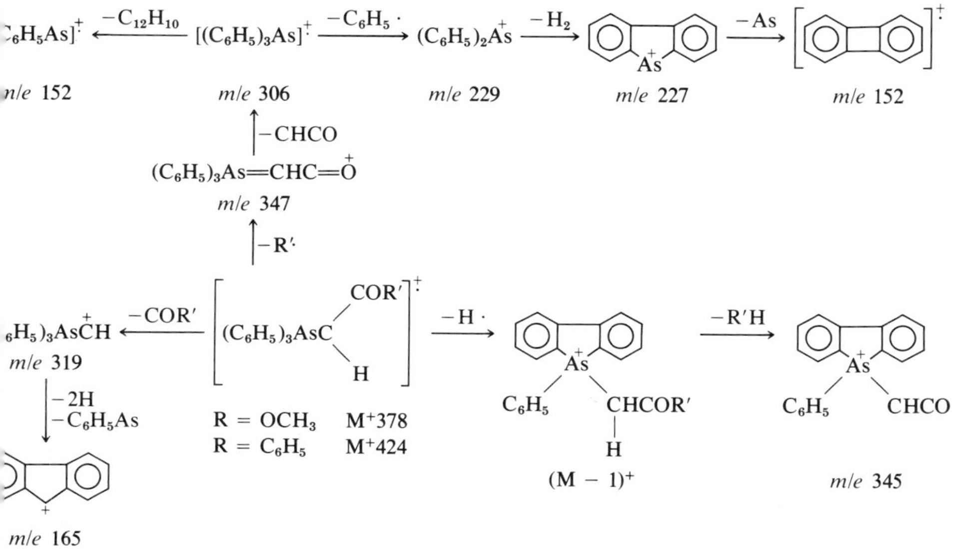

SCHEME 1

We found that:

1. For arsonium ylides 1 and 2 the molecular ion peak and $(M - 1)^+$ peak were characteristic, whereas for the fluorinated arsonium ylides (3 to 9) they were not discernible at all.

2. For compounds 3 and 4 with fluorinated olefinic substituents, as would be predicted, peaks of fluorine rearrangement ions, such as $C_6H_5AsF^+$, $(C_6H_5)_2AsF^+$, and $(C_6H_5)_3AsF^+$ were observed in addition to triphenylarsine and its degradation fragments.

3. For compounds 5 to 11 with fluoroacyl substituents, the base peak was the ion formed by the elimination of the fluoroalkyl group.

4. The fluorene ion peak (m/e 165) with low intensity appeared in the mass spectra of all aforementioned compounds except compounds 3 and 4.

IV

CHEMICAL REACTIONS AND SYNTHETIC APPLICATIONS

A. Reactions of Stabilized Arsonium Ylides with Carbonyl Compounds

Phosphonium ylides that bear an electron-withdrawing substituent (CO_2R, COC_6H_5, CN, etc.) in the alkylidene moiety show considerably reduced nucleophilicity, and in many cases will not react with carbonyl groups. For example, Trippett and Walker (95) reported that neither carbethoxymethylene triphenylphosphorane nor cyanomethylene triphenylphosphorane could react with ketones, whereas Fodor and Tömöskozi (22) claimed that the former reacted with ketones only under forcing conditions, and Ramirez and Dershowitz (77) reported that benzoylmethylene triphenylphosphorane did not react with cyclohexanone at all. The stabilized phosphonium ylides are therefore of little value as reagents in Wittig reactions. As discussed earlier, arsonium ylides are more reactive than phosphonium ylides; the greater reactivity is attributed to their greater polarity which results from less effective $d\pi-p\pi$ bonding to arsenic.

Huang et al. (48) found that the carbomethoxymethylene triphenylarsorane reacted with ketones as well as with aldehydes to give various unsaturated esters in good yield. The by-product, triphenylarsine oxide, can be removed by washing with dilute hydrochloric acid. A comparison

TABLE II

COMPARISON OF YIELDS FROM THE REACTION OF CARBONYL COMPOUNDS WITH
CARBOMETHOXYMETHYLENE TRIPHENYLARSORANE AND WITH THE
CORRESPONDING PHOSPHONIUM YLIDES

Carbonyl compounds	Arsonium ylide (Yield %)	Phosphonium ylide (Yield %)	Reference
C_6H_5CHO	90	77	99
p-$NO_2C_6H_4CHO$	98		
o-ClC_6H_4CHO	96		
p-$CH_3OC_6H_4CHO$	95		
$C_6H_5COC_6H_5$	77	0	96
$C_6H_5COCH_3$	52	0	96
	63	25	96
	32	0, 7	87, 96

of olefin yields from arsonium ylide and phosphonium ylide reactions is shown in Table II.

We found also that carbomethoxymethylene triphenylarsorane reacted with a series of aliphatic ketones under very mild conditions (47). For example, with acetone it reacted at room temperature, whereas with higher aliphatic ketones at their refluxing temperature (boiling point of the aliphatic ketones). The yields of the α,β-unsaturated esters were moderate.

$$Ph_3As=CHCOOCH_3 + RR'C=O \rightarrow RR'C=CHCOOCH_3 + Ph_3AsO \qquad (28)$$
$$R = CH_3; R' = CH_3, C_2H_5, n\text{-}C_3H_7, i\text{-}C_4H_9, n\text{-}C_5H_{11}, n\text{-}C_6H_{13}$$
$$R,R' = CH_2CH_2CH_2CH_2CH_2$$

On the basis of ^{1}H-NMR and gas chromatographic analysis, it was shown that the product from aliphatic ketones in each case was a mixture of E and Z isomers. Changing the solvent, molar ratio of reactants, reaction temperature, reaction time, or the presence of sodium bromide had little effect on the E/Z ratio of reaction products. On the other hand, the structure of the substrate showed a profound effect, especially when a bulky tert-butyl group was adjacent to the carbonyl (see Table III). In this case 99% of E olefinic compound was formed stereospecifically (46).

TABLE III

PROPORTION OF E-ISOMER IN THE PRODUCT

Structure of E-isomer in olefinic product	% of E-isomer
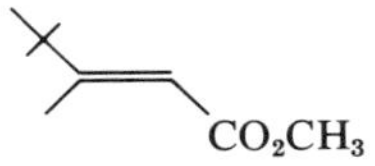	99
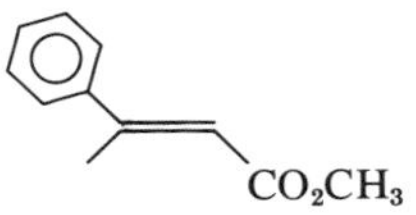	83–95
	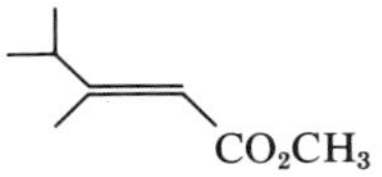 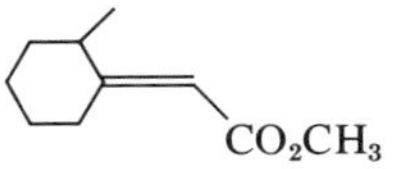 84–93
	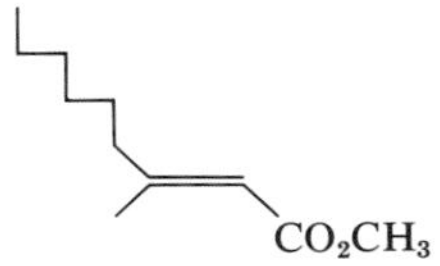 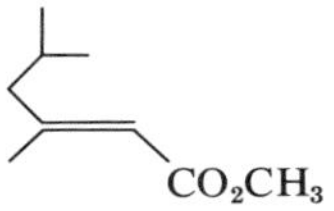 60–70

$$Ph_3As{=}CHCOOCH_3 + \underset{R'}{\overset{R}{\diagdown}}C{=}O \rightarrow \underset{R'}{\overset{R}{\diagdown}}C{=}CHCOOCH_3 + Ph_3AsO \qquad (29)$$

$R = CH_3$; $R' = n\text{-}C_6H_{13}$, $i\text{-}C_4H_9$, iso-C_3H_7, C_6H_5, $t\text{-}C_4H_9$

$R,R' = CH_2CH_2CH_2CH_2C(CH_3)H$

Thus the stabilized arsonium ylide carbomethoxymethylene triphenylarsorane reacted with ketones forming the thermodynamically more stable olefinic compounds predominantly. This could be explained as follows:

The reactants combine to give a betaine with either A or B configuration, the two species being in equilibrium with ylide and ketone. Because of the unfavorable interaction between the negative carbomethoxy group and the oxygen atom in the A isomer, the B isomer should be preferred and so should form more rapidly. Furthermore, the betaine intermediates

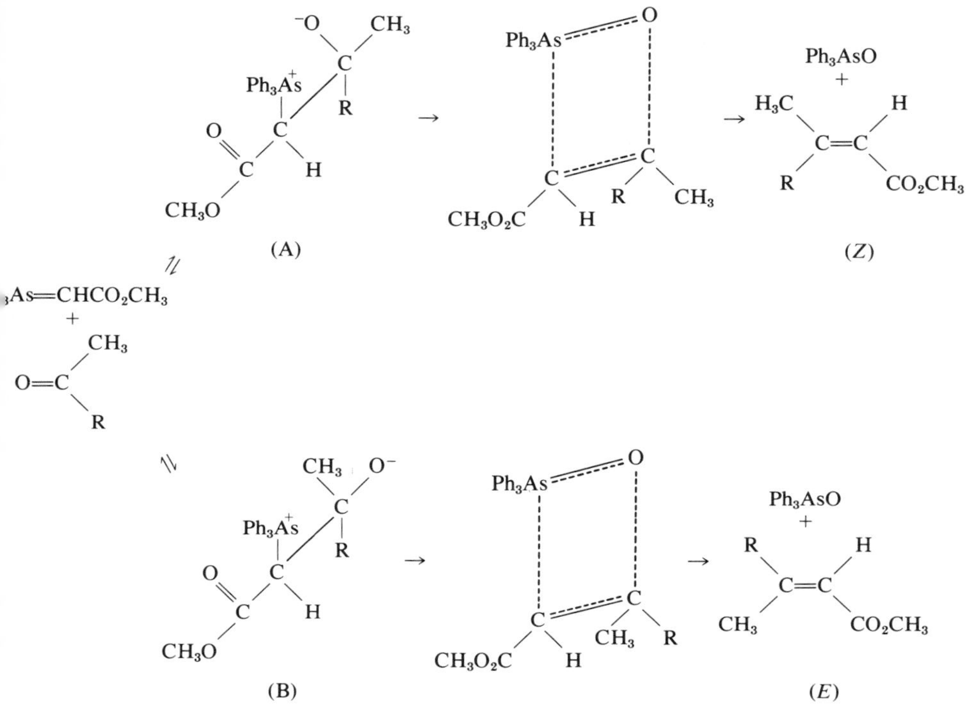

$Ph_3As=CHCO_2CH_3$
+
$O=C$ with CH_3 and R

(A) Ph_3As, ^-O, CH_3, C, R, O, C, C, H, CH_3O

→

$Ph_3As \cdots O$
$C=C$
CH_3O_2C H R CH_3

→

Ph_3AsO
+
H_3C H
$C=C$
R CO_2CH_3
(Z)

(B) Ph_3As, CH_3, O^-, C, R, O, C, C, H, CH_3O

→

$Ph_3As \cdots O$
$C=C$
CH_3O_2C CH_3 R H

→

Ph_3AsO
+
R H
$C=C$
CH_3 CO_2CH_3
(E)

SCHEME 2

eliminate triphenylarsine oxide irreversibly and the E isomer is obtained predominately because of a lower energy transition state and less steric hindrance. Consequently, the B isomer, which is preferentially consumed, is continuously replenished by the mobile equlibrium. Under these circumstances a preference for the B → E route might be expected.

Ramirez and Dershowitz (77) reported that the benzoylmethylene triphenylphosphorane did not react with cyclohexanone and reacted with benzaldehyde only with difficulty. Ting *et al.* found that benzoylmethylene triphenylarsorane reacted with aromatic aldehydes and with cyclohexanone under very mild conditions producing excellent yields of the resulting enones (94). These results, summarized in Table IV, show again that the arsenic reagent is more active than the corresponding phosphorus reagent.

Trippett and Walker (95) reported that cyanomethylene triphenylphosphorane did not react with ketones. In contrast to the phosphorane,

TABLE IV

REACTION OF BENZOYLMETHYLENE TRIPHENYLARSORANE WITH AROMATIC
ALDEHYDES AND CYCLOHEXANONE

Carbonyl compound	Reaction conditions	Product	Yield (%)
p-$NO_2C_6H_4CHO$	Room Temp., Standing	p-$NO_2C_6H_4CH{=}CHCOC_6H_5$	91
p-ClC_6H_4CHO	Refluxing in C_2H_5OH Soln, $\frac{1}{2}$ hr	p-$ClC_6H_4CH{=}CHCOC_6H_5$	80
p-$(CH_3)_2NC_6H_4CHO$	Refluxing in C_6H_6 Soln, $\frac{1}{2}$ hr	p-$(CH_3)_2NC_6H_4CH{=}CHCOC_6H_5$	71
	Refluxing in C_6H_6 Soln, 6 hr	$=CHCOC_6H_5$	73

cyanomethylene triphenylarsorane was found to be reactive toward a number of ketones, giving α,β-unsaturated nitriles in fair to good yields. The products and their yields are shown in Table V (*41*).

Nesmeyanov *et al.* (*74, 75*) reported the reaction of acetylmethylene triphenylarsorane with benzaldehyde. Huang *et al.* (*50*) found that this ylide reacted with ketones as well as aldehydes in moderate to excellent yields.

$$Ph_3As{=}CHCOCH_3 + RR'C{=}O \rightarrow RR'C{=}CHCOCH_3 + Ph_3AsO \qquad (30)$$

R = H: R′ = p-$NO_2C_6H_4$, p-$(CH_3)_2NC_6H_4$, , p-HOC_6H_4

R = CH_3, R′ = C_6H_5; R,R′ = $CH_2CH_2CH_2CH_2CH_2$

Bravo *et al.* reported that the carbonyl-stabilized arsonium ylides reacted with *o*-hydroxybenzaldehydes to give *o*-hydroxychalcones (*12*).

The fact that the stabilized arsonium ylides are more reactive than corresponding phosphonium ylides was further demonstrated by the reaction between dicarbethoxymethylene triphenylarsorane and benzaldehyde which gave diethyl benzylidene malonate in 72% yield (*32*). However, dicyano, benzoylcarbethoxy, acetylcarbethoxy, acetylphenylaminocarbonyl, cyanocarbethoxy, or nitromethylene triphenylarsorane failed to react with benzaldehyde. They were only reactive toward stronger electrophiles such as *p*-nitrobenzaldehyde (*32*).

REACTION PRODUCTS OF CYANOMETHYLENE TRIPHENYL ARSORANE WITH VARIOUS KETONES
AND THEIR YIELDS

Ketone	Product	Yield (%)	Ketone	Product	Yield (%)
CH_3–C(=O)–n-C_3H_7	CH_3, n-C_3H_7 C=CHCN	42	(trimethylcyclohexenyl-propenyl ketone)	(trimethylcyclohexenyl) CH=CH–CHCN	70
cyclohexanone	cyclohexylidene=CHCN	54	1-tetralone	=CHCN (A)[*] + CH_2CN	54
CH_3–C(=O)–C_6H_5	CH_3, C_6H_5 C=CHCN	30	steroid ketone (C_8H_{17})	C_8H_{17} steroid =CHCN (H, CN) (B)[**] + C_8H_{17} steroid =CH (NC, H)	87
C_6H_5–C(=O)–C_6H_5	C_6H_5, C_6H_5 C=CHCN	40			

[a] The isolated yield of (A) was 33%.
[b] The isolated yield of (B) was 32%.

$$Ph_3As = CRR' + p\text{-}O_2NC_6H_4CHO \rightarrow Ph_3AsO + RR'C{=}CHC_6H_4NO_2\text{-}(p) \quad (31)$$

$$R = R' = CN;\ R = COPh, R' = COOC_2H_5;\ R = COCH_3,\ R' = COOC_2H_5,$$

$$R = COCH_3,\ R' = CONHPh;\ R = CN,\ R' = COOC_2H_5;\ R = H,\ R' = NO_2$$

The most strongly delocalized arsonium ylides—37, 38, dibenzoyl, diacetyl, acetylbenzoyl, acetyl-p-nitrobenzoyl, bis(phenylsulfonyl), acetylnitro, and methylene triphenylarsorane—were completely unreactive even toward 2,4-dinitrobenzaldehyde (*32*).

(37)   (38)

Fluorenylidene-(9)- (*51, 55*), carboalkoxymethylene- (*28, 29, 48*), acetylmethylene- (*28, 29, 50*), substituted benzoylmethylene- (*28, 29, 56, 94*) as well as (*x*-nitrobenzylidene)- (*58, 59, 97*) -triphenylarsorane reacted with a variety of aromatic aldehydes to give corresponding olefins and triphenylarsine oxide.

$$Ph_3As{=}CHR + X{-}\langle C_6H_4\rangle{-}CHO \rightarrow Ph_3AsO + R{-}CH{=}CH{-}\langle C_6H_4\rangle{-}X \quad (32)$$

$$R = {-}CO{-}OAkyl,\ {-}COCH_3,\ {-}CO{-}\langle C_6H_4\rangle{-}X,\ {-}\langle C_6H_4\rangle{-}NO_2$$

$$X = H,\ alkyl,\ aryl,\ Hal,\ NO_2$$

The reaction was carried out in alcohol (*28, 48, 58, 59, 97*) benzene (*56, 97*), or ether (*51, 55*) and the olefin formed in each case was trans (*48, 58, 59, 97*).

Cyclopentadiene triphenylarsorane reacted with p-nitrobenzaldehyde in chloroform to give triphenylarsine (10% yield), triphenylarsine oxide (80% yield), and two other unidentified products (*24*). The corresponding 2,3,4-triphenyl- or tetraphenylcyclopentadienylide under similar conditions afforded triphenylarsine oxide and the corresponding substituted fulvene (*25, 68*).

Nesmeyanov *et al.* reported that some stabilized and semistabilized arsonium ylides reacted with a few aldehydes, acetone, and cyclohexanone to form corresponding olefins (*71*).

B. *Reactions of Stabilized Arsonium Ylides with*
   α,β-Unsaturated Esters

Bestmann and Seng (8) reported that the reaction of methylene triphenylphosphorane with crotonic ester gave a cyclopropane derivative. We found that the less active carbomethoxymethylene triphenylphosphorane did not react with crotonic ester at all.

$$(C_6H_5)_3P{=}CH_2 + CH_3CH{=}CHCO_2CH_3 \rightarrow CH_3\underset{\underset{CH_2}{\diagdown\diagup}}{CH}{-}CHCO_2CH_3 \tag{33}$$

$$(C_6H_5)_3P{=}CHCO_2CH_3 + CH_3CH{=}CHCO_2CH_3 \not\rightarrow CH_3\underset{\underset{\underset{CO_2CH_3}{|}}{CH}}{CH}{-}CHCO_2CH_3 \tag{34}$$

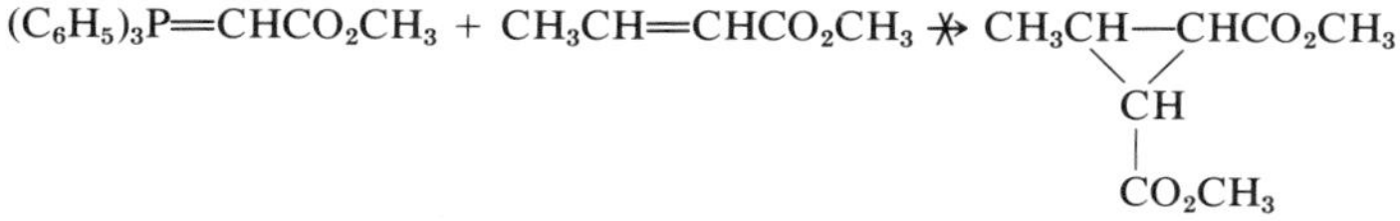

Since carbomethoxymethylene triphenylarsorane and benzoylmethylene triphenylarsorane are more active toward ketones than corresponding phosphorus reagents, we used these arsenic reagents to react with acrylic, crotonic, and α-methylacrylic esters, respectively, and obtained satisfactory results (43). The reactions were stereospecific and the yields of products (39) were moderate to excellent depending upon the esters used. Table VI shows the products and the yields.

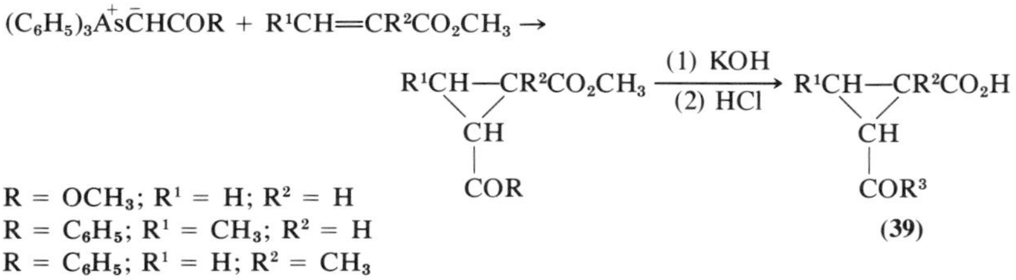

$$(C_6H_5)_3\overset{+}{As}\overset{-}{C}HCOR + R^1CH{=}CR^2CO_2CH_3 \rightarrow$$

$$R^1\underset{\underset{\underset{COR}{|}}{CH}}{CH}{-}CR^2CO_2CH_3 \xrightarrow[\text{(2) HCl}]{\text{(1) KOH}} R^1\underset{\underset{\underset{COR^3}{|}}{CH}}{CH}{-}CR^2CO_2H \tag{35}$$

R = OCH$_3$; R^1 = H; R^2 = H
R = C$_6$H$_5$; R^1 = CH$_3$; R^2 = H
R = C$_6$H$_5$; R^1 = H; R^2 = CH$_3$ (39)

This method is stereospecific. We would like to propose the following reaction mechanism. First, the reaction begins with nucleophilic attack on the β-carbon of the double bond, forming an intermediate.

$$Ph_3As{=}CHCOR + R^1CH{=}CR^2COOCH_3 \rightarrow Ph_3\overset{+}{As}\underset{\underset{R^1CH{-}\overset{-}{C}R^2COOCH_3}{|}}{-}CHCOR \tag{36}$$

Second, the preferred conformation in either transition state, **40** or **41**, whichever has low energy, undergoes subsequent cyclization to give the products.

TABLE VI

PRODUCTS AND YIELDS OF THE REACTIONS OF
CARBOMETHOXYMETHYLENE TRIPHENYLARSORANE AND
BENZOYLMETHYLENE TRIPHENYLARSORANE WITH
α,β-UNSATURATED ESTERS

Product	R^3	R^1	R^2	Yield (%)
39a	OH	H	H	83
39b	OH	CH_3	H	54
39c	OH	H	CH_3	46
39d	C_6H_5	H	H	93
39e	C_6H_5	CH_3	H	47
39f	C_6H_5	H	CH_3	40

From conformational analysis, both **40** and **41** are the preferred conformations:

1. When $R_1 = H$, **40** = **41**, only one product was obtained (**39a**, **39c**, **39d**, and **39f**) in each case. Therefore, their configurations are ascertainable.

2. When $R_1 = CH_3$, $R^2 = H$, and $R^3 = OH$, by rotating 180° along the dotted line, **42** = **43**; therefore also only one product was obtained (**39b**), and its configuration is also ascertainable.

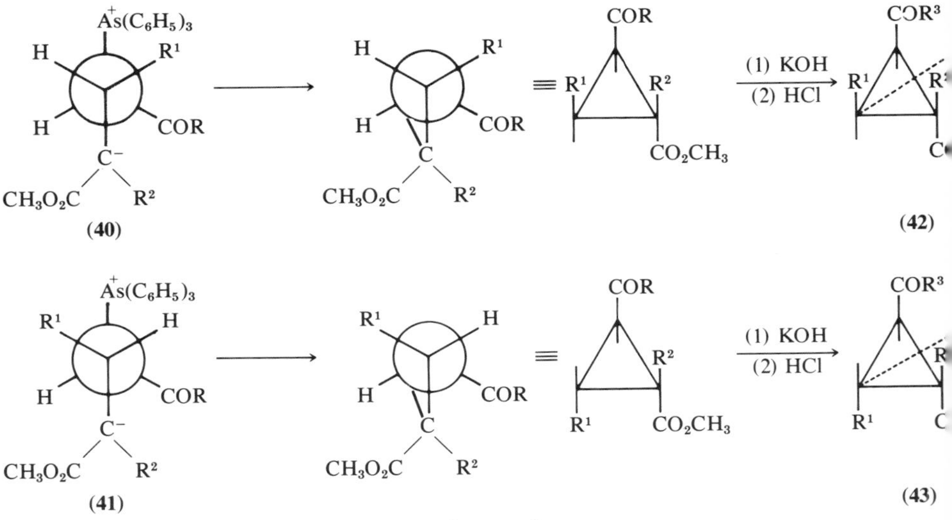

SCHEME 3

3. When $R^1 = CH_3$, $R^2 = H$, and $R^3 = C_6H_5$ according to the above mechanism a mixture of two isomers should be obtained (groups COC_6H_5 and CH_3, cis or trans to each other) the structures of which were verified by ^1H NMR. Table VII shows the configurations of the products.

TABLE VII

CONFIGURATION OF THE CYCLOPROPANE DERIVATIVES AND THEIR MELTING POINTS

Compound	Configuration	mp (°C)	mp (°C, Lit.)	Reference
39a		176–177	177–177.5	69
39b		147–149	148–149	69
39c		168–169	170	69
39d		120–121	120.5–122	3
39e		86–87		
39f		145–146	147–148	57

C. Reactions of Stabilized Arsonium Ylides with α,β-Unsaturated Ketones

Trippett and Walker (95) and Ramirez and Dershowitz (77) claimed, and our experiments demonstrated as well, that carbomethoxymethylene triphenylphosphorane and benzoylmethylene triphenylphosphorane were unreactive toward α,β-unsaturated ketones. Bestmann and Seng (8) reported that carbomethoxymethylene triphenylphosphorane reacted with β-benzoyl acrylic ester to give another ylide.

$$Ph_3\overset{+}{P}-\overset{-}{C}HCO_2CH_3 + C_6H_5COCH=CHCO_2CH_3 \rightarrow \left[\begin{array}{c} Ph_3\overset{+}{P}CH-CO_2CH_3 \\ | \\ H_3CO_2C\overset{-}{C}H-\overset{-}{C}H-COC_6H_5 \end{array} \right] \rightarrow \begin{array}{c} Ph_3\overset{+}{P}-\overset{-}{C}-CO_2CH_3 \\ | \\ H_3CO_2C-\overset{-}{C}HCH_2COC_6H_5 \end{array}$$

However, the arsenic reagents did react with α,β-unsaturated ketones to form either substituted cyclopropanes or the products similar to that of normal Wittig reactions, depending on the groups adjacent to the carbonyl group (45). When a sterically hindered bulky group was adjacent to the carbonyl group, the arsonium ylides added to the double bond to give the substituted cyclopropanes. Otherwise, the arsonium ylides reacted with the carbonyl group to give alkenes. The former reaction was found to be particularly useful in the stereospecific synthesis of 1,2,3-trisubstituted cyclopropanes, for example, 44ba, 45ab, 45ac, 44bc. These compounds would be quite difficult to prepare by other routes. Table VIII shows the yields of various products.

$$Ph_3As=CHC\overset{O}{\underset{R}{\diagdown}} + C_6H_5\overset{O}{\overset{\|}{C}}-CH=CHR^1 \rightarrow C_6H_5-COCH-CHR^1 + Ph_3As \quad (37)$$
$$\underset{CHCOR}{\diagdown\diagup}$$

R = OCH$_3$; R^1 = CH$_3$
R = C$_6$H$_5$; R^1 = C$_6$H$_5$ (44)
R = C$_6$H$_5$; R^1 = CO$_2$CH$_3$

$$44 \xrightarrow[\text{(2) HCl}]{\text{(1) KCH}} C_6H_5COCH-CHR^3 \quad (38)$$
$$\underset{CHCOR^2}{\diagdown\diagup}$$
$$(45)$$

$$Ph_3As=CHC\overset{O}{\underset{R}{\diagdown}} + CH_3-\overset{O}{\overset{\|}{C}}-CH=CR^4R^5 \rightarrow R^4R^5C=CH\overset{CH_3}{\overset{|}{C}}=CHCOR + Ph_3AsO \quad (39)$$
$$(46)$$

R = OCH$_3$; R^4 = H; R^5 = C$_6$H$_5$
R = C$_6$H$_5$; R^4 = CH$_3$; R^5 = CH$_3$

TABLE VIII

YIELDS OF THE PRODUCTS FROM THE REACTIONS OF
CARBOMETHOXYMETHYLENE TRIPHENYLARSORANE AND
BENZOYLMETHYLENE TRIPHENYLARSORANE WITH
α,β-UNSATURATED KETONES

Product	R	R^1	R^2	R^3	Yield %
45aa	—	—	OH	CH_3	40
44ba	C_6H_5	CH_3	—	—	41
44ab	OCH_3	C_6H_5	—	—	31
45ab	—	—	OH	C_6H_5	
44bb	C_6H_5	C_6H_5	—	—	43
45ac	—	—	OH	COOH	32
44bc	C_6H_5	CO_2CH_3	—	—	33

Product	R	R^2	R^4	R^5	Yield %
47aa	—	OH	H	C_6H_5	70
46ba	C_6H_5	—	H	C_6H_5	35
46ab	OCH_3	—	CH_3	CH_3	45

$$46 \xrightarrow[\text{(2) HCl}]{\text{(1) KCH}} R^4R^5C{=}CH{-}\overset{\overset{\displaystyle CH_3}{|}}{C}{=}CHCOR^2 \qquad (40)$$

$$(47)$$

We postulated that the mechanism for the formation of the cyclo-propanes was similar to that for the reaction of arsonium ylides with α,β-unsaturated esters.

1. The reaction begins with nucleophilic attack on the carbon of the double bond forming an intermediate:

$$Ph_3As{=}CHCOR + C_6H_5\overset{\overset{\displaystyle O}{\|}}{C}{-}CH{=}CHR^1 \rightarrow \begin{matrix} Ph_3\overset{+}{A}s{-}CHCOR \\ | \\ H^1CH{-}\overset{-}{C}HCOC_6H_5 \end{matrix} \qquad (41)$$

2. The preferred conformation in the transition state undergoes the subsequent cyclization to give the products. The ^{1}H-NMR spectra of the aforementioned cyclopropanes have been analyzed (85).

The stabilized arsonium ylides reacted either with α,β-unsaturated diketones or diesters to form corresponding cyclopropanes (72).

D. *Reactions of Stabilized Arsonium Ylides with Tropylium Salts*

Cavicchio *et al.* (*14*) studied the reaction of stabilized ylides and related onium salts with tropylium ion and found that the reaction between a tropylium salt and an arsonium ylide led to *trans*-chalcone (**48**) along with triphenylarsine.

R = Ph, OMe

However, in the reaction of carbomethoxymethylene triphenylarsorane in THF, a liquid product was obtained from the nonionic portion after the reaction. On the basis of the mass and ^{1}H-NMR spectra, the product was shown to be a ~1 : 1 mixture of *trans*-methylcinnamate and its α-tropenyl derivative.

E. *Reactions of Semistabilized and Nonstabilized Arsonium Ylides*

Semistabilized ylides such as benzylidene- (*55*) or methylene- (*39*) triphenylarsorane reacted with carbonyl compounds in ether to give olefins as well as oxiranes. For example, reaction of methylene triphenylarsorane with benzophenone gave, besides triphenylarsine oxide and 1,1-diphenylethene (20%), triphenylarsine and diphenylacetaldehyde (69%) (*39*). The benzylidene triphenylarsorane reacted with *p*-nitrobenzaldehyde in ether to give triphenylarsine oxide and *p*-nitrostilbene (21%) as well as triphenylarsine and 3-phenyl-2-(4-nitrophenyl)oxirane (32%) (*55*).

Reaction of benzylidene triphenylarsorane with aromatic aldehydes in alcohol afforded either arsine oxide and olefin or arsine and oxirane (*97*). Thus, from nitro- (*58, 59, 97*) or cyanobenzylidene derivatives (*97*) and aromatic aldehydes, trans-substituted oxiranes were obtained in 50–90% yield (*97*). Similarly, methylene triphenylarsorane reacted with benzaldehyde in alcoholic solution to give triphenylarsine and phenyloxirane (**49**) in 87% yield (*97*).

$$\text{Ar}_3\text{As}{=}\text{CH}-\langle\bigcirc\rangle-\text{R} + \text{R}'-\langle\bigcirc\rangle-\text{CHO} \xrightarrow[-\text{Ar}_3\text{As}]{\text{C}_2\text{H}_5\text{OH, 20°C}} \quad (44)$$

Ar = C$_6$H$_5$, H$_3$CO$-\langle\bigcirc\rangle-$
R = H, Cl
R' = H, Cl, OCH$_3$, NO$_2$, CN

The behavior of nonstabilized arsonium ylides such as methylene or ethylidene triphenylarsorance toward carbonyl compounds are similar to sulfonium ylides (*10, 11, 13, 97*).

Methylene triphenylarsorane (*97*) reacted with α,β-unsaturated ketones to give cyclopropanes with loss of triphenylarsine.

$$\text{Ph}_3\text{As}{=}\text{CH}_2 + \text{ArCO}-\text{CH}{=}\text{CHPh} \xrightarrow{-\text{Ph}_3\text{As}} \quad (45)$$

Ar = Ph, 2,4,6-Me$_3$C$_6$H$_2$

Heating of benzylidene triphenylarsorane in the presence of a catalytic amount of benzyl triphenylarsonium bromide in ether/benzene gave rise to triphenylarsine and stilbene (*76*).

$$2\,\text{Ph}_3\text{As}{=}\text{CHC}_6\text{H}_5 \xrightarrow{(\text{Ph}_3\overset{+}{\text{As}}-\text{CH}_2\text{C}_6\text{H}_5)\text{Br}^-} 2\,\text{Ph}_3\text{As} + \langle\bigcirc\rangle-\text{CH}{=}\text{CH}-\langle\bigcirc\rangle \quad (46)$$

Pyrolysis of carbomethoxymethylene triphenylarsorane afforded 1,2,3-tricarbomethoxycyclopropane (*47*), whereas pyrolysis or irradiation of benzoylmethylene triphenylarsorane gave 1,2,3-tribenzoylcyclopropane (*72*).

The nature of the arsonium substituents also influences the course of the reaction. Gosney *et al.* (*33*) studied the reaction between arsonium salts of type **50** and benzaldehyde in THF using *n*-butyllithium to generate the ylide. The results, summarized in Table IX, clearly show that electron-donating substituents at the arsenic atom promote the formation of alkenes. At one extreme, the reaction of (j) gives a notable yield of stilbene almost to the exclusion of stilbene oxide, whereas at the other, the reaction of (a) yields predominantly stilbene oxide. Table IX shows that electron-donating substituents at arsenic increase the ratio of alkene to epoxide.

$$(X_2Y\overset{+}{As}-CH_2Ph)Br^-$$
$$(50)$$

Trippett and Walker (*97*) found that the reaction of para-substituted benzylidene triphenylarsoranes with carbonyl compounds led to olefins when the benzylidene para substituent is highly electron withdrawing, but otherwise to epoxide. Kumari *et al.* (*64*) prepared two new semistabilized arsonium ylides, *p*-bromo- and *p*-iodobenzylidene triphenylarsorane, which were treated with a series of carbonyl compounds to yield exclusively *trans*-olefins. In no case was an epoxide obtained.

A suitable base for generating *p*-bromobenzylidene triphenylarsorane

TABLE IX

THE INFLUENCE OF THE NATURE OF ARSONIUM SUBSTITUENTS ON THE
COURSE OF THE REACTION[a]

		Yield (%)	
X	Y	cis-trans-Stilbene oxide	cis-trans-Stilbene
(a) Ph	Ph	79	7
(b) p-(CH$_3$O)C$_6$H$_4$	p-(CH$_3$O)C$_6$H$_4$	66	10
(c) C$_6$H$_5$	p-(CH$_3$)$_2$NC$_6$H$_4$	65	11
(d) p-(CH$_3$)$_2$NC$_6$H$_4$	C$_6$H$_5$	57	15
(e) p-(CH$_3$)$_2$NC$_6$H$_4$	p-CH$_3$OC$_6$H$_4$	33	12
(f) p-(CH$_3$)$_2$NC$_6$H$_4$	p-(CH$_3$)$_2$NC$_6$H$_4$	24	25
(g) C$_6$H$_5$	C$_2$H$_5$	47	27
(h) C$_2$H$_5$	C$_6$H$_5$	13	62
(i) n-C$_3$H$_7$	C$_6$H$_5$	18	70
(j) C$_2$H$_5$	C$_2$H$_5$	1	87

[a] From Gosney *et al.* (*33*).

was found to be sodamide in benzene and for p-iodobenzylidene triphenylarsorane it was found to be sodium hydride in THF. Both ylides reacted with substituted benzaldehydes to give exclusively the corresponding substituted *trans*-stilbenes. This contrasts with the results observed for p-chlorobenzylidene triphenylarsorane which gives epoxides (97). The difference may be attributed to the effects of solvent and base on the decomposition of salts from which the ylides were prepared.

The benzylidene and p-chlorobenzylidene triphenylarsorane ylides, when generated from sodium ethoxide in ethanol, react with a series of substituted benzaldehydes to give epoxides regardless of the nature of substituents present on the aromatic aldehyde (88, 97). However, the same ylides generated from sodium hydride in benzene (89), reacted with a series of aldehydes to give olefins. These observations clearly show that the base and solvent, in addition to the nature of substituent present on the ylidic carbanion, play an important role in dictating the exact path of the reaction.

$$X-\langle O\rangle-CH_2\overset{+}{A}sPh_3Br^- \xrightarrow[\text{EtONa/EtOH}]{\text{NaH/C}_6\text{H}_6} X-\langle O\rangle-\overset{-}{C}H-\overset{+}{A}sPh_3 \qquad (47)$$

$$X = H, p\text{-Cl}$$

$$\Big\downarrow \text{ArCHO}$$

$$
\begin{array}{ll}
X = H, \ Ar = 4\text{-NO}_2\text{C}_6\text{H}_4 & X = H, \ Ar = 4\text{-CH}_3\text{C}_6\text{H}_4 \\
X = H, \ Ar = 4\text{-CH}_3\text{OC}_6\text{H}_4 & X = 4\text{-Cl}, \ Ar = 4\text{-NO}_2\text{C}_6\text{H}_4 \\
X = 4\text{-Cl}, \ Ar = 4\text{-NO}_2\text{C}_6\text{H}_4 & X = 4\text{-Cl}, \ Ar = 4\text{-ClC}_6\text{H}_4 \\
X = H, \ Ar = 4\text{-N(CH}_3)_2\text{C}_6\text{H}_4 & X = 4\text{-Cl}, \ Ar = 4\text{-CH}_3\text{OC}_6\text{H}_4
\end{array}
$$

On the other hand, a number of semistabilized arsoranes reacted with the bulky aldehyde 9-anthraldehyde or ketones 9-anthrone and 2-chloro-9-anthrone to form olefins exclusively, regardless of the base or solvent (sodamide–C_6H_6 or sodium methoxide–methanol) used (92).

The semistabilized arsonium ylide, p-acetylbenzylidene triphenylarsorane, generated from its corresponding bromide salt with either NaH–C_6H_6 or NaOCH$_3$–CH$_3$OH, reacted with aromatic aldehydes to form **51** and with p-methylbenzaldehyde to form **52** (90). The above arsorane seems to be more reactive toward benzaldehydes than the corresponding phosphorane (90).

$$H_3C-\overset{\overset{\displaystyle O}{\|}}{C}-\langle\bigcirc\rangle-\bar{C}H\overset{+}{As}Ph$$

(48)

With $X-\langle\bigcirc\rangle-CHO$:

$$H_3C\overset{\overset{\displaystyle O}{\|}}{C}-\langle\bigcirc\rangle-CH=CH-\langle\bigcirc\rangle X$$

(51)

With $H_3C-\langle\bigcirc\rangle-CHO$:

$$H_3C-\langle\bigcirc\rangle-CH=CH-\overset{\overset{\displaystyle O}{\|}}{C}-\langle\bigcirc\rangle-$$
$$CH=CH-\langle\bigcirc\rangle-CH_3$$

(52)

Recently, Indian workers (61) reported that 2-naphthylmethylene triphenylarsorane reacted with substituted benzaldehydes to give the corresponding epoxides exclusively, whereas 1-bromo-2-naphthylmethylene triphenylarsorane reacted with substituted benzaldehydes to give only olefins. In no case were both olefin and epoxide isolated and in all cases only *trans*-epoxides or *trans*-olefins were detected. It has also been reported that reaction of indole-3-carboxaldehyde with semistabilized arsonium ylides followed only the *trans*-carbonyl olefination (91).

Allen *et al.* found that high yields of the dissymmetric *trans*-2,3-diaryloxirans result from the reaction between semistabilized arsonium ylides derived from the benzyl salts of the enantiomers of racemic *o*-phenylenebismethyl phenylarsine and of methyl α-naphthyl-*p*-tolylarsine upon reaction with prochiral aromatic aldehydes. Optical yields of between 4.7 and 38% are obtained using optically active arsonium salts (2).

F. Reactions of Arsonium Ylides Leading to Heterocycles

Toward carbonyl compounds, the behavior of nonstabilized arsonium ylides such as methylene or ethylidene triphenylarsorane is similar to that of sulfonium ylides (10, 11, 13, 97). When an arsonium ylide was reacted with aminoketones in a cold 1:1 DMSO–THF solution a smooth reaction took place and the corresponding 3-substituted indoles were obtained in fair to good yields (11).

Starting from ethylidene triphenylarsorane, the yield was poor (11). Reaction with hydroximinoketones gave 5-hydroxy-4,5-dihydro-1,2-oxazole and 1,2-oxazole (13).

The reactions of benzoyl, acetyl, and carbethoxymethylene triphenylarsorane with diphenylcyclopropenone in benzene afforded the 2-pyrones. The following mechanism has been proposed for the reaction of benzoylmethylene triphenylarsorane with cyclopropenone (32).

$$
\text{R'} \overset{\displaystyle \bigcirc}{\underset{NH_2}{}} COR + H_2C=AsPh_3 \rightarrow \left[\underset{R'}{\overset{\bar{O}\quad R}{\bigcirc\!\!\!\!\nearrow}} \underset{NH_2}{CH_2\overset{+}{A}sPh_3} \rightleftharpoons \underset{R'}{\overset{HO\quad R}{\bigcirc\!\!\!\!\nearrow}} \underset{NH}{CH_2\!\!-\!\!\overset{+}{A}sPh_3} \right]
$$

$$
\begin{array}{c} -Ph_3As \\ -H_2O \end{array} \Big\downarrow \qquad (49)
$$

$$
\underset{R'}{\bigcirc}\overset{R}{\underset{N}{\underset{H}{\diagdown}}}
$$

R = CH$_3$, C$_6$H$_5$, p-CH$_3$C$_6$H$_4$, COOH
R^1 = H, CH$_3$

$$
\begin{array}{c} Ph_3\overset{+}{As}\!\!-\!\!CH \\ \diagdown \\ -\!\!{\overset{\vert}{\underset{\vert}{C}}}\!\!-\!\!R \;+\; O=\!\!\!\triangleleft\!\!\!\begin{array}{c}Ph\\ \\Ph\end{array} \\ \underset{O}{\Vert} \end{array} \rightarrow \underset{Ph}{\overset{Ph_3\overset{+}{As}\!\!-\!\!CH}{\overset{\Vert}{\underset{\bar{O}}{\underset{\triangleleft}{CR}}}}} \xrightarrow{-Ph_3As} \underset{Ph}{\overset{Ph\diagup\diagdown R}{\underset{O}{\overset{\Vert}{\bigcirc}}}} \qquad (50)
$$

The reaction of methylene triphenylarsorane with benzonitrile oxide in DMSO gave rise to 3,5-diphenylisoxazole (**53**) along with the 3-phenyl-4-benzoyl-2-isoxazoline oximes **54** and **55** (*31*). When the reaction took place in ether instead of DMSO, a mixture of 3,5-diphenyl-1,2,4-oxadiazole (**56**) and 3,5-diphenylisoxazole (**53**) was obtained (*31*).

$$
\underset{(53)}{\overset{Ph}{\underset{N\diagdown O\diagup Ph}{\diagdown}}} \qquad \underset{(54)}{\overset{\overset{HON}{\Vert}}{\underset{N\diagdown O}{\overset{Ph}{\diagdown}\!\!C\!\!-\!\!Ph}}} \qquad \underset{(55)}{\overset{\overset{NOH}{\Vert}}{\underset{N\diagdown O}{\overset{Ph-}{}C\!\!-\!\!Ph}}}
$$

$$
\underset{(56)}{\overset{Ph\!-\!\!\overset{N}{\Vert}\!\!-\!\!N}{\underset{N\diagdown O\diagup Ph}{}}}
$$

In contrast to the earlier conclusions reported in connection with sulfonium and pyridinium salts, phenacyl or substituted-phenacyl triphenylarsonium bromide reacted with substituted aniline to give **57**, with α-naphthylamine to give **58**, and with β-naphthylamine to give **59** (4, 5).

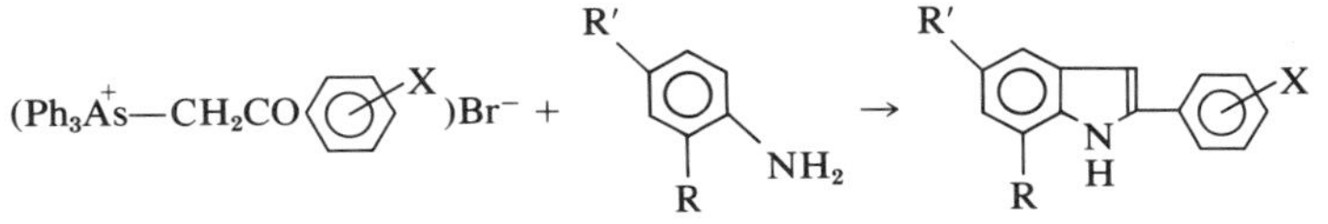

(57)

X = H, *m*-NO$_2$, *p*-CH$_3$ (51)

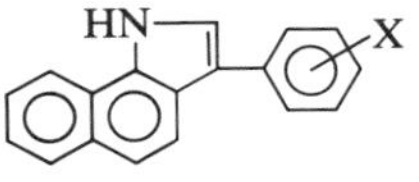

(58) 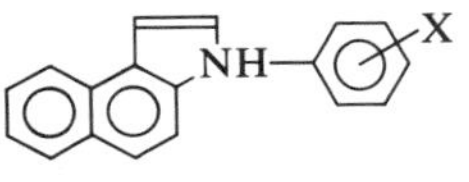   (59)

The reaction of phenacyl or substituted-phenacyl triphenylarsonium bromide with aromatic diazonium salts gave 1,4-dihydro-1,2,4,5-tetrazine (**60**) (6).

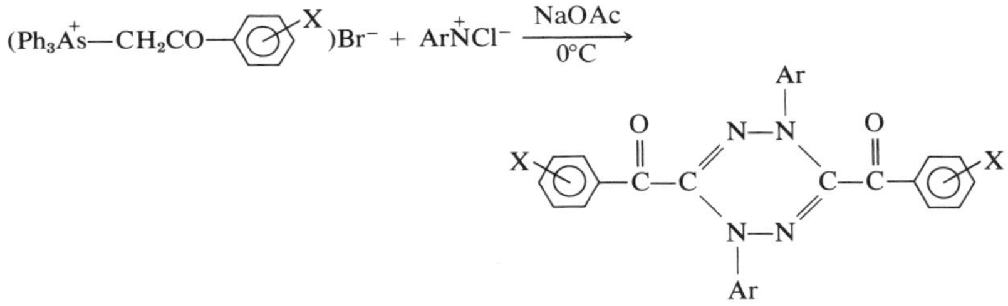

(60)

X = H, *m*-NO$_2$, *p*-CH$_3$

G. *Formation of Metal Complexes*

Along with the ylides of phosphorus, sulfur, and nitrogen, arsonium ylides have also been used as ligands for a variety of metals. When keto-stabilized alkylene arsoranes (Ph$_3$As=CHCOR) are treated with styrene- or benzonitrile–PdCl$_2$, complexes of general formula (Ylide)$_2$PdCl$_2$ are obtained in high yield. Evidence is given for a trans square planar structure in the solid state and for an epimeric equilibrium between trans square planar structures in solution for these new complexes (9).

$$2 \; Ph_3As{=}CHCOR + (PhCN)_2PdCl_2 \rightarrow (RCOCH{=}AsPh_3)_2PdCl_2 + 2 \; PhCN \quad (53)$$

The copper and silver dimethylarsonium bismethylide dimers have been obtained as novel heterocycles by reaction between trimethylarsonium methylide and CuCl or $(CH_3)_2AsAgCl$ in a 2 : 1 molar ratio (82).

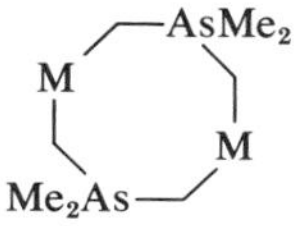

M = Cu
M = Ag

The complexes of palladium and platinum with different heteronium (N, P, As, S) ylides were found to possess the following structures (61):

$$MCl_2LY$$
(61)

M = Pd, Pt; L = $PPh(CH_3)_2$, $P(CH_3)_3$; Y = Ny, Py, Ay, Sy

Ny = p-$CH_3C_5H_4N$=CHCOPh; Py = CH_3Ph_2P=CHCOPh;

Ay = Ph_3As=CHCOPh; Sy = $(CH_3)_2S$=CHCOPh

The configuration of these complexes were established on the basis of IR and ^{1}H-NMR spectra (62). When triphenylarsine was reacted with sulfoxonium ylide complex, pentacarbonyl triphenylarsonium ylide–chromium complex was obtained (98).

$$(CH_3)_2SOCH_2Cr(CO)_5 + Ph_3As \rightarrow Ph_3AsCH_2Cr(CO)_5 + Me_2SO \qquad (54)$$

The reaction of methylene triphenylarsorane with ethylene oxide (79) and $Me_3AsCuCl$ (80) gave the oxarsolane and bis-μ-[dimethylarsonium bis(methylide)digold(I)], respectively.

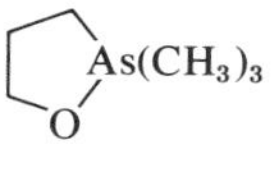

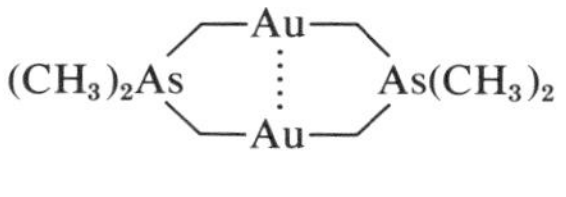

V

CONCLUSION

Phosphonium ylides react with carbonyl compounds to give olefins whereas sulfonium ylides afford epoxides. In their behavior toward carbonyl compounds, the arsonium ylides seem to be intermediate between phosphonium and sulfonium ylides.

Current results indicate that stabilized arsonium ylides such as phenacylide, carbomethoxymethylide, cyanomethylide, fluorenylide, and cyclopentadienylide afford only olefinic products upon reaction with carbonyl compounds. Nonstabilized ylides such as ethylide afford almost exclusively epoxides or rearranged products thereof. However, semi-stabilized arsonium ylides, such as the benzylides, afford approximately equimolar amounts of olefin and epoxide. Obviously, the nature of the carbanion moiety of the arsonium ylide greatly affects the course of the reaction. It is reasonable to suppose that a two-step mechanism is involved in the reaction of heteronium (P, S, and As) ylides with carbonyl compounds (56).

$$X = P, S, As \tag{55}$$

In addition to other considerations, these three factors should be important in determining the course of the reaction: the anionic character of heteronium ylide, the strength of C—X bonding, and the strength of X—O bonding (X = P, As, S). In the case of arsonium ylides, epoxide formation seems to be a normal course of reaction because of the weaker As—O bond. However, when the carbanionic charge in the arsonium ylide is conjugated with another group, such as COC_6H_5, COOR, CN, it leads to the formation of an olefin having a conjugated double bond. Such an olefin would be more stable; thus its formation would compensate for the formation of the weak As—O bond. Therefore, in such a case, path "a" becomes the preferred course of the reaction. In contrast to the stabilized phosphonium ylides which react with ketones sluggishly or not at all, the corresponding arsonium ylides react with ketones smoothly in fair to excellent yield. Thus α, β-unsaturated ketones, esters, and nitriles may be easily prepared by means of arsonium ylides (41, 47, 48, 50).

In comparison to the stabilized phosphonium ylides which fail to react with α, β-unsaturated esters or ketones, the corresponding arsonium ylides react smoothly with α, β-unsaturated esters or ketones (e.g., PhCO—CH=CH—CH₃) to form cyclopropanes in fair to good yields. Furthermore, the reaction is stereospecific (43, 45).

Although in the past two decades, a great number of papers concerning the studies of arsonium ylides have appeared in the literature, many more studies will be required to establish and exploit the new applications of these reactive reagents.

Related to the arsonium ylides, the chemistry of iminoarsorane, to which Froyen (*27, 28*) has made notable contributions, has been reviewed by Bansal and Sharma (*7*) and was therefore not discussed here.

REFERENCES

1. G. Aksnes and J. Songstad, *Acta Chem. Scand.* **18**, 655 (1964).
2. D. G. Allen, N. K. Roberts, and S. B. Wild, *J. Chem. Soc. Chem. Commun.* p. 346 (1978).
3. R. L. Augustine and F. G. Pinto, *J. Org. Chem.* **33**, 1877 (1968).
4. R. K. Bansal and S. K. Sharma, *Tetrahedron Lett.* p. 1923 (1977).
5. R. K. Bansal and S. K. Sharma, *J. Organomet. Chem.* **149**, 309 (1978).
6. R. K. Bansal and S. K. Sharma, *J. Organomet. Chem.* **155**, 293 (1978).
7. R. K. Bansal and S. K. Sharma, *J. Organomet. Chem. Libr.* **9**, 223 (1980).
8. H. J. Bestmann and F. Seng, *Angew. Chem.* **74**, 154 (1962).
9. P. Bravo, G. Fronza, and C. Ticozzi, *J. Organomet. Chem.* **111**, 361 (1976).
10. P. Bravo, G. Gaudiano, P. P. Ponti, and C. Ticozzi, *Tetrahedron* **28**, 3845 (1972).
11. P. Bravo, G. Gaudiano, P. P. Ponti, and M. G. Zubiani, *Tetrahedron Lett.* p. 4535 (1970).
12. P. Bravo, C. Ticozzi, and A. Cezza, *Gazz. Chim. Ital.* **105**, 109 (1975).
13. P. Bravo, G. Gaudiano, and C. Ticozzi, *Gazz. Chim. Ital.* **102**, 395 (1972).
14. G. Cavicchio, M. D'Antonio, G. Gaudiano, V. Marchetti, and P. P. Ponti, *Tetrahedron Lett.* p. 3493 (1977).
15. E. Ciganek, *J. Org. Chem.* **35**, 1725 (1970).
16. A. F. Cook and J. G. Moffatt, *J. Am. Chem. Soc.* **90**, 740 (1968).
17. G. W. Cowell and A. Ledwith, *Q. Rev. Chem. Soc.* **24**, 119 (1970).
18. F. S. Dainton, *Trans. Faraday Soc.* **43**, 244 (1947).
19. A. J. Dale, *Phosphorus* **6**, 81 (1976).
20. A. J. Dale and P. Froyen, *Acta Chem. Scand.* **24**, 3772 (1970); **25**, 1452 (1971).
21. G. Ferguson and D. F. Rendle, *J. Chem. Soc. Chem. Commun.* p. 1647 (1971).
22. G. Fodor and I. Tömöskozi, *Tetrahedron Lett.* p. 579 (1961).
23. B. H. Freeman, G. S. Harris, B. W. Kennedy, and D. Lloyd, *J. Chem. Soc. Chem. Commun.* p. 912 (1972).
24. B. H. Freeman and D. Lloyd, *J. Chem. Soc. C* p. 3164 (1971).
25. B. H. Freeman, D. Lloyd, and M. I. C. Singer, *Tetrahedron* **28**, 343 (1972).
26. G. Fronza, P. Bravo, and C. Ticozzi, *J. Organomet. Chem.* **157**, 299 (1978).
27. P. Froyen, *Acta Chem. Scand.* **23**, 2935 (1969); **25**, 983, 2781 (1971); **27**, 141 (1973).
28. P. Froyen, *Acta Chem. Scand.* **25**, 2541 (1971).
29. P. Froyen, *Phosphorus* **2**, 101 (1972).
30. G. X. Fu, Y. Z. Xu, W. Y. Ting, Y. C. Shen, W. Tsai, and Y. Z. Huang, *Acta Chim. Sinica* (*Hua Hsueh Hsueh Pao*), to be published.
31. G. Gaudiano, C. Ticozzi, A. Umani-Ronchi, and A. Selva, *Chim. Ind.* (*Milan*) **49**, 1343 (1967); *C. A.* **69**, 2941f (1968).
32. I. Gosney and D. Lloyd, *Tetrahedron* **29**, 1697 (1973).
33. I. Gosney, T. J. Lillie, and D. Lloyd, *Angew. Chem.* **89**, 502 (1977); *Angew. Chem., Int. Ed. Engl.* **16**, 487 (1977).

34. S. O. Grim and D. Seyferth, *Chem. Ind. (London)* p. 849 (1959).
35. K. C. Gupta and R. S. Tewari, *Indian J. Chem.* **13**, 864 (1975).
36. G. S. Harris, D. Lloyd, N. W. Preston, and M. I. C. Singer, *Chem. Ind. (London)* p. 1483 (1968).
37. H. Heine and G. D. Wachob, *J. Org. Chem.* **37**, 1049 (1972).
38. J. B. Hendrickson, R. E. Spenger, and J. J. Sims, *Tetrahedron* **19**, 707 (1963).
39. M. C. Henry and G. Wittig, *J. Am. Chem. Soc.* **82**, 563 (1960).
40. L. Horner and H. Oediger, *Chem. Ber.* **91**, 437 (1958).
41. Y. Z. Huang, N. T. Hsing, L. L. Shi, F. L. Ling, and Y. Y. Xu, *Acta Chim. Sinica (Hua Hsueh Hsueh Pao*) **39**, 348 (1981).
42. Y. T. Huang, Y. C. Shen, K. T. Chen, and C. C. Wang, *Acta Chim. Sinica (Hua Hsueh Hsueh Pao)* **37**, 47 (1979).
43. Y. T. Huang, Y. C. Shen, J. J. Ma, and Y. K. Xin, *Acta Chim. Sinica (Hua Hsueh Hsueh Pao)* **38**, 185 (1980).
44. Y. Z. Huang, Y. C. Shen, and Y. K. Xin, *Acta Chim. Sinica (Hua Hsueh Hsueh Pao)*, to be published.
45. Y. Z. Huang, Y. C. Shen, Y. K. Xin, and J. J. Ma, *Chung-kuo K'o Hsueh (Chin. Ed.)* p. 854 (1980); *Sci. Sin (Engl. Ed.)* **23**, 1396 (1980).
46. Y. Z. Huang, L. L. Shi, F. L. Ling, and Y. Y. Xu, *Acta Chim. Sinica (Hua Hsueh Hsueh Pao)*, to be published.
47. Y. T. Huang, H. I. Tai, W. Y. Ting, L. Chen, H. M. Tu, and C. W. Wang, *Acta Chim. Sinica (Hua Hsueh Hsueh Pao)* **36**, 215 (1978).
48. Y. T. Huang, W. Y. Ting, and H. S. Cheng, *Acta Chim. Sinica ("Hua Hsueh Hsueh Pao)* **31**, 38 (1965).
49. Y. Z. Huang, W. Y. Ting, W. Tsai, J. J. Ma, and C. W. Wang, *Chung-kuo K'o Hsueh (Chin. Ed.)* p. 187 (1981); *Sci. Sin. (Engl. Ed.)* **24**, 189 (1981).
50. Y. Z. Huang, Y. Y. Xu, and Z. Li, to be published.
51. A. W. Johnson, *J. Org. Chem.* **25**, 183 (1960).
52. A. W. Johnson, "Ylid Chemistry," pp. 284–299. Academic Press, New York, 1966.
53. A. W. Johnson and R. T. Amel, *Can. J. Chem.* **46**, 461 (1968).
54. A. W. Johnson and R. B. LaCount, *Tetrahedron* **9**, 130 (1960).
55. A. W. Johnson and J. O. Martin, *Chem. Ind. (London)* p. 1726 (1965).
56. A. W. Johnson and H. Schubert, *J. Org. Chem.* **35**, 2678 (1970).
57. V. A. Kalinina, Y. I. Kheruze, and A. A. Petrov, *Zh. Org. Khim.* **3**, 637 (1967).
58. P. S. Kendurkar and S. Tewari, *J. Organomet. Chem.* **60**, 247 (1973).
59. P. S. Kendurkar and S. Tewari, *J. Organomet. Chem.* **85**, 173 (1975).
60. P. S. Kendurkar and S. Tewari, *J. Organomet. Chem.* **102**, 141 (1975).
61. P. S. Kendurkar and S. Tewari, *J. Organomet. Chem.* **108**, 175 (1976).
62. H. Koezuka, G. Matsubayashi, and T. Tanaka, *Inorg. Chem.* **15**, 417 (1976).
63. F. Krohnke, *Chem. Ber.* **83**, 291 (1950).
64. N. Kumari, P. S. Kendurkar, and R. S. Tewari, *J. Organomet. Chem.* **96**, 237 (1975).
65. D. Lloyd and N. W. Preston, *J. Chem. Soc. C* p. 2464 (1969).
66. D. Lloyd and M. I. C. Singer, *Chem. Ind. (London)* p. 510 (1967).
67. D. Lloyd and M. I. C. Singer, *J. Chem. Soc. C* p. 2941 (1971).
68. D. Lloyd and M. I. C. Singer, *Tetrahedron* **28**, 353 (1972).
69. L. L. McCoy, *J. Am. Chem. Soc.* **80**, 6568 (1958).
70. A. Michaelis, *Liebigs Ann. Chem.* **321**, 141 (1902).
71. N. A. Nesmeyanov, E. V. Binshtok, O. A. Rebrova, and D. A. Reutov, *Izv. Akad. Nauk SSSR, Ser. Khim.* p. 2113 (1972).
72. N. A. Nesmeyanov and V. V. Mikulshina, *Zh. Org. Khim.* **7**, 696 (1971).

73. N. A. Nesmeyanov, V. V. Mikulshina, and O. A. Reutov, *J. Organomet. Chem.* **13**, 263 (1968).
74. N. A. Nesmeyanov, V. V. Pravdina, and O. A. Reutov, *Dokl. Akad. Nauk SSSR* **155**, 1364 (1964).
75. N. A. Nesmeyanov, V. V. Pravdina, and O. A. Reutov, *Izv. Akad. Nauk SSSR, Ser. Khim.* p. 1474 (1965).
76. N. A. Nesmeyanov, V. V. Pravdina, and O. A. Reutov, *Zh. Org. Khim.* **3**, 598 (1967).
77. F. Ramirez and S. Dershowitz, *J. Org. Chem.* **22**, 41 (1957).
78. S. Samaan, *in* "Methoden der Organischen Chemie (Houben–Weyl)" (E. Müller, ed.), 4th ed., Vol. XIII/8, pp. 395–441. Thiem, Stuttgart, 1978.
79. H. Schmidbaur and P. Holl, *Chem. Ber.* **109**, 3151 (1976).
80. H. Schmidbaur, J. E. Mandl, W. Richter, V. Bejenke, A. Frank, and G. Huttner, *Chem. Ber.* **110**, 2236 (1977).
81. H. Schmidbaur and W. Richter, *Angew. Chem.* **87**, 204 (1975).
82. H. Schmidbaur and W. Richter, *Chem. Ber.* **108**, 2656 (1975).
83. H. Schmidbaur, W. Richter, W. Wolf and F. H. Köhler, *Chem. Ber.* **108**, 2649 (1975).
84. H. Schmidbaur and W. Tronich, *Inorg. Chem.* **7**, 168 (1968).
85. Y. C. Shen, Y. Z. Huang, Y. K. Xin, and K. Y. Xu, *Acta Chim. Sinica (Hua Hsueh Hsueh Pao)* **39**, 243 (1981).
86. K. A. Starzewski, Ostoja, W. Richter, and H. Schmidbaur, *Chem. Ber.* **109**, 473 (1976).
87. S. Sugasawa and H. Matsuo, *Chem. Pharm. Bull.* **8**, 819 (1960).
88. R. S. Tewari and S. C. Chaturvedi, *Tetrahedron Lett.* p. 3843 (1977).
89. R. S. Tewari and S. C. Chaturvedi, *Indian J. Chem., Sect. B* **18**, 359 (1979).
90. R. S. Tewari and S. C. Chaturvedi, *Synthesis* p. 616 (1978).
91. R. S. Tewari and K. C. Gupta, *Indian J. Chem., Sect. B* **14**, 419 (1976).
92. R. S. Tewari and K. C. Gupta, *J. Organomet. Chem.* **112**, 279 (1976).
93. R. S. Tewari and K. C. Gupta, *Indian J. Chem., Sect. B* **16**, 623 (1978).
94. W. Y. Ting, H. S. Cheng, W. Y. Shen, and Y. T. Huang, *Bull. Nat. Sci. Univ., Chem. Chem. Eng. Sect.* p. 540 (1965).
95. S. Trippett and D. M. Walker, *J. Chem. Soc.* p. 1266 (1961).
96. S. Trippett and D. M. Walker, *Chem. Ind. (London)* p. 990 (1961).
97. S. Trippett and M. A. Walker, *J. Chem. Soc. C* p. 1114 (1971).
98. L. Weber, *J. Organomet. Chem.* **131**, 49 (1977).
99. G. Wittig and W. Haag, *Chem. Ber.* **88**, 1654 (1955).
100. G. Wittig and G. Geissler, *Liebigs Ann. Chem.* **580**, 44 (1953).
101. G. Wittig and H. Laib, *Liebigs Ann. Chem.* **580**, 57 (1953).
102. Y. Yamamoto and H. Schmidbaur, *J. Chem. Soc. Chem. Commun.* p. 668 (1975).

The Methylene Bridge

WOLFGANG A. HERRMANN*

Institut für Anorganische Chemie
Universität Regensburg
Regensburg, Bundesrepublik Deutschland

* Present address: Institut für Anorganische Chemie, Johann Wolfgang Goethe-Universität, Frankfurt am Main, Bundesrepublik Deutschland.

I

INTRODUCTION

Rarely has any field in organometallic chemistry encountered the tremendously rapid recent expansion experienced by the synthesis, spectroscopy, structural chemistry, theory, and reactivity of compounds characterized by terminal carbene (methylene, A) and carbyne (methylidyne, B) functionalities. Fischer and his group pioneered the discovery of both these remarkable classes of compounds (1, 2); many other workers have successfully followed their paths, and several comprehensive reviews are eloquent witness to the amazing plethora of compounds this fascinating area of modern research has yielded (3–9).

Interestingly enough, however, only scattered examples of dinuclear organometallic compounds having methylene or methylidyne bridges (types C and D) were in the literature when their mononuclear counterparts (i.e., compounds belonging to classes A and B) became "famous." Currently, the bridged methylene-compounds (C) are being specifically investigated with vigorous effort in many laboratories. This chapter was originally intended to cover not only methylene complexes in full detail, but also μ-methylidyne derivatives D (Scheme 1). Since many dinuclear-type compounds $L_x\overline{M\text{—}CR\text{—}M}'L'_y$ have been reported (10–32) and since these compounds are expected to display a chemistry basically different from the reactivity of the well-established trinuclear type E members in this series (33–38), a first comprehensive review is certainly justified. However, the past two years have produced so many significant results in the area of the methylene-bridged compounds that the author finally had

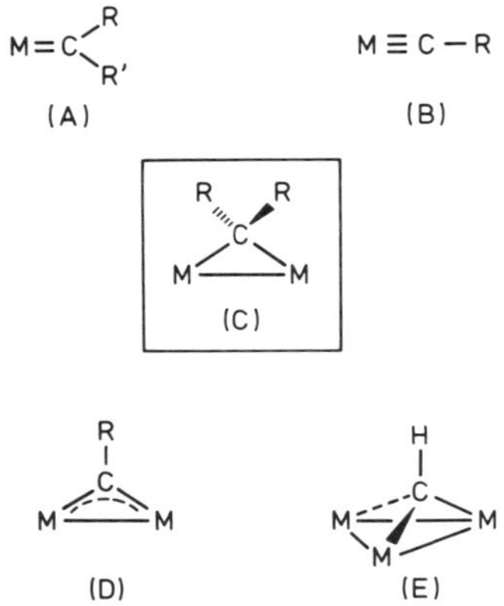

$$M = C\begin{smallmatrix} R \\ \\ R' \end{smallmatrix} \qquad\qquad M \equiv C - R$$

(A)   (B)

(C)

(D)   (E)

SCHEME 1

(F)   (G)

SCHEME 2

to abandon the idea of describing both systems in a single article. Hence both μ-methylidyne (D) and μ-vinylidene complexes (F) will be used only for the sake of comparison with methylene-bridged complexes. Alkanediyl compounds of type G [L_xM—$(CH_2)_x$—$M'L'_y$ ($x > 1$)] will however be included (Section XI) because of their close structural relationship to "true" bridged methylene counterparts (Scheme 2).

Most of the sustained work on the different aspects of μ-methylene complexes has been performed in our own laboratories at Regensburg (*39–61, 296–299*), and by the innovative groups of Knox (Bristol) (*62–67*), Levisalles and Rudler (Paris) (*68–71*), Pettit (Austin) (*72–74*), Shapley (Urbana, Illinois) (*75–83*), and Stone (Bristol) (*84–94*). Other significant contributions originate from the research of Wilkinson (London) (*95–97*), Bergman (Berkeley) (*98*), Tebbe (Du Pont) (*99–101*), Puddephatt (Liverpool) (*102, 103*), and Ziegler (Heidelberg) (*104*).

II

NOMENCLATURE

With increasing numbers of new μ-methylene complexes appearing in the literature, some confusion about their nomenclature has arisen. Both the "carbene" and the "methylene" nomenclature are being used interchangeably, a fact that we find does not contradict the present IUPAC nomenclature (*105, 106*). If a carbene coordinates to just one metal center, the resulting species of type L_xM=CRR' (A) have been shown by a vast number of X-ray and neutron diffraction studies (*3–9*) to possess literally sp^2-hybrized carbene carbon atoms, regardless of the nature of substituents R and R'. The same techniques have been used to assign more or less distorted sp^3-hybridization to the bridgehead carbon of type-C molecules (Scheme 1) (see Section VI).

Basically, the systematic nomenclature of ligands is defined by the constitution (structure) of the respective group; thus a carbon monoxide ligand is called "carbonyl" as a matter of course, no matter how it is bonded to the metal(s); the hybridization state or even the "strength" of the corresponding ligand-to-metal bonds are by no means criteria for nomenclature. The ligands :CRR′, therefore, have to be regarded as "alkylidene" or "methylene" systems; if these groups act as bridges, the "μ" descriptor unambiguously determines the coordination mode [IUPAC rules 7.611 and 7.612 (105)]. For the unsubstituted CH_2 moiety, the term methylene is maintained [rule A 4.1 (106)]; if we are dealing with ligands derived from hydrocarbons by removal of a hydrogen atom, the "alkylidene" nomenclature is used [e.g., "ethylidene" for :CH(CH₃) or "propylidene" for :CH(C₂H₅)]. If two other groups are substituted for both hydrogens, the suffix "methylene" is used [e.g., bis(methoxy-carbonyl)methylene for :C(CO₂CH₃)₂ (106)]. Cyclic carbene (methylene) functionalities are attributed the general ending "ylidene." Consequently, bridged molecules of type G having hydrocarbon chains ($x > 1$) are to be called ω,ω'-alkanediyl complexes. In keeping with the requirements of a uniform nomenclature, compounds containing |CR ligands are classified as methylidyne and alkylidyne systems, respectively, regardless of the coordination mode (terminal or μ_n-bridging, with $n \geqslant 2$). Since the carbyne nomenclature (2, 6, 7) has been widely used, both terms may again be applied interchangeably.

III

HISTORICAL BACKGROUND

The chemistry of μ-methylene complexes did not develop as extensively as that of its mononuclear counterparts, the latter being characterized by terminal carbene ligands. However, there are some significant μ-CRR′ compounds "hidden" in the literature of the late 1950s originating from the research groups of Sternberg and Wender in the United States (107) and of Albanesi in Italy (108–111). These workers synthesized the first heteromethylene compounds, and the prototypal example of an unsubstituted CH_2 bridge was achieved in our laboratories in 1975 (39).

A. The Old "γ-Lactone-Type" Compounds

The reaction of octacarbonyldicobalt with various alkynes has long been known to give alkyne-bridged compounds of the type (μ,η^2-RC≡CR′)Co₂(CO)₆ (112, 113). However, under more drastic conditions, these

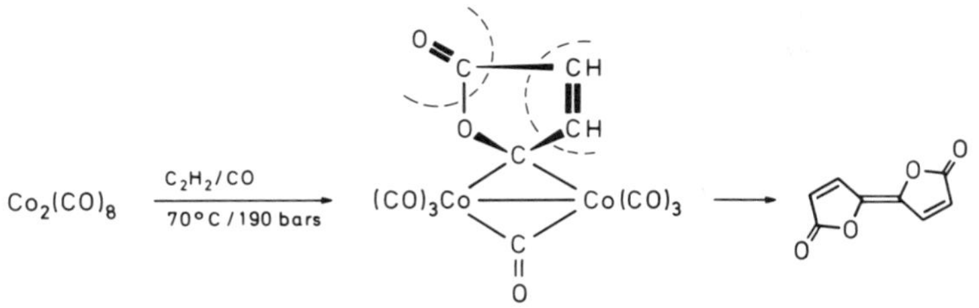

SCHEME 3

products are cleanly converted to dinuclear "γ-lactone" derivatives (1) which contain two carbon monoxide building blocks and one alkyne unit (Scheme 3). These three groups form a planar, bridging 5-oxofuran-2-ylidene ligand symmetrically spanning a metal–metal bond via an essentially sp³-hybridized bridgehead carbon (*114*). The remarkable metal carbonyl-catalyzed formation of diolides directly from CO and acetylene has been shown to occur through sequential CO/alkyne additions within the coordination core of the catalyst (*115–117*).

Compounds 1 exhibit low reactivity in carbonylation (*118*) and cyclo-trimerization (*119*); but they readily react with haloacetylenes to give, with CO elimination and concomitant ligand rearrangement, the mixed μ-oxomethylene–μ-vinylidene derivatives 2 in 60–90% yields. An X-ray diffraction study performed with one of the key products proved the ligands were not bonded to each other. From a preparative point of view, this remarkable reaction afforded μ-dihalovinylidene complexes (e.g., =C=CI₂) for the first time (*120*). Reactions of 1 with other acetylenes have also been reported (*121, 122*). It is of interest to note that the cyclic ligands of γ-lactone-type compounds 1 are prone to carbon–oxygen cleavage reactions induced by strong protic acids such as H₂SO₄. Subsequent esterification yields trinuclear methylidyne clusters belonging to the well-known Co₃(CO)₉(μ-CR) series (Scheme 4) (*123*).

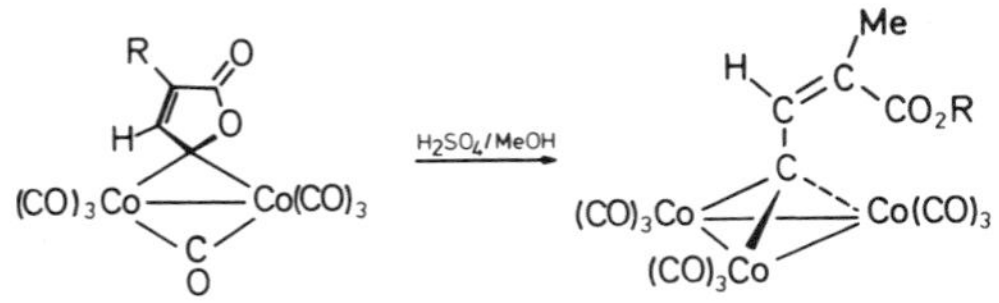

SCHEME 4

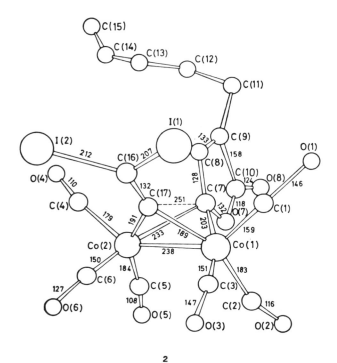

2

B. *The First CH₂ Bridge Complex*

Early in our explorative work on the synthetic potential of diazoalkanes in organometallic chemistry (*8*), we succeeded in preparing the first stable CH_2-bridged transition metal complex. Low-temperature reactions of $(\eta^5\text{-}C_5H_4R)Mn(CO)_2THF$ (R = H, CH_3) with diazomethane did not yield the mononuclear carbene derivatives $(\eta^5\text{-}C_5H_4R)Mn(CO)_2(=CH_2)$ which we had hoped for, based on our experience with the reactivity pattern of other diazoalkanes with the same and similar substrates (*124–126*), but rather the dinuclear compound **3** characterized by its metal-to-metal bond-bridging CH_2 unit (*39, 47, 57*). The mononuclear species is certainly involved in the formation of **3** since crossover experiments have demonstrated that subsequent treatment of $(\eta^5\text{-}C_5H_5)Mn(CO)_2THF$ with first CH_2N_2 (1 : 1, low temperature) and then $(\eta^5\text{-}C_5H_4CH_3)Mn(CO)_2THF$ gives the "mixed" product $(\eta^5\text{-}C_5H_5)(CO)_2Mn\overline{\quad CH_2\quad}Mn(CO)_2(\eta^5\text{-}C_5H_4CH_3)$ (*126*). The thermal stability of these prototopal compounds is quite high (e.g., mp 135–136°C for R = H) and parallels that of many other analogous compounds prepared subsequently. Like the majority of

μ-methylene compunds, the manganese derivatives of type **3** are soluble in most common organic solvents, essentially insensitive to moisture, intensely colored, and displaying only slight sensitivity to oxygen, at least in the crystalline state. Naturally, solutions decompose slowly in air; but, although exceptions are known, no sophisticated techniques are generally required to handle μ-methylene systems.

3a

The structures of these compounds (see Fig. 1; R = H, CH₃) have been established unequivocally by means of X-ray diffraction methods (*47, 57, 61, 127*), as well as by IR and ¹H- and ¹³C-NMR spectroscopy, and will be discussed in Sections VI and VII, respectively.

IV

SYNTHESIS OF μ-METHYLENE TRANSITION METAL COMPLEXES

If we try to establish the major routes that have been used for the preparation of μ-methylene complexes, we find that both direct and indirect methods can be employed. The first approach involves the use of reagents that have had a long history as useful carbene precursors in

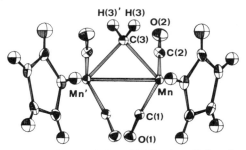

Fig. 1.   Crystal structure of μ-methylene-bis[dicarbonyl(η^5-cyclopentadienyl)manganese](*Mn—Mn*) (**3a**) determined by X-ray diffraction techniques at $-145°C$ (*127*). Identical sets of ligands (C₅H₅/C₅H₅ and (CO)₂/(CO)₂) dispose positions trans to each other.

organic chemistry. Clearly, diazoalkanes represent the starting materials of choice: hundreds of well-characterized members of this class of compound are known and can easily be prepared (*128*), with all kinds of functional groups attached to the diazo carbon center, ranging from electron withdrawing to electron donating, from small to large, from open chained to cyclic, and from saturated to unsaturated. Moreover, different methods of activation are possible for reactions with organometallic substrates: photolysis, thermolysis, and catalysis. There is yet another group of compounds for potential use (namely, the geminal dihaloalkanes CX_2RR') which promise one of the simplest and most general entries into μ-methylene complexes, provided that side reactions originating from halogen transfer can sufficiently be suppressed. Moreover, numerous structurally different examples of these precursors can be made (or purchased), and some of them have already been successfully used for the synthesis of simple alkylidene-bridged compounds (*72, 98*). It should be noted that the isomeric ω,ω'-dihaloalkanes provide an excellent preparative route to alkane-bridged organometals ("alkanediyl" complexes, see Section XI). Finally, there is just one significant example known in which Wittig reagents transfer their ylide functionality to a metal carbonyl giving rise to both alkylidene and vinylidene derivatives (*104*).

Among the indirect routes to bridging methylenes, two methods currently employed will certainly survive the test of time. The first is what may be termed the "carbene path"; it involves addition of electrophilic Fischer-type carbenes to coordinatively unsaturated transition metal complexes. At present, because of the vast number of mononuclear metal carbenes known in the literature (*3–9*), the range of applicability of this straightforward method appears to be limitless. Alternatively, the "acetylene path" is not only of interest to chemists who want to synthesize μ-alkylidene complexes but also to those actively interested in metal-induced reactions of simple hydrocarbons. This second indirect method is characterized by initial insertion of alkynes into metal–ligand functionalities (e.g., M—CO) and subsequent rearrangement of the dimetallacycles thus obtained. This last approach to carbene-bridged compounds has the advantage of also yielding μ-vinylidene compounds in some instances. Organolithium reagents, organomercurials, ketenes, etc., have also been tried as possible starting materials, but they appear to be of rather limited applicability for the construction of methylene bridges. Tebbe's reagent, a molecule exhibiting ylide behavior, represents a particularly good example of a mixed main group element transition metal μ-methylene complex.

Before discussing the various synthetic approaches to μ-methylene complexes, it should be noted that these compounds are prevalent among the Fe, Co, and Ni subgroups. They are also known to involve groups VI and VII elements (W, Mn), but as yet a stable μ-methylene complex has not been encountered with the early transition metals. In isolated cases (titanium, zirconium) methylene bridges have been postulated as intermediates (129, 130). One example of a mixed zirconium/ruthenium μ-methylene complex has been prepared recently [see Addendum (Ref. 315)].

A. Direct Methods

1. The Diazo Method

In recent years, diazoalkanes have proven to be more and more versatile as reagents for a simple synthesis of numerous nitrogeneous and nitrogen-free ligand systems in organometallic systems; a recent review article has surveyed various aspects of their synthetic applicability in transition metal chemistry (8). It now appears that one of the major advantages of diazoalkanes as precursors involves their tendency to undergo metal-induced nitrogen elimination with concomitant carbene transfer to the metal substrate present. The methodological features of metal carbene (methylene) syntheses from diazo precursors are (a) replacement of a less tightly bound two-electron ligand with the carbene unit preformed in the diazoalkane or (b) direct addition of the methylene building block to unsaturated metal–metal bonds (see below). In the first case, subsequent metal–metal bond formation is required in order to achieve di- or polynuclear μ-methylene derivatives, whereas the second route provides methylene-bridged compounds of type C directly (see Section IV,A).

The formation of μ-methylene complexes from diazoalkane precursors generally occurs under very mild conditions, as exemplified by the low-temperature syntheses of complex **3**. High yields are observed if the generation of reactive organometallic intermediates and nitrogen elimination from the diazo substrate occur at roughly the same rate and, in addition, if the mononuclear species thus formed have a lifetime sufficient to allow metal–metal bond formation. Furthermore, the formation of the reactive organometallic species has to occur under conditions that do not cause uncontrolled decomposition or side reactions of the diazo molecules. It follows from this assessment that photolysis cannot be employed with diazoalkanes having pronounced photosensitivity (e.g., CH_2N_2) since light-induced decomposition of the organic precursor will successfully compete with the activation step (e.g., CO elimination) of the organometallic reactant. It is now well appreciated that unsaturated metal–metal systems commonly give very high yields of methylene addition products (see below): first, no activation of these species is necessary since they react with diazoalkanes even at low temperatures and second, the metal–metal bond is present from the very beginning. Diazoalkanes of higher thermal stability, such as the diazomalonates, tolerate more drastic reaction conditions wherein the application of photochemical techniques can be avoided. Thus, bis(trifluoromethyl)diazomethane reacts with oc-tacarbonyldicobalt or bis(cycloocta-1,5-diene)platinum to yield the μ-hexafluoro-2-propylidene derivatives **4** and **5**, respectively (*84, 131, 132*). The analogous μ-CF_2 complex **6** has been obtained by a different route (*133*). The mononuclear precursor dicarbonyl(η^5-cyclopenta-dienyl)cobalt affords methylene-bridged dimers in high yields upon treatment with various diazoalkanes in boiling benzene (*42*).

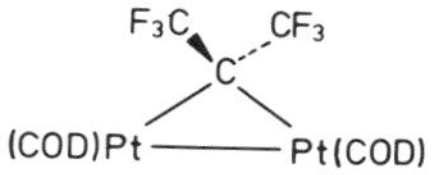

4

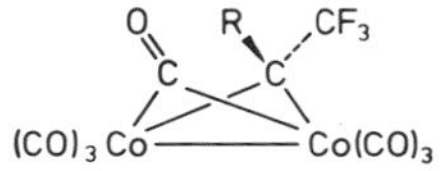

5 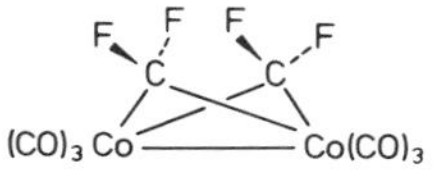   **6**

Photolysis may be used if thermal reactions do not work and if the diazo compound is considerable less photosensitive than the organometallic starting material to be activated for diazoalkane attack. The synthesis of the bis(methoxycarbonyl)methylene complex **7** (Scheme 5) from the mononuclear metal carbonyl (η^5-C_5H_5)Rh(CO)$_2$ is an illustrative example of this particular procedure (*41, 46*). The low-yield synthesis of μ-methyl-ene iron complexes of type $[\mu$-C(H)CO$_2$R](μ-CO)[(η^5-C_5H_5)Fe(CO)]$_2$ (R = C_2H_5, t-C_4H_9) may be regarded as a carbonyl substitution process, although the mechanism is not known (*48*).

The reactivity pattern of diazoalkanes in metal carbonyl chemistry is not only governed by the nature of the starting materials (*8*), but also depends

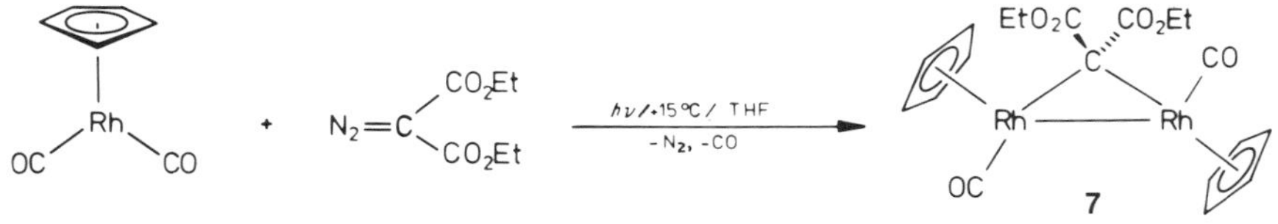

SCHEME 5

markedly on the reaction conditions employed. The half-sandwich complex dicarbonyl (η^5-cyclopentadienyl)cobalt (**8**) is a good example. Under thermal conditions (boiling benzene), a variety of diazoacetates and diazomalonates effect carbonyl substitution, with concomitant loss of nitrogen (Scheme 8). The dinuclear, diamagnetic products (**9**) thus formed in high yield exhibit in bridging positions the respective methylene ligands derived from their diazo precursors, as deduced from their IR, ^{1}H-, and ^{13}C-NMR spectra (*42*) and from X-ray diffraction study (*43*).

Two possible mechanisms have been invoked to explain the formation of these dicobaltacyclopropanes. First, dissociative rupture of a Co–CO bond could initially occur, resulting in the formation of the highly reactive fragment (η^5-C_5H_5)Co(CO), which then adds to the diazoalkane. The latter substrate could, alternatively, directly attack the electrophilic metal center of **8**. In neither case is it known at what stage the diazo nitrogen is released. The cyclic 2-diazo-1,3-dioxoindane reacts the same way, giving the corresponding μ-(1,3-dioxoindane-2-ylidene) complex **10** (*45*). The behavior of diphenyldiazomethane furnishes proof that the formation of μ-methylene derivatives is not restricted to mono- and diacyl diazo substrates (*134*). Compound **11**, having one carbonyl ligand less than the aforementioned products, is formed either in boiling benzene or upon low-temperature photolysis (Scheme 7).

If compound **8** is irradiated ($\lambda > 300$ nm) in the presence of diazoalkanes, the product pattern shifts dramatically for the diazoacetates and -malonates which have oxygen-containing substituents capable of coordination. At temperatures of about -40 to $-20°C$ (tetrahydrofuran), photolysis predominantly yields mononuclear metallacycles **12** which are not yet accessible by any other means (*135, 136*). An X-ray diffraction study revealed the strict planarity of the cyclic system (*136*). The bidentate ligands that wind around the metal arise from metal-mediated carbonylation of the respective carbenes (Scheme 6).

Parallel experiments provide ample evidence that formation of free ketenes is not responsible for the formation of **12**. Rather, the five-membered metallacycles arise from an unprecedented [2+3]-

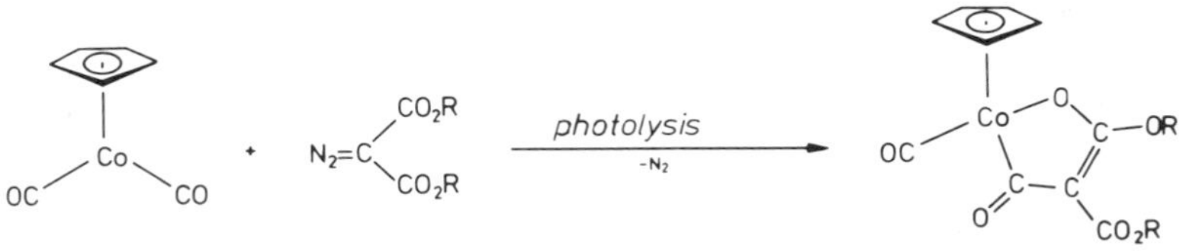

SCHEME 6

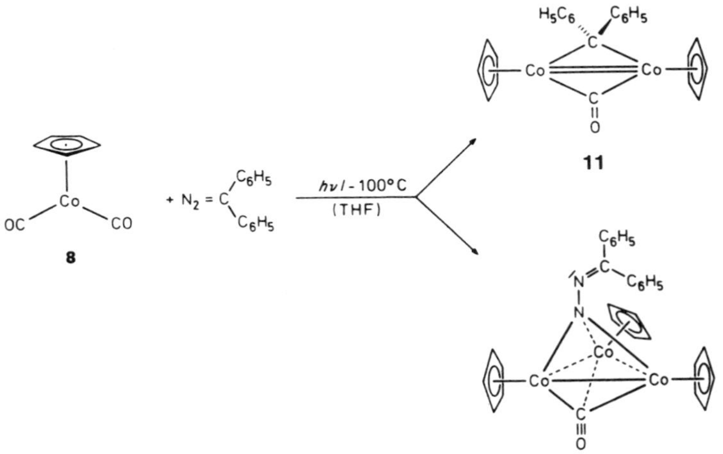

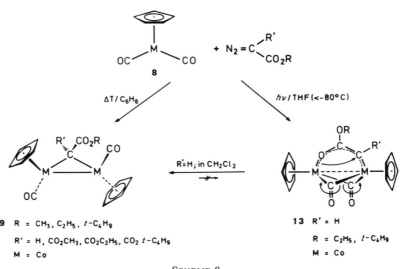

SCHEME 7

cycloaddition of the carbonyl-flanked carbenes onto a Co–C(CO) bond (*136*). The addition of excess (η^5-C_5H_5)Co$(CO)_2$ (**8**) to the metallacycles (boiling benzene), effects rupture of the cobalt–oxygen bond, cleanly yielding the "normal" μ-methylene derivatives (**9** in Scheme 8). Hence there is little doubt that the above-mentioned thermal reactions of **8** with diazoalkanes traverse the mononuclear structures **12.**

9 R = CH_3, C_2H_5, t-C_4H_9
R' = H, CO_2CH_3, $CO_2C_2H_5$, CO_2 t-C_4H_9
M = Co

13 R' = H
R = C_2H_5, t-C_4H_9
M = Co

SCHEME 8

If we lower the reaction temperatures even more (<80°C; tetrahydrofuran), dinuclear derivatives **13** are slowly formed, the structures of which are characterized by doubly bridging alkoxycarbonyl carbenes (*43, 44*). Although the low-temperature products are thermally stable in the solid state, they undergo an interesting irreversible rearrangement in solution (*43, 44*). The products formed during these isomerization are identical to the known μ-methylene comlexes **9** which can be isolated pure without further work-up. By means of IR and ^{1}H-NMR techniques both the carbonyl bridge-opening process and the μ-methylene rearrangement were characterized as intramolecular processes strictly obeying first-order rate laws (e.g., for the rearrangement $[\mu,\eta^2\text{-C(H)CO}_2\text{Et}](\mu\text{-CO})_2[(\eta^5\text{-}C_5H_5)\text{Co}]_2 \rightarrow [\mu\text{-C(H)CO}_2\text{Et}][(\eta^5\text{-}C_5H_5)\text{Co(CO)}]_2$ at T 290.5 $\pm$ 2 K: $k =$ (3.80 $\pm$ 0.03) $\times$ 10^{-4} sec^{-1}; $\tau_{1/2}$ = 30.37 $\pm$ 0.24 min; $\Delta G\ddagger$ = 21.54 $\pm$ 0.20 kcal deg^{-1} mol^{-1}; CH_2Cl_2). The reverse of the rearrangements does not occur (Scheme 8). Diphenyldiazomethane does not possess the structural prerequisites for producing either metallacycles **12** or μ-heterocarbene-type derivatives such as **13**. Thus, it is not surprising that it gives the μ-methylene complex **11** under both photolysis and thermal conditions. It is of some interest that this particular diazo substrate provided the first example of simultaneous stabilization of an intact diazo compound and the corresponding carbene fragment in a transition metal system (*134*), although diphenyldiazomethane ranks among the photolabile members of the diazoalkane series (Scheme 7).

The trinuclear osmium dihydride $(\mu\text{-H})_2\text{Os}_3(\text{CO})_{10}$ illustrates to some extent what we have said about the product-determining influence of hetero substituents in diazoalkanes. Ethyl diazoacetate gives the C,O-bridged compound $[\mu,\eta^2\text{-CH}_2\text{CO}_2\text{Et}](\mu\text{-H})\text{Os}_3(\text{CO})_{10}$ which is structurally related to **13** (*137*), whereas the reaction of the parent diazomethane follows a very different path. Shapley *et al.* (*75*) made the important discovery that bridging methylenes can readily interconvert with methyl ligands: treatment of the trinuclear osmium hydride cluster $(\mu\text{-H})_2\text{Os}_3(\text{CO})_{10}$ with 1 equiv of diazomethane (CH_2Cl_2/Et_2O) at room temperature results in rapid formation of a yellow, air-stable solid of composition "Os$_3$(CO)$_{10}$-CH$_4$" (yield 77%). NMR data showed this product to form a solvent-dependent equilibrium mixture of the two isomers $(\mu\text{-CH}_3)(\mu\text{-H})\text{Os}_3$-(CO)$_{10}$ (**14**) and $(\mu\text{-CH}_2)(\mu\text{-H})_2\text{Os}_3(\text{CO})_{10}$ (**15**) in which the methylene complex is predominant (K = [**15**]/[**14**] = 3.5 $\pm$ 0.1; CD$_2$Cl$_2$, 32°C; calculated from ^{1}H- and ^{13}C-NMR signal intensities) (*75*). The value of the equilibrium constant K decreases in the order acetone > dichloromethane > benzene, but it changes very little with temperature. The interconversion of **14** and **15** (Scheme 9) has been demonstrated in several ways. For example, spin saturation transfer experiments show that irradiating

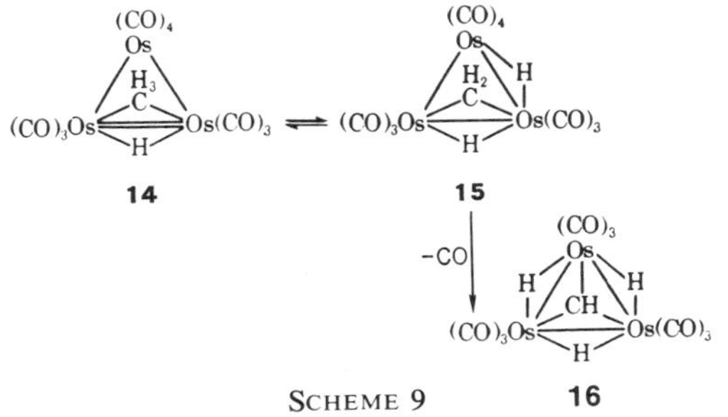

SCHEME 9   16

appropriate signals of either isomer entails significant loss of peak intensity due to the nonirradiated isomer. Furthermore, when diazomethane was added to the osmium cluster at lower temperatures ($-20°C$), only the hydridomethyl compound 14 is observable by ^1H NMR. On raising the temperature, these signals decrease in intensity as those of the methylene isomer grow in strength (k (14 → 15) $\approx 10^{-3}$ sec^{-1}, $\Delta G\ddagger \approx 20$ kcal/mol; 14°C). The equilibrium is shifted to 14 when $P(CH_3)_2C_6H_5$ (1 equiv) is added at 25°C, since only the methyl isomer reacts with the latter substrate. $Os_3(CO)_{10}[P(CH_3)_2C_6H_5]$ and methane are the observed products, whereas no evidence for the expected (known) adduct $HOs_3(CO)_{10}(CH_3)$-$[P(CH_3)_2C_6H_5]$ was found (75).

The methylene isomer 15 eliminates carbon monoxide when heated at 110°C (toluene, xylene; 24 h), and yet another carbon–metal hydrogen shift takes place (75). The methylidyne species 16 formed in this irreversible reaction has been characterized by its NMR spectra [$\delta(C\underline{H})$ 9.36, quartet (76); $\delta(\underline{C}H)$ 118.4 $(CD_3)_2CO$ (77)] which implies a structure analogous to the previously known ethylidyne derivative (μ-H)$_3$(μ-CCH$_3$)-$Os_3(CO)_{10}$ (138). NMR data obtained for partially deuterated samples of the methyl complex 14 have yielded evidence that the methyl ligand occupies an unsymmetrical bridging position between two metal centers, a coordination mode which implies significant C $\cdots$ H $\cdots$ Os interaction (78). This finding nicely models the lability of C—H bonds generally observed with hydrocarbons chemisorbed on metal surfaces (139–141).

Partial deuteration of the methylene isomer 15 provided some insight into the equilibration process. Because of the difference in sign and magnitude between the neutron scattering length of hydrogen and deuterium, it proved possible to determine the abundance of both ligands at each site in the crystalline state by means of neutron diffraction techniques (76). These population data supported by analogous data from solution NMR reflect an equilibrium isotope effect on the H/D distribution that favors the incorporation of deuterium into the methylene bridge and, conversely, of hydrogen into the osmium hydride positions. This effect was interpreted in terms of vibrational zero point energies (76). The lighter nucleus (H) will preferentially be located in the lower frequency site (Os—H—Os) and vice versa, an effect that is general for any case involving exchange between hydrogen and deuterium over carbon and metal bound sites, even if such processes occur on metal surfaces.

The possible configurations of dideuterated species resulting from pairwise H/D interchange (equilibrium constant K_2) in the methylene isomer **15** are depicted in Scheme 10 (76). For the methyl isomer at each level of deuteration there are only two configurations related by the equilibrium constant K_1. Minimization of the difference between experimental and calculated NMR signal intensities yielded $K_1 = 1.74 \pm 0.23$, $K_2 = 1.58 \pm 0.21$, $K_{eq} = 2.45([\mathbf{14}]/[\mathbf{15}])$, and $d_1/d_2 = 0.20$. The values of K_2 and d_1/d_2 agree very well with the site populations determined by neutron diffraction (see Scheme 10) ($K_2 = 2.30 \pm 0.10$; $d_1/d_2 = 0.17$) (76).

Curious and in some way intriguing are the reactions of (η^5-C_5H_5)Rh(CO)$_2$ with alkyl nitrosoureas, common starting materials for the preparation of diazoalkanes (128). It has been demonstrated that nitrosoureas possess very mild nitrosylating power toward the analogous cobalt compound (η^5-C_5H_5)Co(CO)$_2$ (**8**). The latter undergoes only partial nitrosylation when treated with methyl nitrosourea in boiling benzene, yielding the structurally characterized dinuclear product **17** (Scheme 11), even when an excess of the nitroso reagent is employed (142, 143). Surprisingly, no nitrosylation at all occured under exactly the same conditions

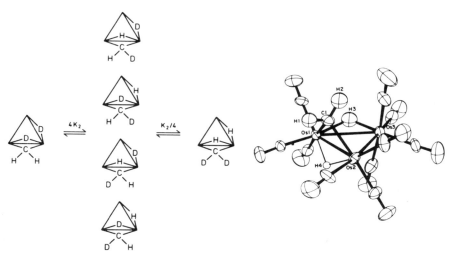

SCHEME 10

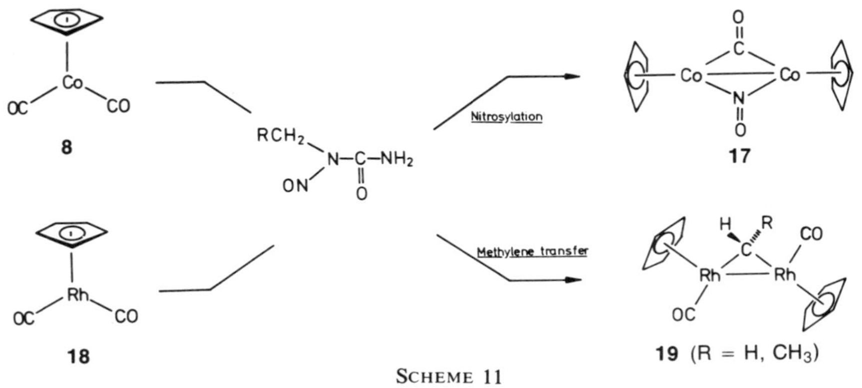

SCHEME 11

when the rhodium compound was used in place of its cobalt homologue. Instead, the μ-methylene derivative **19** was isolated from the reaction mixture (40, 41). The dinuclear complex $(\mu\text{-CO})[(\eta^5\text{-}C_5H_5)Rh(CO)]_2$— formed from $(\eta^5\text{-}C_5H_5)Rh(CO)_2$ (**18**) under thermal or photochemical conditions—was found to be more effective in affording the CH_2 compound **19**, which is now accessible on a 4-g scale (68% yield) directly from the rhodium dimer (144). The same procedure can be used for the synthesis of the ethylidene derivative $[\mu\text{-C(H)CH}_3][(\eta^5\text{-}C_5H_5)Rh(CO)]_2$ (40, 41). The mechanism of this puzzling reaction is not clear, but labeling experiments at least demonstrated that the CH_2 ligand originates from the methyl substituent of the nitroso reagent (41). It therefore appears reasonable to propose alkyl transfer to a reactive mono- or dinuclear fragment derived from $(\mu\text{-CO})[(\eta^5\text{-}C_5H_5)Rh(CO)]_2$ (**20**), with subseqent α-hydrogen abstraction.

Mass spectroscopic evidence for a reversible α-hydrogen shift in $CH_3Rh[P(C_6H_5)_3]_4$ involving intermediate carbene hydrido species has been reported (145). As to the question of what rhodium species might be open to attack by an alkyl group, there remain several possibilities. For instance, the mononuclear fragment $(\eta^5\text{-}C_5H_5)Rh(CO)$ would be a plausible substrate to undergo alkyl transfer; also, the dinuclear starting material itself might methylate under the conditions used. In light of direct carbene additions known for the "double bonded" Rh=Rh dimer $[(\eta^5\text{-}C_5Me_5)Rh(\mu\text{-CO})]_2$ (**21**) (see below), the (unknown) analog $[(\eta^5\text{-}C_5H_5)Rh(\mu\text{-CO})]_2$ may also be invoked as the pivotal intermediate (94).

Another synthetic approach to μ-methylene complexes originates from the very simple idea that there might be an analogy with olefin cyclopropanation (146) in organometallic chemistry if one had starting materials at hand that contain metal–metal double bonds (147). This type of reaction was independently and contemporaneously observed with one and the same metal complex, namely, Maitlis's rhodium dimer $[(\eta^5\text{-}C_5Me_5)Rh(\mu\text{-}CO)]_2$ (*148*), by three different research groups (*54, 81, 94, 149*). Shapley *et al.* (*75*) had already demonstrated a *net* CH_2 addition to an Os=Os double bond. However, as discussed above, the initial step of this reaction is formation of the methyl isomer $(\mu\text{-}CH_3)(\mu\text{-}H)Os_3(CO)_{10}$ (**14**) by insertion of methylene into an Os—H bond. Consequently, an organometallic substrate that no longer had an M—H functionality seemed desirable. The Stone group started from theoretical calculations according to which the rhodium methylene complex $(\mu\text{-}CH_2)[(\eta^5\text{-}C_5H_5)Rh(CO)]_2$ (*92, 94*) can formally be viewed as arising from interaction of an empty π^*-orbital of the (unknown) dimer $(\eta^5\text{-}C_5H_5)_2Rh_2(\mu\text{-}CO)_2$ with CH_2 fragments (*150, 151*). We became interested in carbene addition both because we had prepared the prototypal CH_2 complexes **19** from *mono*nuclear precursors and because we had meanwhile accomplished a straightforward synthesis of the previously hard-to-make compound $[(\eta^5\text{-}C_5H_5)Rh(\mu\text{-}CO)]_2$ (*54, 144, 152*).

Rhodium dimer **21** reacts according to Scheme 12 with a wide range of diazoalkanes to give the triply bridged carbene addition products **22** via nitrogen elimination (*54, 55*). These compounds are generally formed in quantitative yields and can even be isolated in the case of bulky methylene ligands such as tetrabromocyclopentadienylidene or 9-anthrone-10-ylidene (*54*). An X-ray structure analysis of one of the key products has confirmed the triply bridged geometry predicted from spectroscopic data (*54*). Nitrogen elimination prior to formation of **22** is so

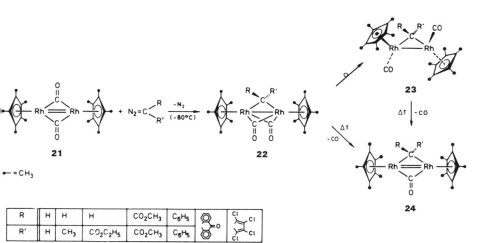

| R | H | H | H | CO$_2$CH$_3$ | C$_6$H$_5$ | | |
| R' | H | CH$_3$ | CO$_2$C$_2$H$_5$ | CO$_2$CH$_3$ | C$_6$H$_5$ | | |

SCHEME 12

rapid, even at $-80°C$ (54), that it could not be traced by means of low-temperature IR spectroscopy (153). The assumption that a five-membered dimetallacycle of the type $\overline{Rh—N\text{=}N—CRR'—Rh}$ is formed in the very first step (55, 94) appears reasonable. The extreme reactivity of **21** is documented by the fact that even the most stable diazoalkanes [e.g., $N_2\text{=}C(CO_2R)_2$] rapidly eliminate dinitrogen in the temperature range $-80–0°C$ when tetrahydrofuran is used as solvent (54). Further reactions of primary products **22** occur. First, carbonyl bridge opening can take place; this isomerization process leads directly to the singly bridged derivatives **23** commonly observed as the final products if (a) about ambient reaction temperatures are employed (81, 94) and (b) if sterically unpretentious carbenes are present (54, 81, 94). By contrast, isomerization does not occur *at all* if we are dealing with bulky methylene bridges (e.g., C_5Br_4, and other α-substituted cyclic carbenes) (54). In the latter case, decarbonylation is observed as an alternative pathway (54). The rhodium–rhodium double bond present in the starting material **21** is restored in the resulting products **24** (54, 81). This process also starts from type-**23** compounds [e.g., $C(C_6H_5)_2$] but it is not yet clear whether a **23** $\rightleftarrows$ **22** equilibrium is the salient prerequisite for decarbonylation, or whether direct loss of CO from **23** occurs.[1] In this context, attention is drawn to a curious difference between the structurally identical compounds $(\mu\text{-}CH_2)[(\eta^5\text{-}C_5H_5)Rh(CO)]_2$ (**19**) and $(\mu\text{-}CH_2)[(\eta^5\text{-}C_5Me_5)Rh(CO)]_2$ (or their CRR' derivatives). The former molecule does not undergo carbonyl exchange up to at least $100°C$ (61, 79, 152, 154), whereas the pentamethyl derivative exhibits dynamic behavior via pairwise scrambling commonly observed for complexes with carbonyl ligands disposed in cis–trans arrangements about metal–metal bonds (54, 94). This process is rapid on the NMR time scale at room temperature and yields "enantiomerization" (54) of the chiral (S,S or R,R, respectively) rhodium centers. ^{13}C-NMR results show that in the unsymmetrically substituted CRR' bridges (e.g., R = H, R' = CF_3), the two carbonyl ligands remain nonequivalent while moving between rhodium atoms (54a). The reported yields for compounds of type **23** range from "almost quantitative" (54, 94) to 5–52% (81). In the latter case, subsequent workup by sublimation (80°C) has most likely resulted in partial decomposition of the products. Clean addition of sulfur dioxide to the metal–metal double bond of **21** has also been accomplished (54, footnote 12).

It seems fitting to stress that intramolecular isomerization of the triply bridged primary products **22** to their singly bridged counterparts **23** is largely controlled by steric factors. If bulky methylene ligands are present, this process is completely suppressed for a very simple reason: carbonyl bridge opening is accompanied by a movement of the C_5Me_5

---

[1] This problem has now been solved (55).

ligand from positions exactly perpendicular to the metal–metal vector (in 22) into positions above or below the Rh_2C plane (in 23; see Section VI). In the course of this ligand movement, the bulky C_5Me_5 rings further approach the methylene bridge. If we now have sterically demanding CRR′ functions from the very beginning, the C_5Me_5/CRR′ approach necessary for CO bridge opening is no longer possible. Comparison of structural data (X-ray) obtained for either class of compounds 22, 23, and 24 clearly demonstrate the correctness of this explanation (54, 155, 156) (Fig. 2).

Elimination of carbon monoxide from the μ-methylene complexes 23 (or 22) with concomitant formation of a rhodium–rhodium double bond implies an interesting synthetic approach to multiply bridged dinuclear species having different methylene bridges. The first example of consecutive carbene addition known started from the Maitlis compound 21 (59). Diphenylmethylene transfer from the diazo compound gives 23a which undergoes decarbonylation either in boiling tetrahydrofuran [95% yield (54)] or upon vacuum sublimation at 130°C (81) to give 24a. Exposure of 24a to excess diazomethane finally produces the mixed $CH_2/C(C_6H_5)_2$ derivative 25. The overall yield of this sequence amounts to as much as 91% (59). Compound 25 constitutes the first known organometallic molecule containing structurally different methylene bridges; its stereochemistry is characterized by R,S (or S,R) configuration of the chiral metal centers ("meso" form) (Scheme 13).

In a logical continuation of this work, carbene addition to an iron–iron double bond has also been exploited for the simple synthesis of the first μ-methylene complex in the nitrosyl series. The readily available μ-nitrosyliron complex $[(\eta^5\text{-}C_5H_5)Fe(\mu\text{-}NO)]_2$ (26) exhibits the same structural features as the rhodium dimer 21 (157) and reacts with diazomethane in the temperature range -80–$25°C$ to give the expected μ-methylene derivative 27 (Scheme 14) as a black, air-stable compound in

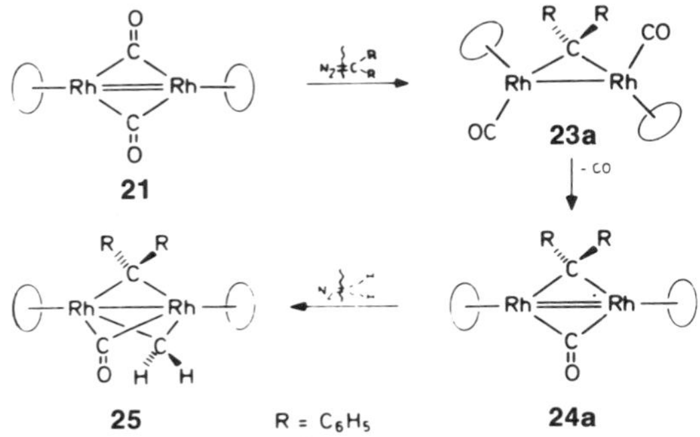

SCHEME 13

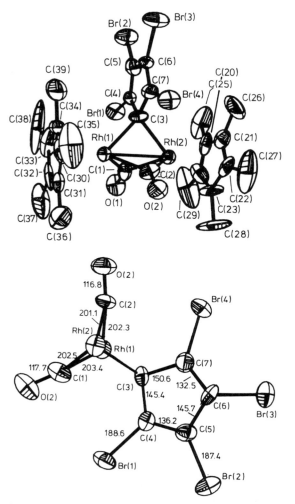

FIG. 2. (a) Crystal structure of μ-tetrabromocyclopentadienylidene-bis[(μ-carbonyl) (η^5-pentamethylcyclopentadienyl)rhodium](Rh—Rh)—an example of sterically crowded μ-methylene complexes. All cyclic ligands are strictly parallel to each other, and the close contacts between Br-1 and Br-4 with the C_5Me_5 rings prevent CO bridge opening isomerization of this molecule (see text). The C_5Br_4 ligand displays alternating C—C bond lengths typical of a 1,3-diolefin (54). (b) Drawing of the $Rh_2(\mu\text{-CO})_2(\mu\text{-C}_5Br_4)$-frame of ($\mu$-tetrabromocyclopentadienylidene)bis[(μ-carbonyl)(η^5-pentamethylcyclopentadienyl)- rhodium](Rh—Rh) showing the alternating C—C bond lengths within the cyclic methylene bridge (55).

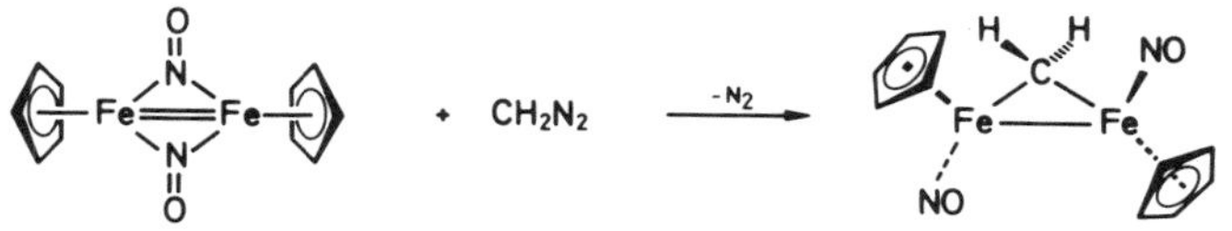

26   **27**

SCHEME 14

yields up to 90% (59). Diazocyclopentadiene forms the cyclopen-
tadienylidene complex of composition $(\mu\text{-}C_5H_4)[(\eta^5\text{-}C_5H_5)Fe(NO)]_2$ (**28**)
(153). As expected, neither molecule isomerizes to nitrosyl-bridged
species nor do they undergo clear thermolysis (153).

If we apply this simple synthetic concept to metal–metal "single"
bonds, their breakdown is expected to be inevitable upon addition of
methylidenes CRR' or equivalent two-electron ligands such as CO or SO$_2$.
Moreover, we have to take into account that the dinuclear species thus
formed is likely to fall apart since any stable μ-methylene known complex
of the type L_xM—CRR'—L_xM exhibits an additional metal-to-metal bond
(see Section VI). In a clever approach, Brown et al. (102, 103) have circumvented this latter problem by using a starting compound $Pt_2Cl_2(\mu\text{-}$
dppm)$_2$ (**29**), the metal centers of which are held in close proximity by
rigid diphosphine bridges. In addition, a platinum–platinum bond is present, as demonstrated by X-ray structural studies (158, 159) as well as by
reversible protonation (160) and CO addition reactions (161). The related
ligand bis-(diphenylarsino)methane (dpam) also gives a similar platinum
carbonyl complex, namely, $Pt_2Cl_2(\mu\text{-}CO)(\mu\text{-}dpam)$ (161). Addition of
diazomethane to **29** in fact gives the stable μ-methylene derivative **30** in
55–65% yield (Scheme 15) (102, 103). ^{1}H- and ^{31}P-NMR spectroscopy not
only demonstrates that the CH$_2$ group acts as a symmetrically bridging
ligand ("A-frame"-type molecule) but further indicates that the Pt—Pt
bond is effectively broken when CH$_2$ and similar ligands (see above) are
added to **29** (103). Low solubility precluded isolation of a crystal for an
X-ray structure determination, but the geometry of the phosphine sug-

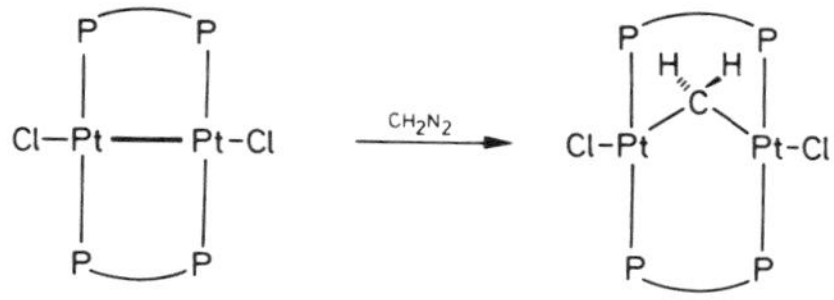

29   **30**

SCHEME 15

gests a fairly large "internal" Pt—CH$_2$-Pt angle (>100°), with the methylene bridge very much resembling the geometry of the "ketonic" carbonyl ligand of the recently reported rhodium complex Rh$_2$Cl$_2$(μ-CO)-(μ-MeO$_2$CC≡CCO$_2$Me(μ-dppm)$_2$[⋪(Rh,CO,Rh) = 116.0(4)°; d(Rh—Rh) = 335.42(9) pm] (*162;* see also Ref. *324*).

Let us now ask what happens if we react diazolkanes with metal–metal triple bonds: Will dimetallacyclopropenes form? They should, provided both substrates bow to our simple formalism. The dimolybdenum compound [(η^5-C$_5$H$_5$)Mo(CO)$_2$]$_2$ (**31**), as one of the nicest examples of (electrophilic) metal–metal triple bonds among metalcarbonyls (*144, 163*), generally displays high reactivity to diazo hydrocarbons. Diazomethane rapidly looses nitrogen even at low temperatures if treated with **31**, but the "messy" reaction successfully resisted isolation of products (*164, 165*). Diphenyldiazomethane, which is generally less reactive in cyclopropanation than the parent compound (*146*), promptly adds to the dimetal frame of **31** (tetrahydrofuran, 25°C) without loss of nitrogen. The diazoalkane complex **32** so obtained very much resembles structurally other (di- and trinuclear) diazo-bridged compounds (*8*). The Mo=Mo double bond expected from formal reasoning has not been retained, mainly due to the fact that the bridging nitrogen ligand acts as a four- rather than a commonly observed two-electron ligand (*8*) via its terminal nitrogen atom. Moreover, a semibridging carbon monoxide group situated on the opposite side of the diazo bridge is evident from X-ray structural data (*164*). On the basis of conventional electron-bookkeeping, nothing more than a Mo—Mo "single" bond has survived the course of this diazoalkane addition to the triple bond. This situation does not change when nitrogen is released at elevated temperatures (60°C), giving the diphenylmethylene complex **33**. Mechanistic studies established an intramolecular rearrangement process when the former (uncomplexed) diazo carbon bends over into a bridging position (*164*). Falling back to one of the phenyl ring's bonding reserves, the "methylene" ligand accomplishes a σ-alkyl/π-allyl type four-electron system (Scheme 16). As a result, the dimetal frame once more escapes maintaining a formal double bond. The ligand is released from coordination upon addition of carbon monoxide under remarkably mild conditions

32

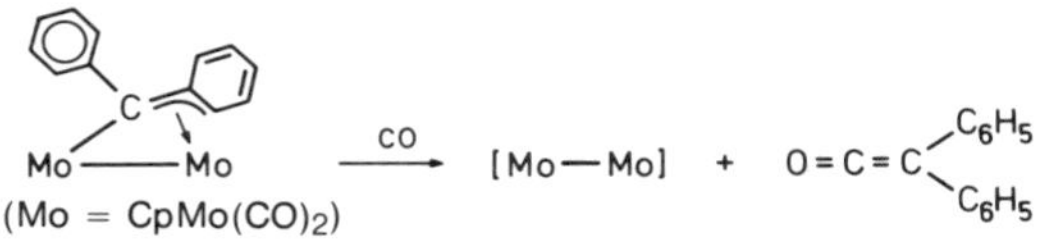

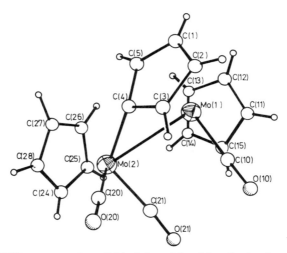

SCHEME 16

(RT, 1 atm), this reaction yielding diphenylketene and $[(\eta^5\text{-}C_5H_5)Mo(CO)_3]_2$ (*164, 164a*).

Diazocyclopentadiene proceeds one step beyond simple carbene addition, in that one CO ligand is replaced with the C_5H_{54} residue from the organic substrate. Here, a compound of composition $(\eta^5\text{-}C_5H_5)_2Mo_2(CO)_3\text{-}(C_5H_4)$ (**34**) is formed in ~40% yield (tetrahydrofuran, 25°C) (*60*). An accurate X-ray structural determination revealed that the cyclopentadienylidene ligand acts as a terminal-type σ-alkyl to the $(\eta^5\text{-}C_5H_5)Mo\text{-}(CO)_2$ fragment and, in addition, as an aromatic η^5-system bound to the remaining $(\eta^5\text{-}C_5H_5)Mo(CO)$ unit (Fig. 3). Exhausting its electronic capabilities, the unique cyclic ligand acts as a six-electron ligand to the dimetal backbone. The latter has a metal–metal separation [309.8(0) pm] compatible with a single-bonded system.

FIG. 3. ORTEP representation of **34**. Selected bond lengths [pm] and angles [deg]: Mo–Mo 309.8(0), Mo(2)–C(4) 211.1(4), Mo(1)–C(4) 222.4(4), Mo(1)–C(1) 230.4(4), Mo(1)–C(2) 231.5(4), Mo(1)–C(3) 225.0(4), Mo(1)–C(5) 226.4(4), Mo(1)–C(11–15) 228.3–232.5, Mo(2)–C(24–28) 233.2–239.1, C(1)–C(2) 139.7(6), C(2)–C(3) 142.4(6), C(3)–C(4) 144.2(6), C(4)–C(5) 145.1(6), C(5)–C(1) 141.7(6); Mo(1)–C(4)–Mo(2) 91.2(1), C(4)–Mo(2)–Mo(1) 45.9(2). For a brief discussion of the structure, see Herrmann *et al.* (*60*).

2. The Dihaloalkane Route

Geminal dihaloalkanes have not as yet been much used for the introduction of alkylidene moieties into organometallic substrates. It is evident from reports dealing with this method that successful suppression of halogen transfer to the metal centers is crucial for the achievement of high yields in methylenated products (see below). There is one prominent case in which even the nature of the counterion controls the effectiveness of the synthesis: replacement of Na^+ with NEt_4^+ in the dinuclear iron complex $M_2[Fe_2(CO)_8]$ drastically lowers the yield of the μ-methylene derivative $(\mu-CH_2)[Fe_2(CO)_8]$ (35) in the reaction with diiodomethane (72). Under the proper conditions, however, 35 is readily available in large (>100 g) quantities (74). Other members of the ω,ω'-dihaloalkane series have been used to extend the synthesis of 35 to the homologous μ-alkylidene congeners (72). The cobalt compound $(\mu-CH_2)[(\eta^5-C_5H_5)Co(CO)]_2$ (36), structurally analogous to its substituted derivatives (42, 43), is obtained from the radical anion $[(\eta^5-C_5H_5)Co(\mu-CO)]_2^{\cdot-}$ and excess CH_2I_2 in 48% yield (98, footnote 6). Likewise, alkanediyl complexes are accessible from ω,ω'-dihaloalkanes (Section XI).

The preparation of dicyanomethylene complexes such as $[\mu-C(CN)_2](\mu-CO)[(\eta^5-C_5H_5)Fe(CO)]_2$ (37) from the highly nucleophilic $[(\eta^5-C_5H_5)Fe(CO)_2]^-$ anion and dibromomalodinitrile suffers greatly from the halogenating tendencies of the organic precursor because of the relatively positive bromine substituents. Indeed, the predominant products of this reaction are the coupling adduct $[(\eta^5-C_5H_5)Fe(CO)_2]_2$ and the bromide $(\eta^5-C_5H_5)Fe(CO)_2Br$, whereas the desired complex 37 (cis configuration) appears in no more than $\sim0.6\%$ yield (166, 167). IR data led to the conclusion that the dicyanomethylene bridge has strong π-acceptor properties similar to the related dicyanovinylidene ligand (166–168), both groups appreciably exceeding the isoelectronic carbon monoxide functionality in this regard. Attempts to achieve a better synthesis of 37 by reaction of tetracyanoethylene oxide with the iron dimer $[(\eta^5-C_5H_5)Fe(CO)_2]_2$ ended up with formation of the known cyanide $(\eta^5-C_5H_5)Fe(CO)_2CN$, again in low yield (7%) (167).

37

If dihaloalkanes are treated with neutral rather than anionic organometals (see above), only small amounts, if any, of the expected

μ-dihalomethylene complexes are observed, even if the metal substrates [e.g., Fe(CO)$_5$, Co$_2$(CO)$_8$] exhibit a pronounced tendency for dehalogenation. This is convincingly demonstrated by the reported photolysis of Fe(CO)$_5$ and of Co(CO)$_3$NO in the presence of CF$_2$Br$_2$ (169, 170). Depending strongly upon the reaction conditions, two new compounds were isolated which are thought to constitute the difluoromethylene-bridged derivatives of composition (μ-CF$_2$)$_2$(μ-CO)[Fe(CO)$_3$]$_2$ (38; mp 104–105°C, dec.) and (μ-CF$_2$)$_2$[Co(CO)$_3$]$_2$ (39; subl 45°C, 760 Torr), respectively, solely based on their mass and infrared spectra. The yields are abysmal. Both compounds decompose to give CO, C$_2$F$_4$, and elemental iron and cobalt (169, 170, 170a), but their structures remain unclear.

3. Synthesis from Wittig Reagents

There is yet another promising synthetic approach to μ-methylene complexes. It was recently discovered by the Ziegler group that a bridging carbonyl ligand can be replaced by C(H)R units originating from Wittig reagents. Thus the iron dimer (η^5-C$_5$H$_5$)$_2$Fe$_2$(CO)$_4$ reacts with (C$_6$H$_5$)$_3$P=CH$_2$ in boiling tetrahydrofuran–dioxane to yield the μ-CH$_2$ complex 40. Both cis and trans isomers were observed and were separated by column chromatography; they equilibrate in solution, with the final equilibrium cis/trans ratio amounting to 3/1 after about 1 h at room temperature. Moreover, minor amounts of the bis(μ-vinylidene) derivative 41 are formed in the course of a typical Wittig reaction between the metal carbonyl and the phosphorylide (104).

cis/trans

40   41

B. Indirect Methods

1. The Carbene Path

A major breakthrough in the synthesis of transition metal methylene and methylidyne complexes has been achieved by Stone and his group; it originates from the simple idea that M=C "double" bonds in Fischer-type carbenes and M≡C "triple" bonds in carbyne systems should add to low valent metal complexes as do C=C and C≡C linkages, respectively. A logical extension of this concept follows from the premise that since Pt^0 readily reacts with alkenes and alkynes, Pt^0 compounds and other low valent, coordinatively unsaturated metal species should also react with M≡M groups. A serendipitous discovery that $Pt(COD)_2$ reacted with perfluoropropene to give the diplatinum complex $Pt_2[\mu\text{-}C(CF_3)_2](COD)_2$ rather than the expected π-complex $Pt(CF_2{=}CFCF_3)\text{-}(COD)$ (*171*) led to the idea that mononuclear metal carbene or carbyne complexes would generally combine with nucleophilic and coordinatively unsaturated metal species M' according to Scheme 17.

In practice, this approach to the syntheses dimetallacyclopropanes or -propenes has been very successful (*27, 86–93*). Complexes **42–47** are

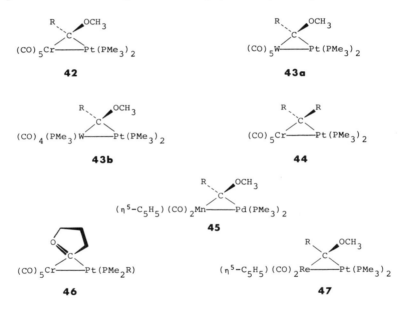

R = C_6H_5 (except for **47** for which R = 4-MeC_6H_4)

representative of those that have been readily prepared in high yields by reacting low-valent metal complexes with either metal carbenes or carbynes (see also Table V). Many key compounds have had structures established by means of X-ray diffraction techniques (Section VI). This novel synthetic approach provides mainly heterobimetallic μ-methylene complexes and undoubtedly involves the methodologically clearest and synthetically cleanest entry to mixed-metal systems that can presently be thought of.

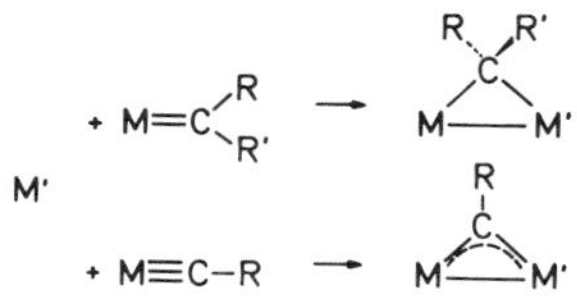

SCHEME 17

Some of the compounds thus obtained display interesting decomposition pathways, which, in turn, provide yet another possibility for synthesizing new μ-methylene complexes not accessible by any other means. For instance, the heterodinuclear compound $(CO)_5Cr[\mu\text{-}C(OCH_3)C_6H_5]$-$Pt(PMe_3)_2$ **(42)** readily decomposes in organic solvents above 80°C to $(CO)_5Cr(PMe_3)$, and the homotrinuclear platinum derivatives of composition $Pt_3[\mu\text{-}C(OCH_3)C_6H_5](\mu\text{-}CO)(PMe_3)_3$ **(48)** and $[Pt\{\mu\text{-}C(OCH_3)C_6H_5\}$-$(PMe_3)]_3$ **(49)**. This pathway reflects migration of the methylene bridge across the Cr—Pt bond with simultaneous transfer of the phosphine to chromium and concomitant rupture of the metal–metal bond (*88*). The process is reminiscent of an earlier observation reported by the Fischer group, according to which $(\eta^5\text{-}C_5H_5)Mo(CO)(NO)[C(OCH_3)C_6H_5]$ transfers its carbene functionality to tetracarbonylnickel, with formation of the trimer $[Ni(\mu\text{-}CO)\{\mu\text{-}C(OCH_3)C_6H_5\}]_3$ **(50)** and $(\eta^5\text{-}C_5H_5)Mo(CO)_2NO$ (*172*). It seems likely that the latter carbene transfer proceeds via the dimetal species $(\eta^5\text{-}C_5H_5)(NO)(CO)Mo[\mu\text{-}C(OCH_3)C_6H_5]Ni(CO)_3$ which escaped isolation under the reaction conditions employed by Fischer and Beck (*172*). A related carbene migration process from one metal center to another is the notable result of the reaction of the mononuclear manganese complex $\overline{Mn(COCH_2CH_2CH_2)}(CO)_4I$ with $Pt(\eta^2\text{-}C_2H_4)_2$-$(P—t\text{-}Bu_2Me)$. The 2-oxocyclopentenylidene ligand originally attached to manganese shows up in a terminal position at platinum of the final product **51**, its metal-to-metal bond being supported by a bridging iodo ligand (*89*).

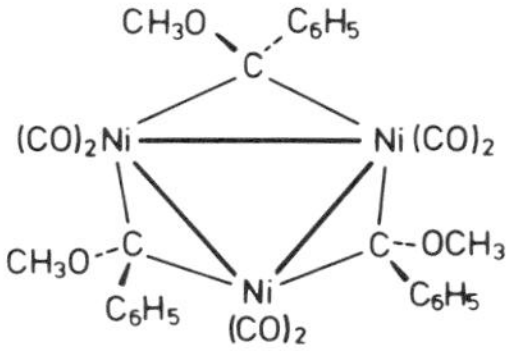

50

Fischer-type carbenes have also been successful precursors for homo- rather than heterodinuclear μ-methylene complexes. Two examples may be quoted here. Upon attempted extension of the Casey method for the synthesis of mononuclear metal carbenes not "stabilized" by a hetero substituent (*173*), a 3-methyl-2-butene-1-ylidene bridge **56** was generated through methyllithium attack at the methoxy(methyl)carbene tungsten complex **52** and subsequent protonation with CF_3CO_2H (*68, 69*). Deuterium labeling and parallel experiments with $LiCD_3$ and Li-*n*-Bu, respectively, support a mechanism that involves synchronous deprotonation of the metal-attached carbene, and CH_3^- addition at the electron-deficient carbene carbon (*68, 69*). The intermediate vinylidene and dimethylcarbene species **53** and **54**, respectively, thus generated may combine via [2+2]-cycloaddition followed by internal rearrangement of the metallacyclobutane **55** (Scheme 18). The olefinic function of **56** becomes coordinated to one of the metal centers upon thermal decarbonylation (room temperature). The possible participation of complexes such as **57** (Fig. 4) in olefin metathesis reactions has been discussed (*70*).

Another significant example of the classical carbene synthesis yielding μ-methylene complexes instead of the corresponding mononuclear compound comes from Fischer's group (Scheme 19). Treatment of either $Re_2(CO)_{10}$ or the preformed metal carbene $Re_2(CO)_9[C(OCH_3)R]$

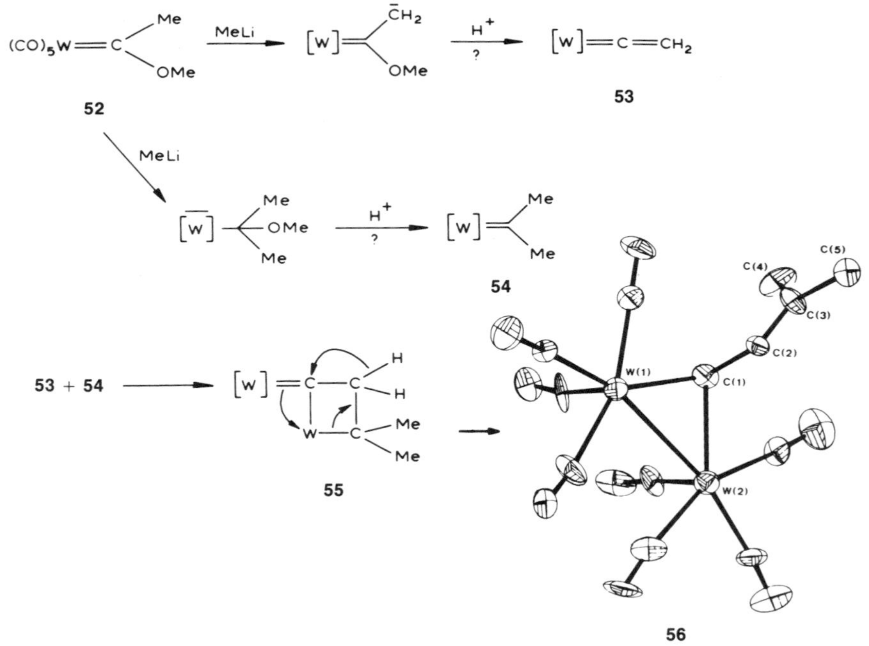

SCHEME 18

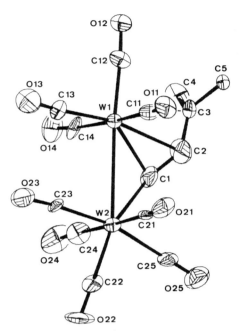

FIG. 4.   ORTEP representation of the μ,η^2-dimethylvinylidene tungsten complex 57. Selected bond lengths (pm) and angles (deg): C(1)–C(2) 124(3), C(2)–C(3) 144(3), W(1)–C(3) 263(1); W(1)–C(1)–C(2) 79.9(14), W(1)–C(2)–C(3) 85.5(13). For further data see Table I.

(R = C_6H_5, C_6H_4-p-CH_3) (*174*) with excess organolithium reagent LiR and subsequent alkylation by means of $(CH_3)_3O^+BF_4^-$ affords the doubly bridged μ-(methoxy)(aryl)methylene system 58 in 38–79% yield (*175*). We take the occurrence of these particular products to be another example in support of the well-documented preference of carbenes to adopt, if possible, bridging rather than terminal positions. In light of this rule of thumb, and keeping in mind the common outcome of Fischer carbene syntheses (*1*, *3–7*), the biscarbene species $[C(OCH_3)R](CO)_4Re$—$Re(CO)_4$-

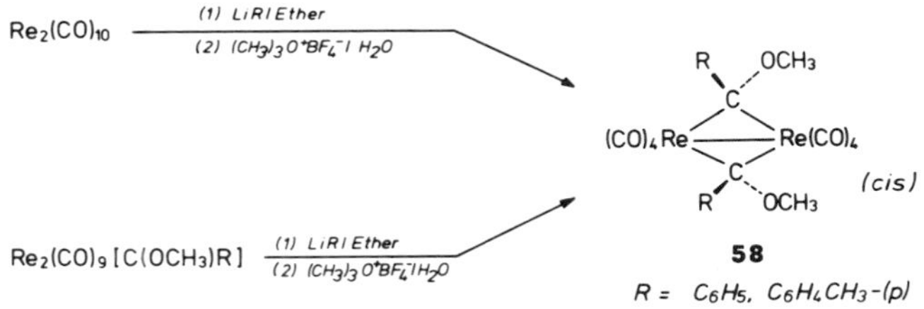

SCHEME 19

[C(OCH$_3$)R] (**59**) are certainly formed via the initial steps of the above reactions. Isomerization of **59** through pairwise formation of carbene bridges should be a low-energy process. Hence it appears that simple dinuclear metal carbonyls in general have all the prerequisites for yielding bis-μ-methylene complexes in the course of the Fischer procedure for metal carbene synthesis.

2. *The Acetylene Path*

The second indirect path leading to μ-methylene complexes was discovered by Knox *et al.* (*176*) The ethyne-derived dimetallacycle **60**, which at up to ~100°C undergoes fluxional breaking and re-forming of the "carbonyl alkyne" carbon–carbon bond, suffers irreversible cleavage of this very bond in toluene (~110°C). This isomerization yields the μ-vinylidene complexes *cis*- and *trans*-(η^5-C$_5$H$_5$)$_2$Ru$_2$(CO)$_2$(μ-CO)(μ-C=CH$_2$) (**61**). Subsequent protonation (HBF$_4$) of **61** gives the corresponding μ-ethylidyne complex *cis*-[(η^5-C$_5$H$_5$)$_2$Ru$_2$(CO)$_2$(μ-CO)(μ-CCH$_3$)]$^+$BF$_4^-$ (**62**) which, in turn, is open to hydride attack (NaBH$_4$) yielding the neutral μ-ethylidene derivative (η^5-C$_5$H$_5$)$_2$Ru$_2$(CO)$_2$(μ-CO)[μ-C(H)CH$_3$] (**63**). The final carbene-bridged compound completes a remarkable sequence in which acetylene is converted, via the above-mentioned dimetallacycle **60**, to μ-C=CH$_2$, then μ-CCH$_3$, and ultimately to μ-C(H)CH$_3$ (*63*) (Scheme 20; R = H).

The same ethylidene ruthenium complex, as well as its iron congener, is alternatively obtained through direct protonation of the dimetallacycles **64a** (M = Fe) and **64b** (M = Ru) (*64*). In this case, the carbonyl "alkyne" carbon–carbon bond is broken irreversibly to give the cationic μ,η^3-vinyl complexes **65a** and **65b**, which undergo nucleophilic attack by hydride (NaBH$_4$) to produce complexes of methylcarbene (**63a,b**) (Scheme 21a). Deuterium-labeling experiments prove that the final compounds arise from initial hydride addition to the β-vinylic carbon of **65**. However, isolation of small amounts of the η^2-ethylene complex **66** indicates that hydride attack can also occur at the α-vinylic carbon (*64*).

A dimetallacycle was employed once more for the net interconversion

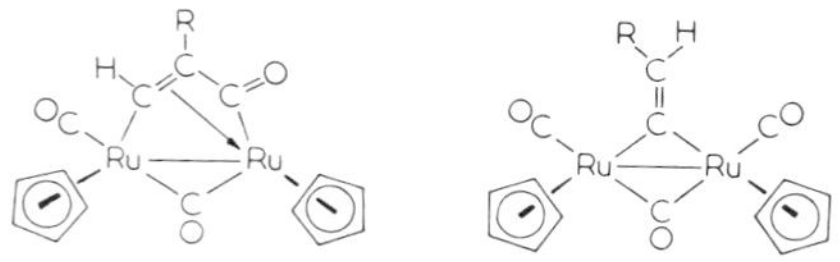

60   **61**

SCHEME 20

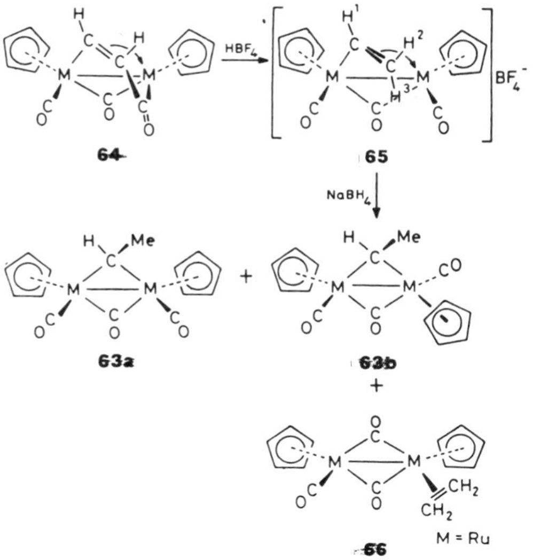

SCHEME 21A

of allene to 2-propylidene along lines similar to those discussed in the previous paragraph (63). The complex $(\eta^5\text{-}C_5H_5)_2Ru_2(\mu\text{-}CO)[\mu\text{-}C(=O)C_2(C_6H_5)_2]$ liberates diphenylacetylene when heated with allene (Scheme 21b). This reaction provides the dinuclear $\mu,\eta^3\text{-}C_3H_4$ complex 67 in quantitative yield. Protonation of the latter compound with HBF$_4$ causes prompt precipitation of the 1-methylvinyl complex 68, which accepts hydride from NaBH$_4$ regiospecifically at the β-vinylic carbon site, producing the dimethylcarbene-bridged complex 69 in fairly good yield. Both cis and trans isomers are again evident from IR spectroscopy, and these are seen in the ^{1}H-NMR spectra (30°C, CDCl$_3$), without interconverting, in a cis/trans ratio of ~2/3 (Scheme 21b).

By contrast, the vinyl-bridged cation $(\eta^5\text{-}C_5H_5)_2Mo_2(CO)_4[\mu,\eta^2\text{—}CH(=CH_2)]^+$ (70) resisted attempted conversion to the expected μ-methylene complex (see above) upon treatment with nucleophiles such as hydride, methoxide, or cyanide. The latter reagents instead regenerate the acetylene complex $(\eta^5\text{-}C_5H_5)_2Mo_2(CO)_4(\mu\text{-}HC\equiv CH)$ (71) which is the immediate precursor of 70 (71 + H$^+$ → 70) (66). The origin of the vinylidene group in 70 is somewhat reminiscent of alkyne isomerization proceeding in the coordination core of the $(\eta^5\text{-}C_5H_5)Mn(CO)_2$ fragment. Thus $(\eta^5\text{-}C_5H_5)Mn(CO)_2(HC\equiv CH)$ gives the dinuclear μ-vinylidene complex $(\mu\text{-}C=CH_2)[(\eta^5\text{-}C_5H_5)Mn(CO)_2]_2$ (72) when reacted with KOH (H$_2$O–THF) (177). The structure of this compound was

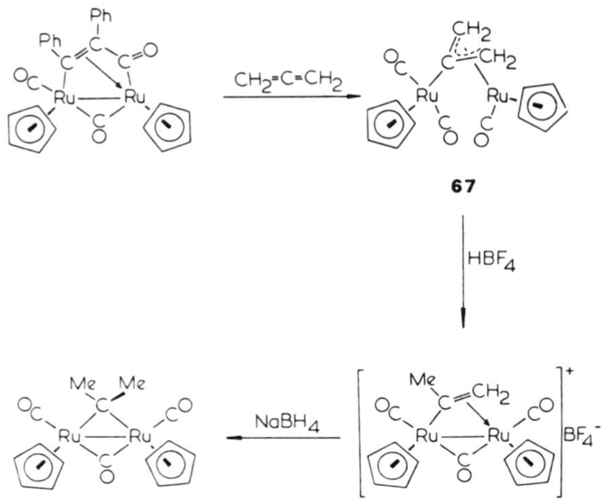

SCHEME 21B

established by X-ray diffraction techniques (177) and very much resembles the geometry of the methylene complex $(\mu\text{-}CH_2)[(\eta^5\text{-}C_5H_5)Mn(CO)_2]_2$ (61, 127). Analogous rearrangement reactions of substituted alkynes were reported earlier by Antonova et al. (178–180).

3. Other Synthetic Approaches

a. Organolithium Reagents. The first μ-methylene complex having hydrogen attached to the bridgehead carbon was discovered by accident. Pentacarbonyliron successfully resisted formation of the expected Fischer-type carbene complexes $(CO)_4Fe[C(OCH_3)R]$ (R = CH_3, C_6H_5) upon treatment with $LiCH_3$ or LiC_6H_5, and subsequent attempted methylation by means of Meerwein's reagent (181–183). Instead, the major product isolated and structurally characterized (X-ray) represents a boat-shaped dimetallacycle, the intermolecular distances of which are consistent with both a phenylferroxycarbene and the alternative benzoyl description (181, 182). The other compound, isolated in 0.5% yield, was spectroscopically recognized as μ-benzylidene complex 73 (183). The two isomers (a and b) that were detected by IR spectroscopy form an equilibrium with each other, which is governed by the temperature and the solvent employed (183) and which thus reflects an interesting parallel with the well-known structural behavior of the isoelectronic carbonyl complex $Co_2(CO)_8$ in solution.

73a 73b

The dinuclear iron compound **74** obtained by treatment of $Fe(CO)_5$ first with $Li[C_6H_3(OCH_3)_2]$–THF and then with $[Et_3O]^+BF_4^-$–CH_2Cl_2 may be considered a borderline case between a true μ-methylene complex and an adduct of the $Fe(CO)_3$ fragment to the conjugated π-system of the mononuclear metal carbene $(CO)_4Fe=C(OEt)Ar$ *(184)*. The stereochemistry of the bridging system is governed mostly by steric constraints which result in a nearly planar configuration of the bridgehead carbon. From structural data, the methylene carbon hybridization is closer to sp^2 than to sp^3 *(184)*.

74

In the hope of obtaining iridium methyl compounds, the dinuclear doubly methylene-bridged derivative of composition $[(\mu\text{-}CH_2)Ir(COD)]_2$ (**75**) appeared as the ultimate product of the reaction of $[(COD)Ir(\mu\text{-}Cl)]_2$ with methyllithium; this contrasts with the formation of $[(\mu\text{-}CH_3)(COD)Rh]_2$ from $LiCH_3$ and $[(1.5\text{-}COD)Rh(\mu\text{-}Cl)]_2$ *(185)*.

A strange and certainly less straightforward method of constructing methylene bridges was reported for the cobalt compound $(\mu\text{-}CH_2)(\mu\text{-}CO)[(\eta^5\text{-}C_5Me_5)Co]_2$ (**76**), the discovery of which stemmed from an attempt to prepare bis(pentamethylcyclopentadienyl)cobalt, via the reaction of $Li^+[C_5Me_5]^-$ (preformed from C_5Me_5H and n-BuLi) with anhydrous $CoCl_2$ in tetrahydrofuran. The final product isolated in $\leqslant 22\%$ yield was shown not to be the desired cobalt sandwich but rather the novel μ-methylene complex **76** *(186)*. There is good evidence *(186)* that the CH_2 group as well as the CO group originate from the lithium enolate of acetaldehyde $[Li^+CH_2C(=O)H^-]$, which, in turn, arises from a known *(187)* side reaction between butyllithium and the solvent.

b. Dimethylmagnesium. A single, but significant, case is known in which dimethylmagnesium functions as a methylene transfer reagent *(95, 96)*. If added in excess to either $Ru_2(O_2CMe)_4Cl$ or the μ_3-oxo-centered acetate complex $Ru_3O(O_2CMe)_6(H_2O)_3(O_2CMe)$, in the presence of trimethylphosphine, the tris-μ-methylene complex 77 is formed in yields of up to 27%. The observed liberation of methane clearly demonstrates that methyl (hydrido) intermediates originating from α-hydrogen abstraction are mechanistically involved.

77

c. Organomercurials. Neither for the synthesis of carbene complexes nor for the preparation of their μ-methylene relatives have organomercurials of type $Hg(CX_3)R$ been successful starting materials. Moreover, dihalocarbene ligands are scarce (*188, 189*) compared to the wealth of other carbene ligands. For instance, the reaction of octacarbonyldicobalt with the dichlorocarbene precursor (*190*) $Hg(CCl_3)C_6H_5$ (hexane, 30–60°C) affords the μ_3-chloromethylidyne ["chlorofred" (*34*)] complex (μ_3-CCl)[Co_3(CO)_9] (**78**) in ~50% yield (*191*). Trapping experiments using cyclohexene have demonstrated that dichlorocarbene is in fact formed, and from this finding it has been suggested that the dinuclear intermediate $(\mu$-$CCl_2)(\mu$-$CO)[Co(CO)_3]_2$ (**79**) plays a pivotal role in the formation of the final product (*191*). Compound **78** is also formed when CCl_4, $CBrCl_3$, or CBr_2Cl_2 are used instead of organomercurials under the same conditions. A radical mechanism has been proposed for these reactions (*192*).

d. Ketenes. In some isolated cases, ketenes have been employed for the preparation of μ-methylene complexes. The reaction of $[Rh(CO)_2(\mu$-$Cl)]_2$ with excess diphenylketene involves decarbonylation of the organic precursor (*192–196*), primarily yielding a polymeric product (structure not yet established) of composition $[Rh_2\{C(C_6H_5)_2\}(\mu$-$CO)(\mu$-$Cl)]_x$ (**80**), which, upon treatment with NaC_5H_5, gives the dimeric derivative (η^5-$C_5H_5)_2Rh_2[\mu$-$(C_6H_5)_2]_2(\mu$-$CO)$ (**81**). Other compounds arising from subsequent reactions of **80** were originally thought to contain terminal diphenylcarbene ligands (*192, 193*), but were later reformulated as $Rh_2[\mu$-$C(C_6H_5)_2]_2(\mu$-$CO)Cl_2py_2$ (**82**) (py = pyridine) and (η^5-$C_5H_5)_2Rh_2[\mu$-$C(C_6H_5)_2]_2$ (**83**), respectively. Complex **83** is reportedly also formed from $[Rh(CO)_2(\mu$-$Cl)]_2$ and diphenyldiazomethane (*192–194*). Unfortunately, ketenes tend either to produce stable η^2-complexes (*197*) or, alternatively, to undergo deoxygenation reactions with metal carbonyls (*198*); such side paths of course limit the use of ketenes for any general synthesis of μ-methylene complexes. Beyond that, in just one case a mononuclear carbene was formed from a heterocumulene precursor (Scheme 22). Triethylphosphite induces clean CO elimination from the η^3-diphenyl-ketene iron complex **84** in which the π-bonded, bent heterocumulene acts as a four-electron rather than a common two-electron ligand; other phosphines give quite different products (*199*).

e. Strained Hydrocarbons. A model system for the fragmentation of a metallacycle to carbene and olefin, as occurs in olefin metathesis (*200*), was established through metal-induced degradation of a strained hydrocar-bon (*201*). Low-temperature photolysis of Fe(CO)₅ in the presence of tricyclo[3.2.0.0²,⁷]hept-3-ene primarily yields the thermolabile olefin com-plex **85** of type Fe(CO)₄L (Scheme 23). Prolonged photolysis induces carbonyl elimination from **85** with concomitant C—C bond cleavage. The σ/π-allyl derivative **86** thus obtained in high yield carbonylates at 20°C (1 atm) to give the ketone complex **87**, whereas addition of Fe(CO)₄C₆H₁₀ (C₆H₁₀ = cyclohexene) at $-20°$ produces the dinuclear μ-alkylidene com-plex **88** (22%). Independent trapping experiments suggest that the mononuclear carbene intermediate **89** occurs prior to the final products **87** and **88** [Scheme 23; M = Fe(CO)₃].

f. Rearrangement Reactions. Hydrogen transfer from a metal site to the β-carbon of a μ,η^2-vinyl complex may also yield edge bridging al-kylidenes (Scheme 24). The only report of such an internal rearrangement

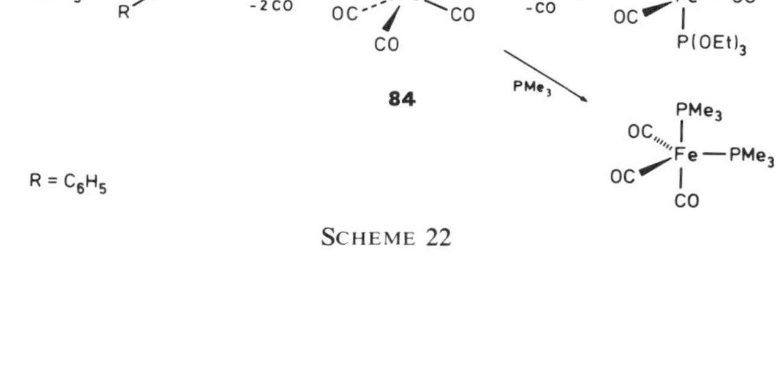

SCHEME 22

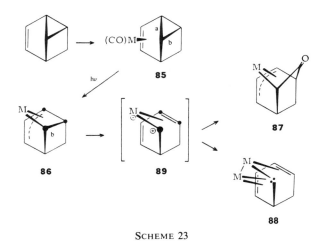

SCHEME 23

process comes from Lewis *et al.* (*202*). The hydrido clusters $(\mu\text{-}H)_3(\mu,\eta^2\text{-}CR\!\!=\!\!CHR')Os_4(CO)_{11}$ [**90**; R = H, R' = CH$_3$ or C$_6$H$_5$; R = R' = C$_6$H$_5$; R + R' = —(CH$_2$)$_4$—] are cleanly converted (yields 80%) to $(\mu\text{-}H)_2(\mu\text{-}CR\text{—}CH_2R')Os_4(CO)_{12}$ (**91**) by uptake of carbon monoxide at elevated temperatures.

Another intriguing isomerization leading to a μ-perfluoroethylidene ligand starts from the dinuclear species $(\mu\text{-}C_2F_4)(\mu\text{-}SCH_3)_2[Fe(CO)_3]_2$ (**92**) which, in turn, is accessible by photochemically induced insertion of tetrafluoroethylene into the Fe—Fe bond of $(\mu\text{-}SCH_3)_2[Fe(CO)_3]_2$. Under remarkably mild conditions (boiling pentane), compound **92** undergoes fluorine migration generating the structurally characterized μ-C(F)CF$_3$ isomer of composition $[\mu\text{-}C(F)CF_3](\mu\text{-}SCH_3)_2[Fe(CO)_3]_2$ (**93**) in 40% yield [(*203*); see also Table I]. It has been proposed (*203*) that this isomerization process involves an ion-pair-type intermediate, the σ,π-bonded vinyl group of which is considered to be open for nucleophilic attack by the fluoride counterion at the β-position, as it is, for instance, in Os$_3$H(HC≡CH$_2$)(CO)$_{10}$ (*204*). Experiments involving phosphine derivatives of **92** strongly support this mechanism (*203*). An analogous reaction was previously known to occur between C$_2$F$_4$ and Co$_2$(CO)$_8$ under thermal conditions (*133*).

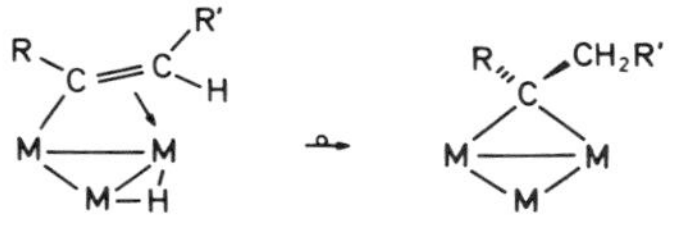

SCHEME 24

4. Tebbe's Reagent

The first μ-methylene complexes having both a main group element and a transition metal attached to CH_2 groups were reported by Tebbe et al. in 1978 (100). The straightforward reaction of 2 equiv of $Al(CH_3)_3$ with (η^5-$C_5H_5)_2TiCl_2$ in toluene (60 h, 25°C) produces the heterodinuclear compound 94[2] in 80–90% yield. The aluminium alkyl inhibits the catastrophic decomposition of (η^5-$C_5H_5)_2Ti(CH_3)_2$ and dictates the abstraction of hydrogen from methyl rather than cyclopentadienyl groups. Practically unlimited amounts of this highly air-sensitive, crystalline, reddish compound can be made in a single run. Compound 94 is also formed from (η^5-$C_5H_5)_2$-$Ti(CH_3)_2$ and $ClAl(CH_3)_2$. Dimethylzinc and either (η^5-$C_5H_5)_2Ti(CH_3)_2$ or (η^5-$C_5H_5)_2TiCl_2$ react to yield again methane and products believed to contain $TiCH_2Zn$ moieties, based on the appearance of diagnostic (8, 47) low-field ^{1}H-NMR resonances (τ 1.32 and τ 1.69, respectively, C_6D_6) and on their reactivity toward ketones (see below).

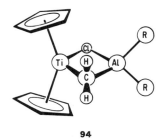

94

The methyl groups in 94 exchange with certain aluminium alkyls and halides (Scheme 25a), but the methylene group remains unreactive (100). Treatment of 94 with $Al(CD_3)_3$ results in statistical scrambling of methyl groups between free and titanium-bound aluminium alkyl, with no deuterium incorporation in the cp or CH_2 positions. It is likely that reactions between AlY_3 and 94 [or the analogous compound (η^5-$C_5H_5)_2Ti$-(μ-CH_2)(μ-CH_3)$Al(CH_3)_2$ similarly obtained from (η^5-$C_5H_5)Ti(CH_3)_2$ and $Al(CH_3)_3$ (100)] proceed via ring opening of the four-membered heterocycle involving cleavage of the CH_2—Al bond, a process which probably gives the transient species (η^5-$C_5H_5)_2Ti(=CH_2)ClAl(CH_3)_2$. Consistent with this interpretation, compound 94 undergoes olefin homologation. It reacts with ethylene at room temperature (toluene) to form propylene (32%, 18 h), and with propylene to form isobutylene (59%), a trace of

[2] Reprinted with permission from F. N. Tebbe, G. W. Parshall, and G. S. Reddy, J. Am. Chem. Soc. 100, 3611 (1978). Copyright 1978 American Chemical Society.

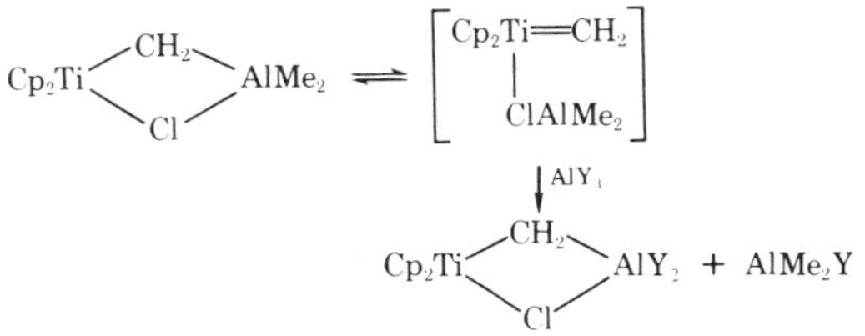

SCHEME 25A. Reprinted with permission from F. N. Tebbe, G. W. Parshall, and G. S. Reddy, *J. Am. Chem. Soc.* **100**, 3611 (1978). Copyright 1978 American Chemical Society.

methylcyclopropane, and other C_4 hydrocarbons. Isobutylene is relatively unreactive alone, but when tetrahydrofuran or trimethylamine is added, 1,1-dimethylcyclopropane and another C_5H_{10} compound are each formed in 2% yield. Most of the isobutylene is recovered. The bridging methylene unit and the isobutylene methylene group were found by ^{13}C labeling to exchange as a unit in the degenerate olefin metathesis reaction **94** (CH_2) + $^{13}CH_2$=$C(CH_3)_2$ → **94** $(^{13}CH_2)$ + CH_2=$C(CH_3)_2$ (205).

Addition of triethylamine to the ethylene reaction leads to formation of a minor amount of cyclopropane, a product *not* detected in the absence of the amine. The reactions seem to involve transfer of the methylene group specifically because $(\eta^5\text{-}C_5H_5)_2Ti(\mu\text{-}CH_2)(\mu\text{-}Cl)Al(CD_3)_2$ reacts with C_2H_4 to form propylene-d_0. On the other hand, reaction of C_2D_4 with unlabeled **94** gives CD_3CD=CH_2 and CH_2DCD=CD_2 in a 1.4 : 1 ratio. Taking these labeling experiments into account and considering the analogy with olefin metathesis mechanisms based on transient aluminium-stabilized tungsten carbene complexes (206), it appears reasonable to believe that the metallacycle (**95**) arising from addition of ethylene across the Ti—CH_2 bond acts as an intermediate in ethylene homologation (Scheme 25b). The failure of deuterium scrambling to occur throughout the olefin rules out a metathesis process operating prior to propylene formation. However, the formation of cyclopropanes does reflect a second mode of reaction of **94** in which reductive coupling occurs.

Another interesting feature of Tebbe's reagent is its capability for converting ketones to terminal olefins. For example, reaction with cyclo-

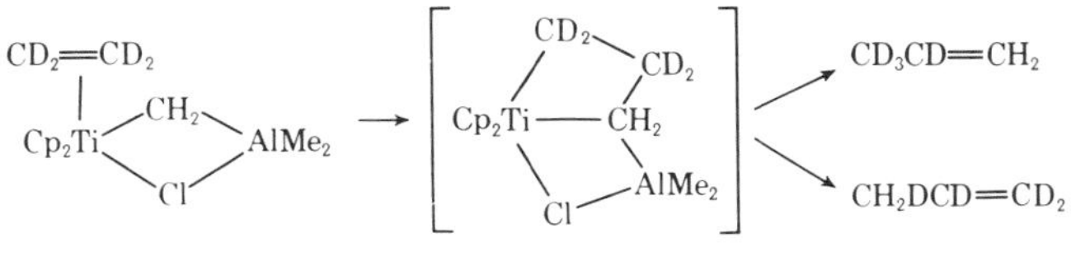

95

SCHEME 25B

hexanone (toluene, -15–$25°C$) produces methylenecyclohexane in 65% yield. The reactivity of **94** to organic carbonyls is comparable to that of $(Me_3CCH_2)_3Ta{=}CHCMe_3$, which produces *tert*-butyl substituted olefins (*9, 207*). Attempts to synthesize $(\eta^5\text{-}C_5H_5)_3Zr{-}CH_2{-}CH_2{-}AlEt_2$ from $(C_5H_5)_4Zr$ and $AlEt_3$ failed. Instead, the hydrido-bridged compound $(\eta^5\text{-}C_5H_5)_3Zr{-}H{-}AlEt_3$ was isolated and structurally characterized (*208*).

V

THEORY AND BONDING IN SIMPLE DIMETALLACYCLOPROPANE COMPLEXES

Considering the structures of μ-methylene complexes, the important question arises as to whether the bridging CRR' ligands are to be regarded essentially as carbenes or, alternatively, as being derived from alkanes. Albright, Hofmann, Hoffmann, and Pinhas have carried out detailed extended Hückel-type MO calculations for the rhodium compound $(\mu\text{-}CH_2)[(\eta^5\text{-}C_5H_5)Rh(CO)]_2$ (**19a**) and have compared this system with isolobal derivatives with $Rh(CO)_2^-$, $CH^{+,-}$, or cpRh(CO) in place of CH_2 (*150, 151*). It appears from this work that the methylene compound may best be viewed as composed of a CH_2 unit formally added to the hypothetical $cp_2Rh_2(CO)_2$ dimer having a $Rh{=}Rh$ double bond. Apart from the σ-type bonding between the methylene group and the organometallic frame, the qualitative interaction scheme demonstrates the existence of a situation exactly corresponding to the Walsh description of cyclopropane, derived from an ethylene and a methylene structural unit. Therefore, μ-methylene complexes very much resemble "dimetallacyclopropanes," which, in turn, are intermediate members of the series of olefin complexes ("metallacyclopropanes"), μ-methylene complexes ("dimetallacyclopropanes"), and trinuclear metal clusters ("trimetallacyclopropanes"). Of importance to the bonding characteristics of the CH_2 unit is an electron transfer from the Rh—Rh σ-bond to an orbital containing a π^*_{xy} component and then to the carbon atom. The formal metal–metal double bond thus becomes a "single bond." The NMR data (*47, 61*) suggest that the carbon bridge exhibits far greater charge density than the carbene carbon of electrophilic Fischer-type carbene complexes $L_xM{=}CRR'$.

Lichtenberger *et al.* (*209*) have reported the He(I) photoelectron spectrum of the manganese compound $(\mu\text{-}CH_2)[(\eta^5\text{-}C_5H_4CH_3)Mn(CO)_2]_2$ (**3b**) in the ionization energy range below 11 eV and have compared the results with the ionizations of $(\eta^5\text{-}C_5H_4CH_3)Mn(CO)_3$ and $(\eta^5\text{-}C_5H_4CH_3)Mn(CO)_2$-$(\eta^2\text{-}C_2H_4)$. Excellent agreement was found between the observed ionizations and the predictions of parameter-free molecular orbital calculations. Again, the bonding scheme was constructed from the CH_2 orbitals and those of the hypothetical $cp_2Mn_2(CO)_4$ dimer. In the first step of this analysis, the frontier orbitals of each $cpMn(CO)_2$ portion simply combine into the corresponding bonding and antibonding combinations. The strongest combination occurs between the d_{z^2}-type orbitals that are each directed toward the same region. As in the bonding of carbenes with single atoms metal, the methylene group has a filled orbital (a_1) that can act as a sigma donor to the system. It also has an empty p_π-orbital (b_1) that can accept π-electron density from the metal(s). The interaction of these orbitals with those of the dinuclear $cp_2Mn_2(CO)_4$ fragment are shown in Fig. 5. The LUMO of the dinuclear fragment has the correct symmetry, is relatively low in energy, and has a nearly ideal spatial distribution for accepting electron density from the donor orbital of the CH_2 group. Similarly, the HOMO of the dinuclear framework is well situated for donating electron density to the empty p_π-orbital of the methylene group. The μ-CH_2 ligand was shown to act overall as a better acceptor than donor in this system. Fenske–Hall calculations place the originally empty p_π-

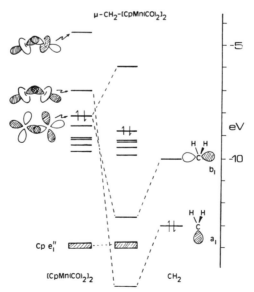

FIG. 5. MO diagram of μ-methylenebis[dicarbonyl(η^5-cyclopentadienyl)manganese] (Mn—Mn) (3a) according to Lichtenberger *et al.* (209).

orbital of the CH_2 group slightly below the filled metal levels. Because of this ordering, the donation into this p_π-orbital should be effectively described as a charge transfer from the HOMO of the dinuclear fragment to the methylene group. This results in the high negative charge on the CH_2 carbon; the calculated electron-density value amounts to -0.526 e which is in reasonable agreement with the numerical result obtained from an experimental electron density determination (-0.78 e) performed by Clemente *et al.* (*127*) (Fig. 6). Both electron donation from the methylene unit and its electron acceptor interaction tend to produce a net metal–metal bond.

A comparative study of the unknown monomeric mononuclear species $(\eta^5\text{-}C_5H_5)Mn(CO)_2(=CH_2)$ indicates that increased stability is gained by the additional metal–metal bond formation, which, in turn, seems to be the driving force for bimetallic systems to be favored over the mononuclear systems (*209*). In agreement with the molecular orbital description given for $(\mu\text{-}CH_2)[(\eta^5\text{-}C_5H_5)Rh(CO)]_2$ (*150, 151*) and $(\mu\text{-}CH_2)[Os_3(CO)_{10}(\mu\text{-}H)_2]$ (*80*), the bonding in the Mn—C—Mn triangle can be summarized as a six-electron, three-center frame similar to cyclopropane. Hofmann has emphasized the close analogy between the bonding scheme of dimetallacyclopropanes and related molecules having other π-acceptor ligands in place of the CRR′ bridge (e.g., carbon monoxide, vinylidenes) (*150*). In light of this work, the suggested representation of the vinylidene homolog $(\mu$—C$=CH_2)[(\eta^5\text{-}C_5H_5)Mn(CO)_2]_2$ of complex **3a**, as a geminally disubstituted olefin rather than a carbenoid species is certainly not mean-

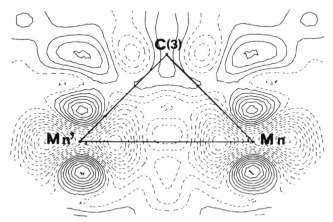

FIG. 6. Section of the X–X deformation density through the Mn–C(3)–Mn′ plane of $(\mu\text{-}CH_2)[(\eta^5\text{-}C_5H_5)Mn(CO)_2]_2$ (**3a**; see Fig. 1). Contour interval 0.1 e$\mathring{A}^{-3}$. Dashed contours for negative, bold line for zero deformation density. Resolution $2 \sin \vartheta_{max}/\lambda = 1.41$ $\mathring{A}^{-1}$ ($\vartheta_{max} = 30°$). Average e.s.d. of the observed density in general position is 0.07 $e\mathring{A}^{-3}$ (*127*).

ingful, not least because the arguments in favor of this bonding mode suffer from questionably validity (*177*).

Hoffmann *et al.* (*210*) have discussed the relative stability of carbonyl versus methylene bridges in dimers of type M_2L_{10}. Both ligands are similar in their calculations, except that CH_2 is both a better σ-donor and a better π-acceptor ("two-electron acceptor bridge"). It has been concluded that in d^7–d^7 systems the μ-CR_2 compounds will be more stable than their μ-CO analogues. If one takes into account that a better donor metal strengthens the acceptor bridging, then it becomes understandable why all known μ-CO d^7–d^7 cases belong to the type $cp_2M_2(\mu\text{-CO})_2$, whereas the μ-CR_2 cases are of the $M_2(CO)_8(\mu\text{-}CR_2)_2$ type (e.g., **58**). Replacement of three CO molecules by cp makes the metal center a better donor, which is needed in order to form strong bridging bonds with the inferior (relative to CR_2) acceptor CO. The d^7–d^7 $cp_2M_2(CO)_2(\mu\text{-}CR_2)_2$-type compounds are consequently predicted to be even more stable than their $M_2(CO)_8(\mu\text{-}CR_2)_2$ counterparts. In general, CO bridging will require a better electron-donor metal than does CR_2; the CO fluxionality in $(\mu\text{-}CH_2)[(\eta^5\text{-}C_5Me_5)Rh(CO)]_2$ is in excellent agreement with this statement (*54, 54a, 94*), whereas the C_5H_5 compound exists exclusively in the nonbridged form [(*40, 41, 61a*); see Section VIII,A and Table IV, footnote *e*].

VI
STRUCTURAL FEATURES OF μ-METHYLENE COMPLEXES

The discussion in the previous section has emphasized the necessity of metal–metal interactions to provide enhanced stability to the methylene acceptor bridging. It is a fact that the entire set of μ-methylene transition metal complexes belonging to Class C (Section I) reveals quite short metal–metal separations, covering the range 245–320 pm [M = Cr (*27*), Mo (*165*), W (*68–70, 86, 88, 90*), Mn (*47, 57, 61, 87, 89, 127, 211*), Re (*175, 211, 212*), Fe (*48, 64, 72, 104, 184, 213, 214*), Ru (*96, 97*), Os (*80, 202, 215, 217*), Co (*43, 45, 114, 120, 186*), Rh (*40, 41, 54, 56, 155, 194, 196*), Pt (*27, 86–90, 92*)]. The single exception to this rule is Puddephatt's platinum CH_2 complex **30** (see Section IV); we may learn from this example that a M—CH_2—M linkage can survive if chelating ligands span both metals, thus preventing the bridging system from falling apart. Ligand bridging, possibly via semibridging CO molecules may also account for the exceedingly great stability of the radical cation $[(\mu\text{-}CH_2)\{(\eta^5\text{-}C_5H_5)Mn(CO)_2\}_2]^+$ cleanly obtained by electrochemical oxidation of the neutral precursor **3a** in acetonitrile, in a fully reversible process. An ESR study of the cation has shown that the metal–metal bond present in **3a** (Table I) is not maintained upon oxidation since the unpaired electron is localized at just one metal center (221). A systematic survey of all known structures (Table I) permits the following general statements with regard to the common structural features of dimetallacyclopropane-type molecules:

1. Symmetrical methylene (alkylidene) bridges represent the predominant geometry. This is documented by isosceles triangular frameworks defined by the two metal centers and the methylene carbon atom. Asymmetric bridging has been encountered in heterodinuclear μ-methylene complexes. For example, the differences observed in M—C(methylene) and M'—C (methylene) bond lengths in compounds **43a** and **43b** are far greater than one would expect from the difference in the corresponding covalent metal radii (Table I). Carbonyl semibridging also occurs, along with occasional asymmetric methylene bridging (27, 86, 93). It is generally observed that the metal–metal distances are slightly longer (averaging ~15 pm) than the values predicted from the addition of Pauling covalent radii for two "single-bonded" metal atoms.

2. The "internal" angles α around the methylene carbon atoms (**96**) are generally very acute and show little variation with the nature of the metal and the carbene substituents R and R' involved (43). Typically, these angles fall in the range 76–81°, and no case is known in which the values are smaller or greater than 81 ± 7°. This means that the bonding metal–metal interaction is great enough to distort the angle to a significantly more acute one than expected for the tetrahedral result (109°28'). It is obvious that the metal–metal distance required for a bonding interaction determines the magnitude of α (e.g., (324); Table I, footnote r).

96

3. The "external" angles β defined by the bridgehead carbon and the nonmetal atoms directly attached to it are far greater than the corresponding M,C,M angles. The 104–110° range is typical of dimetallacyclopropanes, but considerable variations within 105 ± 13° are not unusual (Table I). Such changes in the external angle β cannot be readily correlated with (a) changes in the internal angle α, (b) the nature of the groups R and R' (e.g., steric effects, such as crowding, are not logically related to variations in α or β), (c) the nature of the metals and the remainder ligands surrounding them, or (d) obvious features of the metals themselves, such as the d-electron count. However, such a lack of correlation between α

TABLE I

STRUCTURAL DATA FOR μ-METHYLENE TRANSITION METAL COMPLEXES BELONGING TO THE DINUCLEAR TYPE $[\mu\text{-CRR'}]_{1x}[ML_y]_2$ AND OF RELATED ONE-CARBON BRIDGED MOLECULES

ML_y	CRR'	x	$d(M{-}M)/$ $d(M{-}M')$ [pm]	$d(M{-}CRR')/$ $d(M'{-}CRR')$ [pm]	⟨(M,C,M)/ ⟨(M,C,M') [deg]	⟨(R,C,R') [deg]	Reference
Class A: Compounds with metal–metal bonds							
$(\eta^5\text{-C}_5\text{H}_5)\text{Mo(CO)}_2(\eta^5\text{-C}_5\text{H}_5)\text{Mo(CO)}$	C_2H_4	1	309.8(0)	211.1(4) 222.4(4)	91.2(1)	101.8(3)	165
$\text{W(CO)}_4\text{Re(CO)}_3(\text{PMe}_3)(\mu\text{-CO})$	$\text{C}(p\text{-CH}_3\text{C}_6\text{H}_4)(\text{PMe}_3)$	1	296.8(2)	230(4) [M = Re(?)] 234(4)[M = W(?)]	79.6(12)	105.9(25)	211, 211a
$\text{W(CO)}_4/\text{W(CO)}_5$	$\eta^3\text{-C(H)CH=CMe}_2$	1	318.9(1)	218(2) 220(2)	93.3(7)	—[a]	69
W(CO)_4	$\eta^1\text{-C(H)CH=CMe}_2$	1	315.7(1)	230(2) 234(2)	85.8(5)	—[a]	68, 70
$\text{W(CO)}_4/\text{W(CO)}_5$	$\eta^2\text{-C(H)C(H)CMe}_2(\text{PMe}_3)$	1	314.0(1)	222(2) 226(2)	88.9(6)	—[a]	71
$(\eta^5\text{-C}_5\text{H}_5)\text{Mn(CO)}_2$	CH_2	1	279.9(0)	202.6(1)	87.4(1)	108.8(1.5)	61, 127
$(\eta^5\text{-C}_5\text{H}_4\text{CH}_3)\text{Mn(CO)}_2$	CH_2	1	277.9(1) 278.6(1)	201.3(5) 201.5(5)	87.3(2) 87.2(2)	92(8) 116(3)	47, 57 304a
Re(CO)_4	$\text{C(OCH}_3)\text{C}_6\text{H}_5$	2	281.0(3)	230(4) 222(3)	77(1)	115(3)	175
$(\eta^5\text{-C}_5\text{H}_5)\text{Mn(CO)}_2/\text{Re(CO)}_4$	$\text{C(CO)C}_6\text{H}_5$	1	281.7(3)	221(3) [Mn] 224(3) [Re]	79(1)	111(3)	212
Fe(CO)_4	CH_2	1	250.7(1)	—	—[a]	—[a]	72
$(\eta^5\text{-C}_5\text{H}_5)\text{Fe(CO)}$	CH_2	1	249.9(9)	192(6) 200(5)	77(2)	—[a]	104
$(\eta^5\text{-C}_5\text{H}_5)\text{Fe(CO)}$	C(H)CH_3	1	252.0(1)	198.6(3)	78.8(1)	112(2.5)	64
$(\eta^5\text{C}_5\text{H}_5)\text{Fe(CO)}$	$\text{C(H)CO}_2\text{-}t\text{-Bu}$	1	254.0(5)	179.4(8) 192.6(8)	79.4–85.3	113(5)	48

Complex							
$Fe(CO)_3$	C_7H_8	2	263(−)	197.7(3), 206.4(3)	81.4	—[a]	*201*
$[\mu\text{-}C(H)\text{-}C(NHR)C_6H_5][\mu\text{-}P(C_6H_5)_2][Fe(CO)_3]_2$	$C(H)C(NHR)C_6H_5$	1	257.6(1)	207.7(2), 204.4(2)	77.4(0)	108(1)	*213*
$[\mu\text{-}C(H)\text{-}C(NEt_2)C_6H_5][\mu\text{-}P(C_6H_5)_2][Fe(CO)_3]_2$	$C(H)C(NEt_2)C_6H_5$	1	254.8(1)	204.2(4)	76.2(0)	—[a]	*214*
$Fe(CO)_4$	$\eta^2\text{-}C(OEt)Aryl$[b]	1	253.5(2)	192.0(8), 200.0(7)	80.6(3)	117.2(6)	*184*
$Ru(PMe_3)_3$	CH_2	3	265.0(1)	210.3(6)–211.2(4)	77.7(1)–78.1(3)	—[a]	*96*
$[Ru(PMe_3)_3]^+$	CH_2	2	273.2(1)	209.0(7), 210.9(7)	80.8(4), 81.6(4)	—[a]	*96*
$[(Me_3P)_4Ru(\mu\text{-}CH_2)_2Ru(\mu\text{-}CH_2)_2\text{-}Ru(PMe_3)_4][BF_4]_2$	CH_2	2/2	263.7(1)	197.9(12), 198.3(8)	80.7(4)	—[a]	*97*
$(\mu\text{-}CH_2)(\mu\text{-}H)_2[Os_3(CO)_{10}]$[c]	CH_2	1	282.4(3)	215.1(5), 215.0(6)	82.1(2)	106.0(8)	*80*
$[\mu\text{-}C(H)CH_2C_6H_5](\mu\text{-}H)[Os_4(CO)_{12}]$	$C(H)CH_2C_6H_5$	1	280.1–296.4	211(2), 215(2)	80.6(7)		*202*
$[\mu\text{-}C(H)CH{=}NEt_2](\mu\text{-}H)[Os_3(CO)_{10}]$	$C(H)CH{=}NEt_2$	1	278.5(2)	215(3), 216(3)	80.6(9)	110.2(−)	*215, 216*
$[\mu\text{-}C(H)CH_2PMe_2\phi][(\mu{-}H)][Os_3(CO)_{10}]$	$C(H)CH_2PR_3$	1	280.02(6)	216.1(17)	80.8(3)		*217*
$(\eta^5\text{-}C_5H_5)Co(CO)$	$C(H)CO_2Et$	1	249.5(1)	194.4(7), 195.1(6)	79.6(2)	92.6(2)	*43*
$(\eta^5\text{-}C_5H_5)Co(CO)$	$C_9H_4O_2$[d]	1	247.5(1)	191.6(4), 197.3(4)	79.0(1)	104.3(3)	*45*
$(\eta^5\text{-}C_5Me_5)Co/(\eta^5\text{-}C_5Me_5)\text{-}Co(\mu\text{-}CO)$	CH_2	1	232.0(1)[e]	190.9(9)	74(−)[f]	—[a]	*186*
$Co(CO)_3/Co(CO)_3(\mu\text{-}CO)$	$C_4H_2O_2$[g]	1	245(1)	202(8) (av.), 202.9(4), 205.5(4)	75(−)	108(−)	*114*
$(\eta^5\text{-}C_5H_5)Rh(CO)$	CH_2	1	266.49(4)		81.7(1)	115.9(4)	*40, 41*
$(\eta^5\text{-}C_5H_5)Rh(CO)$	CH_2	1	266.2(1)	205.0(1), 207.6(6)	81.0(1)	110.4(1)	*56*[c]
$(\eta^5\text{-}C_5Me_5)Rh(CO)$	$C(CO_2Me)_2$	1	266.3(1)	208.3(6)	79.6(2)	108.2(5)	*54, 55*

(continues)

TABLE I (Continued)

ML_y	CRR'	x	$d(M{-}M)/$ $d(M{-}M')$ [pm]	$d(M{-}CRR')/$ $d(M'{-}CRR')$ [pm]	$\sphericalangle(M,C,M)/$ $\sphericalangle(M,C,M')$ [deg]	$\sphericalangle(R,C,R')$ [deg]	Reference
$(\eta^5\text{-}C_5Me_5)Rh(CO)$	CH_2	1	267.2(1)	202.6(8)	82.5(4)	109.4(10)	54a
$(\eta^5\text{-}C_5Me_5)Rh(CO)$	$C_9H_4O_2{}^d$	1	266.3(2)	205.3(2)	79.9(2)	105.3(4)	155, 296
$(\eta^5\text{-}C_5Me_5)Rh(CO)$	$C(H)CH_3$	1	265.8(2)	205.6(17)	80.6(8)	—^{a}	155
$(\eta^5\text{-}C_5Me_5)Rh(\mu\text{-}CO)$	C_5Br_4	1	261.2(2)	210.1(15)	76.5(4)	101.6(9)	54, 55
$(\eta^5\text{-}C_5Me_5)Rh(CO)$	$C(C_6H_5)_2$	1	264.2(5)	209.3(6) 210.5(8)	79.9(3)	104.3(6)	155
$(\eta^5\text{-}C_5H_5)Rh$	$C(C_6H_5)_2$	2	246–248	196–208	—^{a}	—^{a}	194, 195
$(\eta^5\text{-}C_5H_5)Rh/$ $(\eta^1\text{-}C_5H_5)Rh(\mu\text{-}CO)$	$C(C_6H_5)_2$	2	254.8(1)	209.4(6)	75.1(2)	111.1(5)	196
$(\eta^1\text{-}C_5H_5N)RhCl^i$	$C(C_6H_5)_2$	2	251	—^{a}	—^{a}	—^{a}	195
$Pt(PMe_3)_2/W(CO)_4(PMe_3)^j$	$C(OMe)C_6H_4\text{-}p\text{-}Me$	1	282.5(1)	203(1) [Pt] 237(1) [W]	79.4(4)	104.6(9)	90
$Pt(PMe_3)_2/W(CO)_5{}^j$	$C(OMe)C_6H_5$	1	286.1(1)	204(1) [Pt] 248(1) [W]	77.8(3)	106.6(9)	86
$[Pt(P\text{-}t\text{-}Bu_2Me)(CO)]_2[W(CO)_4]^k$	$C(OMe)C_6H_5$	1	262.7(1) [Pt–Pt]	204(2) 207(2)	79.5(8)	102(2)	88
$Pt(PMe_3)_2/Cr(CO)_3{-}$ $(\mu\text{-}CO)(PMe_3)^j$	$C(CO_2Me)C_6H_5$	1	264.6(7)	198(3) [Pt] 227(4) [Cr]	77(1)	110(3)	27
$[(\eta^5\text{-}C_5H_5)Fe(\mu,\eta^1{:}\eta^3\text{-}C_5H_4)\text{-}$ $Au_2\{P(C_6H_5)_3\}_2]BF_4{}^j$	C_5H_4	1	276.8(3) [Au–Au]	215(4) 277(4)	78(1)	—^{a}	235
$Au[P(C_6H_5)_3]$	$C(CN)_2$	1	291.2(1)	210(1)	87.7(3)	—^{a}	304b

204

Class B: Compounds without metal–metal bonds

$(\eta^5\text{-}C_5H_5)Fe(CO)_2$	$(CH_2)_3$	1	—[b]	208(1)	115(1)[m]	—[a]	218
$(\eta^5\text{-}C_5H_5)Fe(CO)_2$	$(CH_2)_4$	1	—[b]	208(1)	113(1)[n]	—[a]	218
$Fe(CO)_3f(\mu\text{-}SCH_3)^o$	$C(F)CF_3$	1	296.3(6)[f]	203.7(7)	93.4(1)	100.7(6)	203
$Re(CO)_5^p$	$(CH_2)_2$	1	—[b]	199(1)	121(1)[q]	—[a]	219
$Rh_2Cl_2(\mu\text{-}CF_3C{\equiv}CCF_3)(\mu\text{-}dpm)_2$	CH_2	1	346.4(1)[l]	205(1) 207(1)	114.5(5)	—[a]	324[r]

[a] Not given.

[b] Aryl = 3,5-bis(methoxy)phenyl; see comment on page 191 (Cpd. 74).

[c] Neutron diffraction.

[d] 1,3-Dioxoindaneylid-2-ene.

[e] Bond order 2 according to the EAN Rule.

[f] Calculated from the reported distances.

[g] 5-Oxo-2(5H)-furaneylid-2-ene.

[h] Note that the values independently obtained from X-ray and neutron diffraction do not agree within the experimental errors.

[i] $[\{\mu\text{-}C(C_6H_5)_2\}_2\{\mu\text{-}CO\}\{(\eta^1\text{-}C_5H_5N)RhCl\}_2 \cdot 2\ CH_2Cl_2$.

[j] Asymmetric methylene bridge.

[k] $[\mu\text{-}C(OCH_3)C_6H_5)][\mu\text{-}W(CO)_4]\{Pt(PR_3)CO\}_2$.

[l] No metal–metal bond.

[m] $\not< (Fe,C,C) = 115(1)°$.

[n] $\not< (Fe,C,C) = 113(9)°$.

[o] Average values; two independent molecules in the unit cell; no metal–metal bond according to the EAN Rule.

[p] The bis-metallated ethane adopts a staggered conformation.

[q] $\not< (Re,C,C) = 121(1)°$.

[r] These data confirm the assessment on p. 201; in accord with this, the magnitude of the metal–metal separation clearly determines the acuteness of the "internal" angle α.

and β is not very surprising in light of the fact that there is no correlation between such angles in organic methylene (R^1—CH_2—R^2) fragments *(220)*, which, *cum grano salis* may be compared with the quasitetrahedral methylene bridges in dimetallacyclopropanes. Note that even in the case of the smallest substituents (hydrogen) β angles can approach the upper limit of the aforementioned range (e.g., compounds **14** and **19**; neutron diffraction; Table I).

4. The dimensions of the dimetallacyclopropane rings are susceptible to changes in the nature of the peripheral ligands. In accord with Hoffmann's prediction *(210)*, the metal–carbon distances become shorter (due to strengthening of the acceptor bonding) as the metal becomes a better electron donor. For example, the μ-C—W distance [237(1) pm] in the PMe_3-substituted complex **43b** is markedly shorter than that in the corresponding CO derivative **43a** [248(1) pm; Table I]. This tightening of the ring system by substitution of PMe_3 for CO on tungsten is reflected in greater thermal and oxidative stability *(90, 92)*.

5. Substitution of methylene bridges by carbon monoxide hardly effects the geometry of the three-membered framework; Table II illustrates this statement. Slight differences between these two systems are consistent with the following general trends: (a) M,CO,M angles are greater (2–10°C) than M,CRR',M angles; (b) M—CO distances are somewhat shorter (2–12 pm) than M—CRR' distances; (c) metal-to-metal bonds are slightly longer (<10 pm) when bridged by CO rather than by a methylene function. This structural analogy between CO and methylenes can even be extended to bridging one-carbon iminium ions such as the 1,3-dipolar $>$C⁻—H— CH=N⁺Et₂ ligand (Table I) despite the fact that these fragments formally act as three- rather than two-electron acceptor bridges in strictly symmetrical or slightly asymmetrical positions. The same is true for Kreißl's phosphoniomethanidyl ligand present in compound **97** *(211)*. The analogy

$$H_5C_6\text{'''}\underset{\diagdown}{\overset{PMe_3}{\diagup}}C$$
$$(CO)_4Re\diagup\diagdown W(CO)_4$$
$$\underset{\|}{C}$$
$$O$$

97

still holds for apparently even more remote systems. Barely any structural difference is found along the isoelectronic series $(\mu$-X$)[(\eta^5$-$C_5H_5)$Rh(CO)$]_2$ [X = CH_2, CO, SO_2 *(53)*], with the small deviations observed for SO_2 being caused to a large measure by the greater atomic radius of sulfur as compared to carbon (Table II, Fig. 7). It is no longer surprising then that

TABLE II

Structural Comparison of Corresponding μ-Methylene and μ-Carbonyl Transition Metal Complexes

Compound	d(M—M) [pm]	d(M—CRR') [pm]	d(M—CO) [pm]	∢(M,CRR',M) [deg]	∢(M,CO,M) [deg]	Reference
[μ-C(H(CH₃)(μ-CO)[(η⁵-C₅H₅)Fe(CO)]₂ (cis)	252.0(1)	198.6(3)	190.2(3)	78.8(1)	83.0(1)	64
(μ-CH₂)(μ-CO)[(η⁵-C₅H₅)Fe(CO)]₂ (cis)	249.9(9)	192(6) 200(5)	197(4)	77(2)	84(2)	104
(μ-CO)₂[(η⁵-C₅H₅)Fe(CO)]₂	253.1(2)	—	192(1)	—	82.6(2)	224
(μ-C₄H₂O₂)(μ-CO)[Co(CO)₃]₂	245(1)	199(8) 205(8)	190(8) 196(8)	75(–)	79(–)	114
(μ-CH₂)(μ-CO)[(η⁵-C₅Me₅)Co]₂	232.0(1)	190.9(9)	184.9(8)	—	—	186
(μ-CO)₂[(η⁵-C₅Me₅)Co]₂	233.8(2)	—	184.1–186.0	—	80(–)	225, 226
(μ-CH₂)[(η⁵-C₅H₅)Rh(CO)]₂ (trans)	266.49(4)	202.9(4) 205.5(4)	—	81.7(1)	—	40, 41
(μ-CO)[(η⁵-C₅H₅)Rh(CO)]₂ (trans)	268.1(2)	—	199.0(17) 201.7(17)	—	84.0(6)	227
(μ-SO₂)[(η⁵-C₅H₅)Rh(CO)]₂ (trans)	271.1(1)	225.2(1)	—	74.0(0)	—	53
[μ-C(C₆H₅)₂]₂(μ-CO)[(η⁵-C₅H₅)Rh]₂	254.8(1)	209.4(6)	199.5(9)	75.1(2)	79.4(3)	194, 195
(μ-C₃Br₄)(μ-CO)₂[(η⁵-C₅Me₅)Rh]₂	261.2(2)	210.3(15) 211.8(11)	202.3(14) 202.5(14)	76.5(4)	80.1(5) 80.7(7)	54
[μ-C(CO₂Me)₂][(η⁵-C₅Me₅)Rh(CO)]₂	266.3(1)	207.6(6) 208.3(6)	—	79.6(4)	—	54, 55
(μ-CO)[(η⁵-C₅Me₅)Rh(CO)]₂	274.3(1)	—	196.2(8) 197.5(11)	—	88.3(4)	54a

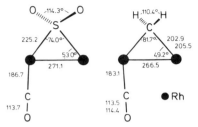

FIG. 7. Structural comparison of the triangular frames of $(\mu\text{-}X)[(\eta^5\text{-}C_5H_5)Rh(CO)]_2$ (X = SO$_2$, left; X = CH$_2$, right) (53). Bond lengths in picometers.

μ-vinylidenes, which have an additional π^*-acceptor orbital orthogonal to the plane of the three-membered ring, hardly depart from the μ-methylene geometry (Table III). With both these complexes the M—M and M—C distances are a little shorter, an effect which results in practically identical "internal" framework angles α.

TABLE III

Structural Comparison of Corresponding μ-Methylene and μ-Vinylidene Manganese, Iron, and Ruthenium Complexes and Geometrically Related Molecules (X-Ray Diffraction Data)

Compound	$d(M\text{—}M)$ [pm]	$d(M\text{—}C)/$ $d(M\text{—}X)$ [pm]	$\sphericalangle(M,C,M)/$ $\sphericalangle(M,X,M)$ [deg]	Reference
$(\mu\text{-}CH_2)[(\eta^5\text{-}C_5H_5)Mn(CO)_2]_2$ (*trans*)	279.9(0)	202.6(1)	87.4(1)	61, 127
$(\mu\text{-}CH_2)[(\eta^5\text{-}C_5H_4CH_3)Mn(CO)_2]_2$ (*trans*)	277.9(1)	201.3(5)	87.3(2)	47, 57
$(\mu\text{-}C\!\!=\!\!CH_2)[(\eta^5\text{-}C_5H_5)Mn(CO)_2]_2$ (*trans*)	275.9(2)	197.9(7) 197.1(6)	88.6	177
$[\mu\text{-}C\!\!=\!\!C(H)C_6H_5][(\eta^5\text{-}C_5H_5)Mn(CO)_2]_2$ (*trans*)	273.4(2)	194(1) 199(1)	88.0(5)	179
$[\mu\text{-}S(C_2H_5)][(\eta^5\text{-}C_5H_4CH_3)\text{-}Mn(CO)_2]_2{}^+ClO_4{}^-$ (*trans*)	293.0(1)	224.2(2) 227.0(2)	81.0(1)	229
$[\mu\text{-}As(CH_3)_2][(\eta^5\text{-}C_5H_5)Mn_2(CO)_6]$	291.2(4)	235.0(4) 236.2(4)	76.3	230
$[\mu\text{-}C\!\!=\!\!C(CN)_2](\mu\text{-}CO)[(\eta^5\text{-}C_5H_5)Fe(CO)]_2$ (*cis*)^a	250.9(4) 251.2(4)	184(2)^b 190(2)^b	85(1)^c 83.9(9)^c	168
$[\mu\text{-}C\!\!=\!\!C(C_6H_5)_2][Fe(CO)_4]_2$	263.5(3)	198(1)	83(1)	231, 232
$(\mu\text{-}C\!\!=\!\!CH_2)(\mu\text{-}CO)[(\eta^5\text{-}C_5H_5)Ru(CO)]_2$ (*cis*)	269.5(1)	202.5(7) 203.3(7)	83.2(3)^d	62

^{a} Two independent molecules in the unit cell.
^{b} Average values.
^{c} $\sphericalangle(Fe,CO,Fe)$ 80(1)° and 80.9(9)°.
^{d} $\sphericalangle(Ru,CO,Ru)$ 82.6(2)°.

VII
SPECTROSCOPIC CHARACTERIZATION OF
μ-METHYLENE METAL COMPLEXES

A. ¹H- and ¹³C-NMR Spectroscopy

How can we discriminate between terminal and bridging methylene groups (types A and C, respectively) if X-ray structural data are not available? Moreover, is spectroscopy a reliable tool to establish whether a metal–metal bond is present in addition to the methylene bridges (types C and G, respectively)? A survey of ¹H- and ¹³C-NMR data (Table IV) has revealed a fairly safe NMR diagnosis (8) which, in most cases, allows unambiguous structural assignments. The pictorial representation given in Fig. 8 summarizes the following general observations:

1. The metal-bound carbene or alkylidene carbons of the aforementioned classes of compounds resonate in ranges nicely separated from each other (¹³C; δ ppm):

$$M\!=\!\underline{C}RR'\ (A) \qquad 240\text{–}370$$
$$M\!-\!\underline{C}RR'\!-\!M\ (C) \qquad 100\text{–}210$$
$$M\!-\!(\underline{C}RR')_x\!-\!M\ (G) \qquad 0\text{–}10$$

The only apparent exception to this rule is the hydrido-bridged osmium cluster $(\mu\text{-}CH_2)(\mu\text{-}H)_2[Os_3(CO)_{10}]$ (75) (Table IV). The alkanediyl derivatives (type G) may even be distinguished from the closely related metal alkyls. The chemical shifts of the latter compounds typically cover the

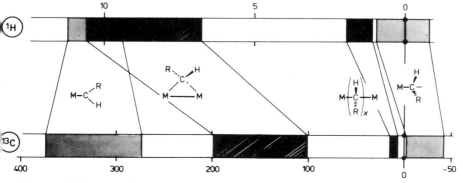

FIG. 8. NMR diagnosis of μ-methylene complexes. The diagram indicates the chemical shift ranges (δC$\underline{H}$, above; δ$\underline{C}$H, below) found for 95% of all metalcarbenes (type A), μ-methylene complexes (type C), ω,ω′-alkanediyl complexes (type G), and metal alkyls belonging to the metal carbonyl series (see text).

TABLE IV

^{13}C- AND ^{1}H-NMR DATA OF μ-METHYLENE TRANSITION METAL COMPLEXES BELONGING TO TYPE $[\mu\text{-}CRR']_x[ML_y]_{12}$[a]

ML_y	CRR'	x	$\delta CRR'$ (^{13}C)	solvent	$\delta C\underline{H}$ (^{1}H)	solvent	Reference
Class A: Compounds with metal–metal bonds[b]							
$(\eta^5\text{-}C_5H_5)_2Ti/AlMe_2$[b]	CH_2	1	188	C_6D_6	8.49	C_6D_6	99
$W(CO)_4$	$C(H)CH{=}CMe_2$	1	145.1	$CDCl_3$	10.16	$CDCl_3$	69
$(\eta^5\text{-}C_5H_5)Mn(CO)_2$	CH_2	1	150.03	CD_2Cl_2	8.65	C_6D_6	39, 126
$(\eta^5\text{-}C_5H_4CH_3)Mn(CO)_2$	$C(OCH_3)C_6H_5$	1	153.10	C_6D_6	8.82	C_6D_6	39
$Re(CO)_4$		2	217.5	THF-d_8	—		175
$(\eta^5\text{-}C_5H_5)Fe(NO)$	CH_2	1	127.38	THF-d_8	8.09	THF-d_8	59
$(\eta^5\text{-}C_5H_5)Fe(NO)$	C_5H_4	1	130.10	THF-d_8	—		59a
$(\eta^5\text{-}C_5H_5)Fe(\mu\text{-}NO)$	CBr_4	1	154.12	THF-d_8	—		153
$(\eta^5\text{-}C_5H_5)Fe(\mu\text{-}NO)$	$C(C_6H_5)_2$	1	151.12	THF-d_8	—		59a
$Fe(CO)_4$	CH_2	1		NR	5.45	Acetone-d_6	72
$Fe(CO)_4$	$C(H)C_6H_5$	1		NR	6.96	CCl_4	183
$(\eta^5\text{-}C_5H_5)Fe(CO)$[c]	$C(H)CO_2C_2H_5$	1		NR	10.27 / 10.25	DMF-d_7 / THF-d_8	48
$(\eta^5\text{-}C_5H_5)Fe(CO)$[c]	$C(H)CO_2{-}t{-}C_4H_9$	1		NR	10.36 / 10.33	DMF-d_7 / THF-d_8	48
$(\eta^5\text{-}C_5H_5)Fe(CO)$[c]	CH_2	1	(cis) 172.9 / (trans) 173.5	Pyridine-d_5	(cis) 10.29, 8.38 / (trans) 9.54	$CDCl_3$ / $CDCl_3$	104
$(\eta^5\text{-}C_5H_5)Fe(CO)$[c]	$C(H)CH_3$	1		NR	(cis) 11.60 / (trans) 10.62	Pyridine-d_5	64
$[(PMe_3)_3Ru(CH_2)_2Ru(CH_2)_3Ru(PMe_3)_3](BF_4)_2$	CH_2	2/2		NR	6.67	CD_3NO_2	97
$Ru(PMe_3)_3$	CH_2	3	131.26	C_6D_6	5.22	C_6D_6	96
$[Ru(PMe_3)_3]^+$[d]	CH_2	2	139.3	C_6D_6	8.41	C_6D_6	96
$(\mu\text{-}CH_2)(\mu\text{-}H)_2Os_3(CO)_{10}$	CH_2	1	25.8	$CDCl_3$	5.12, 4.32 [$^2J(H,H) = 5.9$ Hz]	$CDCl_3$	75
	$C(H)CF_3$	1	—		4.57 [$^2J(F,H) = 18.0$ Hz]	CD_2Cl_2	299
$(\eta^5\text{-}C_5H_5)Co(CO)$	CH_2	1		NR	7.10	CD_2Cl_2	98
$(\eta^5\text{-}C_5Me_3)Co/(\eta^5\text{-}C_5Me_5)Co(\mu\text{-}CO)$	CH_2	1	190.21	C_6D_6	10.68	C_6D_6	186
$(\eta^5\text{-}C_5H_5)Rh(CO)$	CH_2	1	102.25 / 100.9	Toluene-d_8 / $CDCl_3$	7.07	Acetone-d_6	41, 152
$(\eta^5\text{-}C_5H_5)Rh(CO)$	$C(H)CH_3$	1		NR	7.68	Acetone-d_6	41
$(\eta^5\text{-}C_5Me_3)Rh(CO)$[e]	CH_2	1	111.36 [$^1J(Rh,C) = 29$ Hz]	$CDCl_3$	5.97 [$^2J(Rh,H) = 0.76$ Hz]	$CDCl_3$	54, 54a

Compound	C	n	δ(¹H)	δ(¹³C)	Solvent	Solvent	Ref.
$(\eta^5\text{-}C_5Me_5)Rh(CO)$	$C(H)CH_3$	1	7.09	NR		$CDCl_3$	54
$(\eta^5\text{-}C_5Me_5)Rh(CO)$[c,f]	$C(H)CO_2C_2H_5$	1	5.78	106.3 [$^1J(Rh,C) = 32$ Hz]		$CDCl_3$	54, 94
$(\eta^5\text{-}C_5Me_5)Rh(CO)$[g]	$C_9H_4O_2$	1		125.4	CD_2Cl_2	—	153
$(\eta^5\text{-}C_5H_5)Rh$[h]	$C(C_6H_5)_2$	2		188.2	CD_2Cl_2	—	195
$(\eta^5\text{-}C_5H_5)Rh$[i]	$C(C_6H_5)_2$	2		156.0	CD_2Cl_2	—	195
$(\eta^5\text{-}C_5H_5N)RhCl$[k]	$C(C_6H_5)_2$	2		185.2	CD_2Cl_2	—	195
$Pt(PMe_3)_2/W(CO)_5$[l]	$C(C_6H_4CH_3)OCH_3$	1		204	?	—	86, 87
$Pt(PMe_3)_2/(\eta^5\text{-}C_5H_5)Mn(CO)_2$	$C(C_6H_4CH_3)OCH_3$	1		193.6 ($^2J(P,C) = 73$ Hz) [$^2J(P,C) = 1.5$ Hz]	$CDCl_3$	—	89
$Pt(PMe_3)_2/(\eta^5\text{-}C_5H_5)Re(CO)_2$	$C(C_6H_4CH_3)OCH_3$	1		166.0 [$^2J(P,C) = 78$ Hz]	$CDCl_3$	—	27
$Pt(cod)/(\eta^5\text{-}C_5H_5)Mn(CO)_2$	$C(C_6H_5)OCH_3$	1		198	CD_2Cl_2	—	87
$Pt(cod)/W(CO)_5$	$C(C_6H_5)OCH_3$	1		197	$CH_2Cl_2/CDCl_3$	—	87
$(cod)Ir$	CH_2	2	7.75	—	?		316
Class B: Compounds without metal–metal bonds[m]							
$(\eta^5\text{-}C_5H_5)Fe(CO)_2$	$(CH_2)_3$	1	1.42	7.8		$CDCl_3$	218
$(\eta^5\text{-}C_5H_5)Fe(CO)_2$	$(CH_2)_4$	1	1.35	3.8		$CDCl_3$	218
$PtCl\text{-}\mu\text{-}dppm$[n]	CH_2	1	1.08	NR		$C_2D_2Cl_4$	103
$Re(CO)_5$	$(CH_2)_2$	1	2.02	6.4 [$^1J(C,H) = 133.3$ Hz; $^2J(C,H) = 6.4$ Hz]		CD_2Cl_2	219
$(\eta^5\text{-}C_5H_5)W(CO)_2[Pt(C_2H_5)_3]$	$(CH_2)_2$	1	2.56	NR		CD_2Cl_2	219

---

[a] Spectra recorded at about 30°C unless otherwise noted; δ values [ppm] referenced against TMS. NR = Not reported.

[b] $(\eta^5\text{-}C_5H_5)_2Ti(\mu\text{-}CH_2\text{-}\mu\text{-}Cl)AlMe_2$.

[c] Additional CO bridge present; molecules of type $(\mu\text{-}CRR')(\mu\text{-}CO)[(\eta^5\text{-}C_5H_5)Fe(CO)]_2$.

[d] $[(\mu\text{-}CH_2)_2Ru(PMe_3)_3][BF_4]_2$.

[e] CO equilibration at 32°C ($\delta\underline{C}O$ 197.8; $^1J(Rh,C) = 44.5$ Hz; $^1J(Rh,C) = 89.7$ Hz) (54, 153). Intramolecular CO exchange had also been observed for the $\mu\text{-}C(H)CO_2Et$ derivative (94).

[f] For comparison: $(\mu\text{-}CH_2)[\mu\text{-}C(C_6H_5)_2](\mu\text{-}CO)[(\eta^5\text{-}C_5Me_5)Rh]_2$, $\delta\underline{C}H_2$ 5.03 (CDCl₃); $\delta\underline{C}H_2$ 4.22 Hz (59); $(\mu\text{-}C_7H_8)[Fe_2(CO)_6]$ (Cpd. 88); $\delta\underline{C}H$ 5.64 [$^3J_{H,H} = 8.0$ Hz] (R. Aumann, H. Wörmann, and C. Krüger, Chem. Ber. 110, 1442 (1977)). $(\mu\text{-}C_7H_8)[Os_3(CO)_{11}]$, $\delta\underline{C}H_2$ 62.5 [$J_{C,H^\alpha} = 147$ Hz; $J_{C,H^\beta} = 144$ Hz; CD₂Cl₂]; $\delta\underline{C}H_2$ 6.47, 7.75 [$^3J_{H,H} = 7.2$ Hz; CD₂Cl₂; CDCl₃] (300).

[g] $C_9H_4O_2 = $ 1,3-dioxoindanylid-2-ene.

[h] $[\mu\text{-}C(C_6H_5)_2][\mu\text{-}CO][(\eta^5\text{-}C_5H_5)Rh]_2$.

[i] $[\mu\text{-}C(C_6H_5)_2]_2[(\eta^5\text{-}C_5H_5)Rh]_2$.

[k] $[\mu\text{-}C(C_6H_5)_2][\mu\text{-}CO][(\eta^5\text{-}C_5H_4N)RhCl]_2$.

[l] For comparison: $(CO)_4Mn\text{-}C_5H_6O\text{-}Pt(PMe_3)_2$, $\delta(\mu\text{-}\underline{C})$ 226 (CDCl₃); 2-oxocyclopentene-1-ylid-1-ene as bridging ligand (87). $(CO)_4Mn(\mu\text{-}1)Pt(C_5H_6O)[P(t\text{-}Bu)Me]$, $\delta(\mu\text{-}\underline{C})$ 261 (CDCl₃); $C_5H_6O = $ 2-oxocyclopentaneylid-1-ene as terminal ligand (89).

[m] For comparison: $(\eta^5\text{-}C_5H_5)(CO)Co\text{-}CH_2\text{-}CH_2\text{-}Co(CO)(\eta^5\text{-}C_5H_5)$, $\delta\underline{C}H_2$ 31.9, 20.5 (C₆D₆); $\delta\underline{C}H_2$ 1.66, 0.68 (C₆D₆); see comment on page 239f. (Cpd. 125) (98).

[n] 80°C.

δ/CH_2R range -30 to -5. An average upfield shift of 100–200 pm is found for the carbene–carbon nuclei upon coordination of a mononuclear metal-carbene to another metal fragment (90, 222). Related terminal–bridging carbonyl and vinylidene systems (Table V) show the same behavior, but methylidyne groups (Table V) do not always follow this trend. Note that the *nucleophilic* carbene ligand in $(\eta^5\text{-}C_5H_5)_2Ta(\!=\!CH_2)(CH_3)$ ($\delta\underline{C}H_2$ 228; $\delta C\underline{H}_2$ -0.22) is one of the very few which gives rise to a ^{13}C chemical shift similar to those of μ-methylene compounds (129a).

2. If the bridging methylene functions bear hydrogen(s), ^{1}H-NMR spectroscopy enables us to establish whether an additional metal–metal bond is present or not. Alkanediyl compounds belonging to type G (Class B, Table IV) once again very much resemble metal alkyls with respect to their $C\underline{H}_x$ chemical shift range. Mononuclear metalcarbenes (A) and μ-methylene complexes (C), respectively, cannot be identified reliably on the basis of their ^{1}H-NMR data, although methylene bridge hydrogens generally experience significant upfield shifts. The following $\delta C\underline{H}R$ pattern (ppm) should be considered along with the ^{13}C-NMR data (see above):

M=C$\underline{H}$R (A)	9–11
M—C$\underline{H}$R—M (C)	5–11
M—(C$\underline{H}$R)$_x$—M (G)	1–3
M—C$\underline{H}$RR'	1––1

The fact that both the methylene carbon and the μ-alkylidene α-hydrogen resonances are shifted upfield from the corresponding mononuclear metal carbene M=CHR resonance reflects, in a rough approximation, much higher electron density in bridging carbene groups (see Section V). Accordingly, additional upfield shifts of alkylidene α-hydrogen resonances are observed when electron-donating substituents are introduced into the peripheral ligands of dimetallacyclopropanes [e.g., substitution of C_5Me_5 for C_5H_5 (54, 81, 94)]. The magnitude of hydrogen–hydrogen coupling [$^2J(H,H)$] in CH_2-bridged compounds is reminiscent of cyclopropanes [e.g., $(\mu\text{-}CH_2)(\mu\text{-}H)_2Os_3(CO)_{10}$, 5.9 Hz (75); $(\mu\text{-}CH_2)[\mu\text{-}C(C_6H_5)_2]$-$[(\eta^5\text{-}C_5Me_5)Rh(CO)]_2$, 4.2 Hz (59)], but no trends are apparent from the hitherto limited set of data. The same is true for carbon–hydrogen coupling [$^1J(C,H)$] which was found to amount to ~145 Hz in some μ-CH$_2$ rhodium and osmium complexes (54a, 75, 79, 94).

It may be noted in this context that methylene bridges have never been observed to undergo fluxional behavior of their organometallic frames. See, however, Addendum, section 14 (Ref. 309). Furthermore, methylene bridge opening in the absence of other reagents (H$_2$, ethylene, etc.) is as yet an unknown process. This has been demon-

TABLE V

^{1}H- AND ^{13}C-NMR DATA [ppm] FOR SELECTED VINYLIDENE AND METHYLIDYNE COMPLEXES[a]

Compound	$\delta C\underline{H}R$	$\delta\underline{C}=C(H)R/\delta\underline{C}$-R	$\delta C=\underline{C}(H)R$	Reference
Vinylidene complexes				
$(\mu\text{-}C=CH_2)[(\eta^5\text{-}C_5H_5)Mn(CO)_2]_2$	6.77 (C_6D_6)	—[b]	—[b]	177
$(\mu\text{-}C=CHC_6H_5)[(\eta^5\text{-}C_5H_5)Mn(CO)_2]_2$	8.46 (CS_2)	284.16 (CH_2Cl_2)	125.22 (CH_2Cl_2) $[^1J(C,H) = 150$ Hz]	180
$(\eta^5\text{-}C_5H_5)Mn(CO)_2(C=CHC_6H_5)$	6.91 (CS_2)	379.54 (CH_2Cl_2)	123.54 (CH_2Cl_2)	180
$(\mu\text{-}C=CH_2)(\mu\text{-}CO)[(\eta^5\text{-}C_5H_5)Fe(CO)]_2$	6.90 (C_6D_6)	277.2 (CD_2Cl_2)	125.6 (CD_2Cl_2)	91
$(\mu\text{-}C=CH_2)(\mu\text{-}CO)[(\eta^5\text{-}C_5H_5)Ru(CO)]_2$	6.27 ($CDCl_3$)	249.1 (CD_2Cl_2)	122.7 (CD_2Cl_2)	62
Methylidyne complexes[c]				
$(\eta^5\text{-}C_5H_5)(CO)_2W(\equiv CAr)$	—	300.1 (CD_2Cl_2)	—	292
$(\eta^5\text{-}C_5H_5)(CO)_2W(\mu\text{-}CAr)Pt(PMe_3)_2$	—	338 ($CDCl_3$)	—	293
$[(\eta^5\text{-}C_5H_5)(CO)_2Mn(\equiv CC_6H_5)]BCl_4$	—	356.9 (acetone-d_6)	—	27
$[(\eta^5\text{-}C_5H_5)(CO)_2Mn(\mu\text{-}CAr)Pt(PMe_3)_2]BF_4$	—	338.1 ($CDCl_3$)	—	27

[a] See Ref. 291.
[b] Not reported.
[c] Ar = C_6H_4— $p\text{-}CH_3$.

strated, for instance, in the case of the rhodium dimer $(\mu\text{-}CH_2)[(\eta^5\text{-}C_5Me_5)Rh(CO)]_2$, which undergoes racemization (54, 94) arising from intramolecular carbonyl exchange (Table IV, footnote e). The μ-methylene–rhodium coupling [$^1J(Rh,C) = 29$ Hz] does not change when CO equilibration is frozen at about $-20°C$ [$^2H_8$]THF (153).

NMR spectroscopy is also useful for detecting chirality. Dimetallacyclo-propanes maintain the gross stereochemical features of cyclopropanes and thus exhibit chiral metal centers provided that different sets of periph-eral ligands are in positions above and below the M,C,M plane. In case of trans-oriented sets of corresponding ligands, the metals are assigned R,R (or S,S) configurations (pair of enantiomers), whereas cis configuration leads to R,S (or S,R) ("meso" form). A methylcyclopentadienyl ligand represents a good probe for metal chirality since the atoms in α- and those in β-positions with respect to the methyl label become diastereotopic. This effect was first observed for $(\mu\text{-}CH_2)[(\eta^5\text{-}C_5H_4CH_3)Mn(CO)_2]_2$ (3b) by means of both ^1H- and ^{13}C-NMR spectroscopy (Figs. 9 and 10).

B. *Mass Spectra*

The electron impact-induced decomposition pathways of several struc-turally related μ-methylene complexes of cobalt, rhodium, manganese, and iron have been elucidated by high resolution measurements, analysis of metastable transition (DADI; linked scan), and ^2H labeling (46). Termi-nal carbonyl ligands are generally lost prior to further fragmentation of the three-membered frameworks. Subsequent rearrangement reactions of the

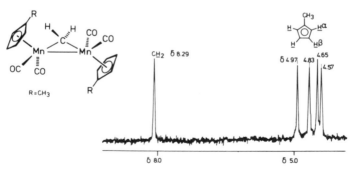

FIG. 9. C_5H_4 and CH_2 region of the ^1H-NMR spectrum of 3b (240 MHz; $-90°C$; acetone-D$_6$).

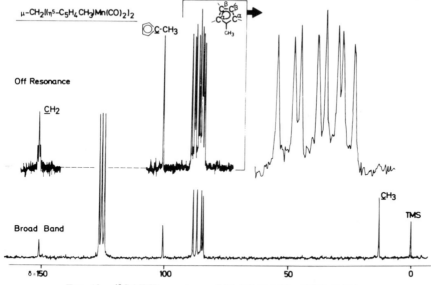

FIG. 10. ^{13}C-NMR spectrum of **3b** (22.63 MHz; 34°C; C_6D_6).

dinuclear fragment ions $cp_2M_2(CH_2)^+$ depend strongly upon the nature of the metals. In the case of manganese, synchronous loss of Mn $[\rightarrow cp_2Mn(CH_2)^+]$, $MnCH_2^-$ $(\rightarrow cp_2Mn^+)$, and $MnCH^-$ $(\rightarrow cp_2MnH^+)$ was observed; these fragmentations include cyclopentadienyl transfer across the dimetal frame (*39, 46*). In contrast, fragmentation of the rhodium compound $(\mu\text{-}CH_2)[(\eta^5\text{-}C_5H_5)Rh(CO)]_2$ (**19**) branches out one step earlier. The fairly abundant ion $cp_2Rh_2(CH_2)(CO)^+$ looses (a) another CO group $[\rightarrow cp_2Rh_2(CH_2)^+]$, (b) C_2H_2O $(\rightarrow cp_2Rh_2^+)$, (c) $Rh(C_2H_2O)$ $(\rightarrow cp_2Rh^+)$, and (d) CH_2O $(\rightarrow cp_2Rh_2C^+)$. Metal-induced combination of CH_2 and CO ligands $(\rightarrow CH_2\!\!=\!\!C\!\!=\!\!O$?) very likely accounts for the direct formation of the ion $cp_2Rh_2^+$. The analogous cobalt compounds display yet another framentation pathway, namely, elimination of C_6H_6 from the key fragments $cp_2Co_2(CH_2)^+$ The latter species is formed even if the original μ-methylene ligand does not bear any hydrogen(s) directly attached to the bridgehead carbon. The carbyne hydrido species $cp_2Co(\equiv CH)\!\!-\!\!Co(H)cp^+$ resulting from C—H bond activation may be invoked as precursors responsible for metal-centered formation and subsequent elimination of benzene (C_5H_5 + CH $\rightarrow C_6H_6$) (*46*). In neither case has direct elimination of the methylene unit been observed, be it from the parent ion or from any fragment ion

(46). On the other hand, formation of carbido species such as $cp_2M_2C^+$ or cp_2MC^+ appears to be a very common process. μ-Alkylidene ligands frequently rearrange within the metal core and subsequently eliminate as olefins (e.g., $C(H)CH_3 \rightarrow C_2H_4$). μ-Methylene complexes of the nitrosyl series (e.g., $(\mu\text{-}CH_2)[(\eta^5\text{-}C_5H_5)Fe(NO)]_2$) follow the same pathways upon electron impact-induced fragmentation. The fragment $cp_2FeCH_2^+$ occurs as the first ion, and $cpFeCH^+$ is also seen (rel. int. 34%) (153). The remarkable formation of methane from pentamethylcyclopentadienyl derivatives such as $(\mu\text{-}CH_2)[(\eta^5\text{-}C_5Me_5)Rh(CO)]_2$ merits notice among the major steps of fragmentation (153).

VIII

REACTIVITY OF METHYLENE-BRIDGED TRANSITION METAL COMPLEXES

Stability against thermolysis and photolysis is one of the striking properties of dimetallacyclopropanes, especially of those involving carbonyl and cyclopentadienyl ligands on the metals. For example, the rhodium methylene complex $(\mu\text{-}CH_2)[(\eta^5\text{-}C_5H_5)Rh(CO)]_2$ resists temperatures of 80°C over a two-day period (boiling benzene) or prolonged UV irradiation and is thus quite robust compared with the carbonyl analogue $(\mu\text{-}CO)$-$[(\eta^5\text{-}C_5H_5)Rh(CO)]_2$. This and many other examples reflect our belief (61a) that any μ-methylene complex should be accessible and stable if the corresponding μ-carbonyl is known. There is so far not a single exception to the rule that a μ-methylene complex is at least as stable as its μ-carbonyl counterpart. The systems $(\mu\text{-}CH_2)[(\eta^5\text{-}C_5H_5)Mn(CO)_2]_2$ (stable)/$(\mu\text{-}CO)$-$[(\eta^5\text{-}C_5H_5)Mn(CO)_2]_2$ (unknown), $(\mu\text{-}CRR')_2[Re(CO)_4]_2$ (stable)/$(\mu\text{-}CO)_2$-$[Re(CO)_4]_2$ (unknown), $(\mu\text{-}CH_2)[(\eta^5\text{-}C_5H_5)Co(CO)]_2$ (stable)/$(\mu\text{-}CO)[(\eta^5\text{-}C_5H_5)Co(CO)]_2$ (unstable), $(\mu\text{-}CH_2)[(\eta^5\text{-}C_5Me_5)Rh(CO)]_2$ (stable)/$(\mu\text{-}CO)$-$[(\eta^5\text{-}C_5Me_5)Rh(CO)]_2$ [photo- and thermolabile (233)] suffice to illustrate this general trend. As pointed out earlier, methylene bridging is favored over terminal coordination if electron-donating peripheral ligands are present.

Despite their thermodynamic stability, dimetallacyclopropanes proved to be quite reactive when treated with appropriate reagents, specifically, unsaturated hydrocarbons and protic acids. Again, the nature of the metals and coligands distinctly influence the reactivity.

A. Intermolecular Carbene Exchange and Intramolecular Carbonyl Exchange

Bergman has shown that intermolecular methylene bridge exchange is a facile process in the cobalt series. Crossover experiments using (μ-CH$_2$)-[(η^5-C$_5$H$_5$)Co(CO)]$_2$ and its C$_5$H$_4$CH$_3$ derivative resulted in formation of the corresponding mixed CH$_2$ complex (cp/cp') in benzene at 65°C (234). The failure of the rhodium homologs (Rh in place of Co) to give methylene exchange (235a) reflects the higher stability of the dimetallacyclopropane frameworks for the heavier elements within a given group of the periodic table, since cleavage of the three-membered ring systems clearly constitutes the salient prerequisite of CH$_2$ transfer. Consistent with this interpretation the mixed Co–Rh system does not form upon attempted crossover synthesis from the Co–Co and Rh–Rh dimers (235a).

We have already emphasized that bridging methylene ligands do not participate in intramolecular ligand scrambling [See, however, Addendum, section 14 (Ref. 309)]. If CO exchange occurs, then a pairwise bridge–terminal site exchange operates (54, 94). Loss of carbon monoxide has also been established (thermolysis, photolysis), but neither process destroys the methylene bridge (54, 81, 183). Metal-centered methylene(hydrido)/methyl tautomerism—relevant to catalysis—and further reaction to methylidyne species (75–78) will certainly prove to be a characteristic of dimetallacyclopropanes having metal-bound hydrogen ligands.

B. Coordination of Peripheral μ-Methylene Functions and CO Substitution

The 3,3-dimethyl-2-propene-1-ylidene bridged tungsten complex [μ-C(H)—C(H)=C(CH$_3$)$_2$][W(CO)$_4$]$_2$ is a curious example of dimetallacyclopropanes in that it does not conform to the EAN rule if the unsaturated hydrocarbon group is counted as a two-electron acceptor bridge. Bond lengths and angles of the W,C,W triangle are normal, and there is neither carbonyl semibridging nor any close contact of the methylene substituents to the metals apparent from the X-ray data (68, 69). It is therefore not surprising that the molecule escapes this situation by additional metal coordination of the β-vinylic function with concomitant CO addition (see Section IV,B,1) (70). The reverse of this process occurs upon addition of certain phosphines (71), a reaction which also results in phosphine addition to the hydrocarbon bridge (71).

C. Substitution at the Methylene Bridge

Substitution reactions at the methylene bridge have been observed, albeit the yields were quite low suggesting that major side reactions had occurred. Addition of Meerwein's salt to the heterodinuclear complex **42** with subsequent treatment of the presumed intermediate $[(CO)_5\text{-}Cr[\mu\text{-}C(C_6H_5)]Pt(PMe_3)_2]^+$ with lithium base $(LiC_6H_4\text{-}p\text{-}Me)$ gave the μ-diarylmethylene derivative **44** (27). The phosphine-substituted derivative of **42**, $(Me_3P)(CO)_4Cr[\mu\text{-}C(OCH_3)C_6H_5]Pt(PMe_3)_2$ (**98a**) displays a more robust metal-to-metal bond and consequently reacts more cleanly with $Me_3O^+BF_4^-$ to give the cation $[(PMe_3)(CO)_4Cr[\mu\text{-}C(C_6H_5)]Pt(PMe_3)_2]^+$ as its BF_4^- salt which, on treatment with sodium methoxide, affords the stable μ-methoxycarbonyl(phenyl)methylene complex **98b** in ~60% yield (27). A CO group evidently migrates from the chromium site to the phenylmethylidyne bridge, by a mechanism as yet unclear, and is subsequently attacked by the methoxide reagent.

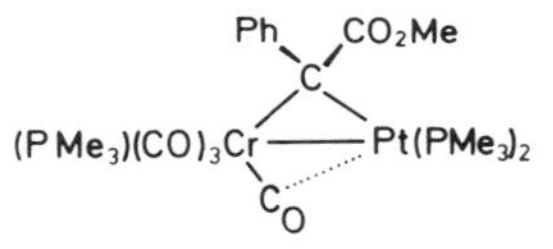

98b

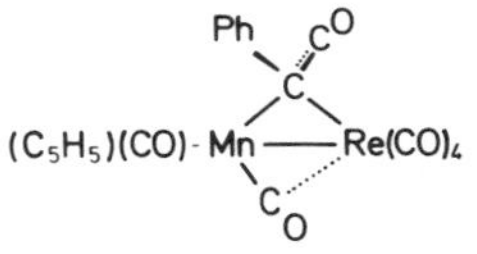

99

Fischer et al. (212) and Stone et al. (93) encountered a very similar CO migration when they attempted the synthesis of a μ-phenylmethylidyne

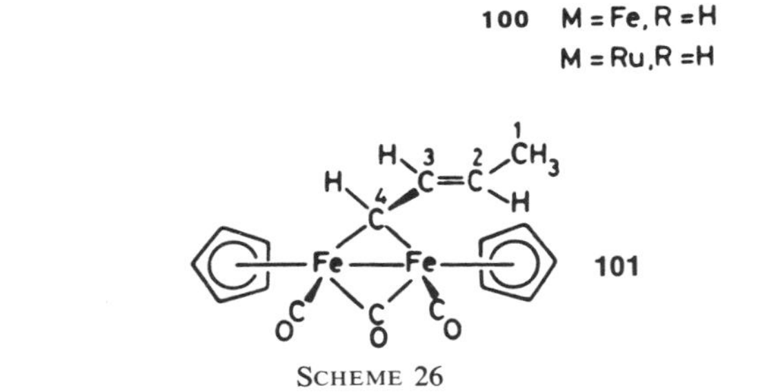

63

100   M = Fe, R = H
         M = Ru, R = H

101

SCHEME 26

complex via addition of the $[Re(CO)_5]^-$ anion (Na salt) to the pronouncedly electron-deficient carbyne unit of $[(\eta^5\text{-}C_5H_5)Mn(CO)_2(\equiv CC_6H_5)]^+$. Instead, the rather curious heterodinuclear compound **99** having a three-electron phenylketenyl ligand as part of a dimetallacyclopropane-type framework was isolated as the final product and structurally characterized. Closely related to true μ-methylene complexes (see Section IV,B), the bridgehead carbon forms the center of a distorted tetrahedron (Table I). The above-mentioned transformation of $\overline{M[\mu\text{-}C(OR)R']M'}$ into $\overline{M(\mu\text{-}CR')M'}$ systems will inevitably have important consequences regarding further reactions of the latter compounds.

D. Reactions with Hydrogen and Unsaturated Hydrocarbons

1. Alkynes

High reactivity of bridging carbenes toward alkynes has been nicely demonstrated by Knox et al. for some iron and ruthenium compounds (65). Under photochemical conditions, the μ-ethylidene complexes (η^5-$C_5H_5)_2M_2(CO)_2(\mu\text{-}CO)[\mu\text{-}C(H)CH_3]$ (**63**; M = Fe, Ru) insert acetylenes into the original metal–carbon bridge, thus producing a three-carbon bridging system **100**. From X-ray as well as ^{1}H-NMR data, it appears that these products are best viewed structurally as μ,η^3-vinylcarbene complexes (Fig. 11). The alternative description as π-allyl-type compounds receives some support from the similarity of the carbon–carbon distances within the bridge. When **100** is subjected to 100 atm of CO (50°C, acetone, ~17 h), the μ,η^1-vinylcarbene complex **101** appears in high yield (cis–trans isomers). Reversible decarbonylation is effected by either heating **101** to 100°C or subjecting the compound to UV irradiation (Scheme 26) (65).

The titanium–aluminium methylene complex **94** ["Tebbe's reagent" (99)] also proved to be very reactive toward alkynes. Thus reaction with bis(trimethylsilyl)acetylene or diphenylacetylene in the presence of tetrahydrofuran gives the titanacyclobutenes **102** and **103**, respectively (Scheme 27). The former compound was shown by X-ray diffraction methods to involve an entirely planar four-membered cycle having different carbon–carbon distances within the ranges expected for C—C double and single bonds, respectively [134.4(6) versus 153.7(6) pm]. A series of elegant experiments yielded ample evidence that these heterocycles preserve the highly reactive (and thus quite unstable!) $(\eta^5\text{-}C_5H_5)_2Ti(\equiv CH_2)$

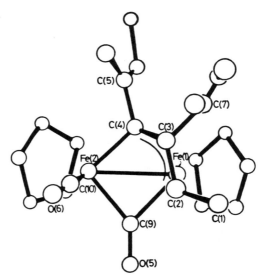

FIG. 11. Molecular structure of the μ,η^3-vinylcarbene iron complex **100a**. Selected bond lengths [pm] and angles [deg]: Fe(1)–Fe(2) 254.0(2), Fe(1)–C(2) 211.1(13), Fe(1)–C(3) 204.2(12), Fe(1)–C(4) 196.9(12), Fe(2)–C(4) 196.8(12), C(2)–C(3) 143.3(18), C(3)–C(4) 145.7(17); C(1)–C(2)–C(3) 118.5(1.1), C(2)–C(3)–C(4) 119.3(1.1), C(3)–C(4)–C(5) 117.9(1.0); R_F 0.10 (from Ref. *65*).

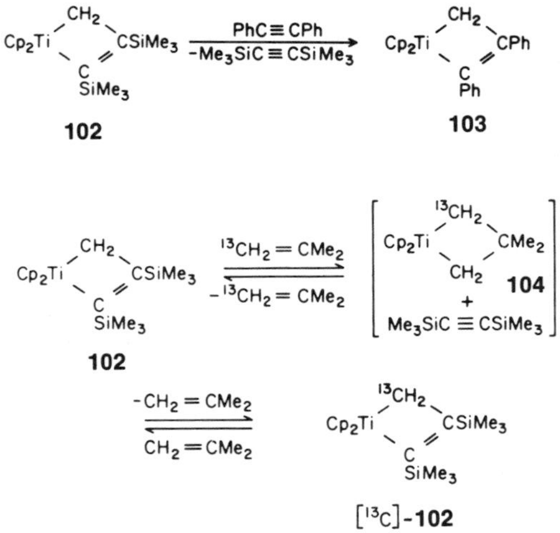

SCHEME 27

group for controlled reactions with a variety of unsaturated hydrocarbons. Acetylene or olefin substitution of the $Me_3SiC{\equiv}CSiMe_3$ fragment in **102** occurs at elevated temperatures; for example, tolane converts **102** to **103** (36%, 85°C, 6 h), whereas the reverse reaction does not occur with any significance. Furthermore, [1-^{13}C]-2-methylpropene and **102** react to yield [^{13}C]-**102** with enrichment in the methylene (C-3) position. Methylene equilibration between the titanacycle and olefin occurs within a period of 30–40 h at 85°C.

One important mechanistic step of this exchange reaction certainly involves substitution of acetylene in **102** by olefin to construct the metallacyclobutane **104**; this intermediate, as yet not isolated or detected, very likely accounts for CH_2 exchange between $(\eta^5\text{-}C_5H_5)_2Ti(\mu\text{-}CH_2)(\mu\text{-}Cl)AlMe_2$ (**94**) and isobutene as well (*101*). These experiments also conveyed the information that a titanacyclobut*ene* (e.g., **102**) is considerably more stable than the corresponding titanacyclobut*ane* (**104**). Tebbe's reagent is recovered if the four-membered metallacycle **102** is treated with dimethylaluminium chloride in the absence of THF. Hence uncomplexed $ClAlMe_2$ seems to compete effectively with $Me_3SiC{\equiv}CSiMe_3$ for the reactive $(\eta^5\text{-}C_5H_5)_2Ti({=}CH_2)$ intermediate, but $ClAlMe_2$–THF is conceivably not competitive at all (*100*).

Direct methylenation of esters and lactones surely involves one of the most attractive synthetic applications of Tebbe's reagent. For example, ethyl acetate $MeC({=}O)OEt$ reacts with the electrophilic μ-methylene complex **94** to give the corresponding vinyl ether $MeC({=}CH_2)OEt$ (*100*). Grubbs *et al.* (*236*) have discovered the general applicability of titanium-mediated methylene-transfer reactions to an incredibly wide range of esters and lactones. In all cases, the methylenation products were the only compounds observed. These high-yield (79–96%) syntheses are easy to perform (toluene solutions of **94** can be stored on the shelf over 2 months without loss in "active reagent titer") and tolerate both ketal and olefin functionalities. Moreover, the olefin positional integrity is maintained throughout the entire reaction. No evidence was found for any olefin isomerization. If γ-ketoesters are employed, both the ketone and the ester functions undergo methylenation upon addition of doubled quantities (1 : 2) of Tebbe's reagent. Competition experiments revealed that ketone methylenation proceeds roughly four times faster than ester methylenation. Beyond that, the presence of a conjugated olefin does not interfere with the methylene-transfer process, and the substrates do not suffer loss of stereochemistry, as demonstrated by methylenation of both E and Z cinnamate esters (*236*). This straightforward method closes a synthetic gap, especially because of the failure of the less electrophilic phosphorus ylides to provide a generally useful operation for the above-mentioned purpose. It should be mentioned in this context that Schrock's mononuclear ylide-type carbene complexes [e.g., $R_3Ta{=}C(H)CH_3$] have also been used to effect alkylidene transfer to carboxylic acid derivatives (*207*).

Grubbs *et al.* (237) succeeded in trapping the reactive $cp_2Ti({=}CH_2)$ fragment derived from Tebbe's reagent with *t*-Bu—CH=CH$_2$ in the presence of Lewis bases (e.g., pyridine). Type-**104** titanacyclobutanes thus formed in high yields react with alkynes to form the titanacyclobutenes $(\eta^5$-$C_5H_5)_2Ti$—CH_2—$CR{=}CR$. Metallacycle exchange with olefin was also observed, and a metalcarbene–metallacycle equilibrium has been proposed to operate in these reactions.

A notable double insertion of methylacetylene into the methylene bridge has been observed with the classical γ-lactone-type cobalt complexes **1**. The heterocyclic methylene ligand (CRR') originally attached to both metal centers is displaced by the new alkylidene ligand μ-C(CH$_3$)— C(H)=C(CH$_3$)—C(H)=CRR' (*116, 122*) (Scheme 28a; other ligands omitted for clarity). This example also points to the possibility of methylene homologation in the coordination core of a catalyst (see Section IX).

2. *Alkenes and Hydrogen*

Ethylene insertion into the methylene bridge and hydrogen addition to latent coordination sites may also proceed under relatively mild conditions. When the iron compound $(\mu$-$CH_2)[Fe_2(CO)_8]$ (**35**) is treated with hydrogen (11 atm) at 60°C in benzene, methane (81%) and minor amounts of acetaldehyde (5%) are produced along with $Fe_3(CO)_{12}$. Similar reactions with mixtures of hydrogen (14 atm) and carbon monoxide (28 atm) yield the same organic products in roughly the same ratio, but at a markedly decreased rate. This suggests that formation of CH_4 involves prior loss of CO with formation of $(\mu$-$CH_2)[Fe_2(CO)_7]$ which then undergoes oxidative addition of H_2 to give, via $(\mu$-$CH_2)[Fe_2H_2(CO)_7]$, the unstable methyl (hydrido) complex of composition $H(CO)_4Fe$—$Fe(CO)_4CH_3$ as a possible precursor of methane.

Treatment of **35** with ethylene (20 atm, benzene, 55°C) produces propylene ($\geq 90\%$) and $(C_2H_4)Fe(CO)_4$ in a reaction which, once again, is inhibited by CO. The same primary step as above, namely, addition of ethylene (instead of H_2), is conceivable for this process, followed by

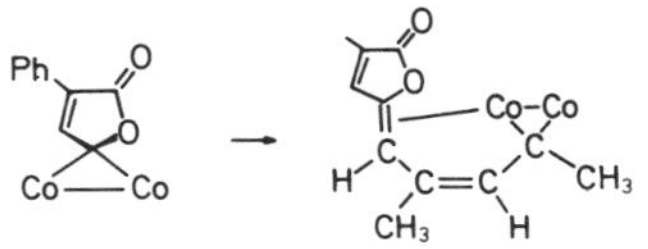

SCHEME 28A

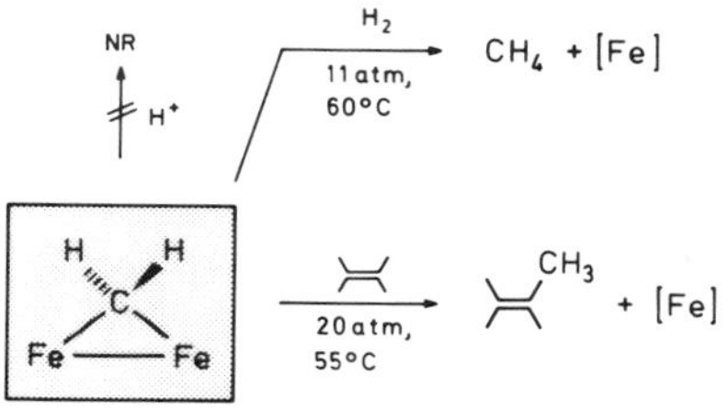

SCHEME 28B

metal-centered ethylene–methylene combination yielding a diferracyclo-pentane frame (72; see also 98). β-Hydrogen elimination from the metal-lacycle and reductive elimination of propylene are thought to account for the final outcome of this reaction (Scheme 28b). Both the mono- and dica-tions [Ru$_2$(μ-CH$_2$)$_2$(μ-CH$_3$)(PMe$_3$)$_6$]$^+$ and [Ru$_2$(μ-CH$_2$)$_2$(PMe$_3$)$_6$]$^{2+}$ catalyze the hydrogenation of 1-hexene in methanol, giving good yields of hexane (90%) at room temperature after several hours (97, 240).

E. Protonation

The most striking differences within this class of compounds (μ-methylene complexes) are encountered upon attempted protonation. First-row metal derivatives such as (μ-CH$_2$)[(η^5-C$_5$H$_5$)Mn(CO)$_2$]$_2$ (3a) and (μ-CH$_2$)[Fe$_2$(CO)$_8$] (35) prove to be completely resistant to even strong protic acids (e.g., HBF$_4$, CF$_3$SO$_3$H),[3] whereas second-row rhodium gives rise to some intriguing chemistry, both synthetically and mechanistically, when attacked by a proton (49, 51, 52, 61).

From Scheme 29, we first learn that the dimetallacyclopropane-type rhodium complex 19 reacts in THF solution with excess tetrafluoroboric acid and other strong Brønsted acids unexpectedly to give trinuclear, cationic cluster compounds 105 in essentially quantitative yields. The structure of the complex cation [(η^5-C$_5$H$_5$)$_3$Rh$_3$(μ-CO)$_2$(μ_3-CH)]$^+$ centers around a nearly equilateral rhodium triangle which is symmetrically span-ned by a methylidyne functionality. In addition, the pronounced asym-metric posture of the "noninnocent" edge-bridging carbonyl ligands is of interest because of the dynamic behavior of the metal carbonyl backbone in solution. Temperature-dependent NMR spectroscopy performed with

---

[3] With the μ-methylene iron complex (μ-C$_7$H$_8$)Fe$_2$(CO)$_6$, we encounter an apparent excep-tion of this rule. Extensive H/D exchange occurs upon treatment of this compound with CF$_3$CO$_2$D, indicating formation of a rapidly equilibrating intermediate cationic species [R. Aumann, H. Wörmann, and C. Krüger, Chem. Ber. 110, 1442 (1977)].

PROTON - INDUCED METHYLENE / METHYL / METHYLIDINE INTERCONVERSION

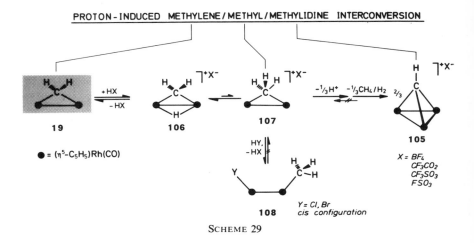

SCHEME 29

highly ^{13}CO-labeled samples clearly demonstrates this solid-state structure to represent a "stopped-action" picture of rapid CO scrambling occuring in solution. Since the face-bridging carbyne ligand does not share
the dynamic behavior of the carbon monoxide groups, the only way to
rationalize this process is as a "merry-go-round" type movement, with
the carbonyl groups "chasing" each other around the triangular metal
frame. We believe strongly that CO equilibration in molecules like **105**
originates from their tendency to counter-balance the electronic requirements of all metal atoms sharing this process and, consequently, to
meet the EAN rule (*52*).

The proton-induced synthesis of the first rhodium compound containing
unsubstituted methylidyne (CH) as a ligand from the methylene (CH$_2$)
precursor looks quite paradoxical and contrasts sharply with the generation of stable cationic methyl bridges upon protonation of a related compound (Scheme 30). As was recently reported by Wilkinson and his asso-
ciates (*95, 96*), the triply methylene-bridged ruthenium phosphine complex
(μ-CH$_2$)$_3$[Ru(PMe$_3$)$_3$]$_2$ promptly undergoes protonation yielding mono- or
dipositively charged methyl species. This molecule successfully resisted
hydride abstraction. It had been hoped this would afford a methylidyne
(CH) group. Instead, even with trityl tetrafluoroborate very high yields
of the protonated products were obtained in tetrahydrofuran, this curious
reaction being initiated by proton transfer from the organic substrate
(*97*). The observation that both methane and hydrogen (X = BF$_4$) are
produced in the course of the acid-induced transformation starting from
19 and leading to **105** (in THF), whereas not even traces of other plausible
by-products except (η^5-C$_5$H$_5$)Rh(CO)$_2$ could be detected, initiated the fol-

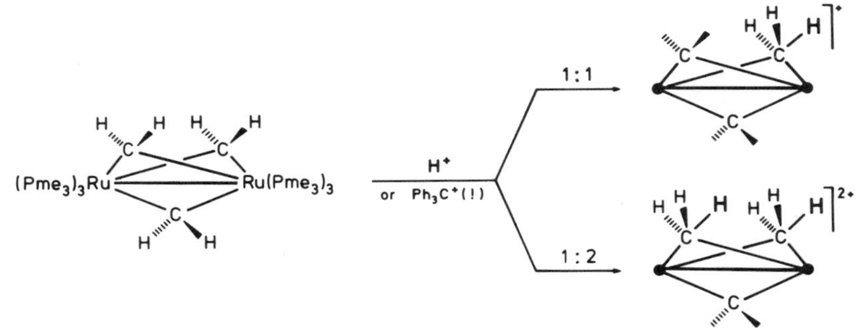

SCHEME 30

lowing experiments which were able to identify some key intermediates responsible for the final methylidyne products (see Scheme 29).

1. If Brønsted acids such as HBF_4 are added to solutions of **19** in diethyl ether rather than tetrahydrofuran at temperatures around $-20°C$, yellow precipitates separate immediately, and quantitatively, without evolution of gas. Conductivity measurements and total elemental analyses as well as IR and NMR data convincingly prove that this sensitive compound **106** constitutes the expected 1 : 1 protonation product. The extra proton exclusively occupies the metal site and does not convert the methylene bridge to a methyl ligand. We are thus dealing with hydrido (methylene) complexes. Suffice to say that the μ-methylene metal frame largely maintains its integrity up to this point.

2. Deprotonation of **106** is effected by even weakly basic solvents such as THF, DMSO, or methanol, yielding the authentic neutral precursor. That rapid proton exchange over the methylene and metal hydrido sites occurs is proved by labeling experiments. Thus the ionic complex deutero-**106,** obtained by deuteration of methylene complex **19,** instantaneously and quantitatively gives its labeled precursor upon treatment with sodium methoxide in methanol. The starting compound formed in this way displays a roughly statistical deuterium distribution over the methylene proton sites. Accordingly, highly deuterium-labeled samples of the methylidyne cluster **105** ($>80\%$ d_1) are formed if an excess of deutero-trifluoroacetic acid reacts with **19** in THF. These results are fully consistent with the observed production of methane (and hydrogen) and clearly demonstrate that organometallic intermediates containing methyl groups participate in the synthesis of the final methylidyne compounds. Since the methyl tautomers escaped direct characterization, formula **107** may not be correct in detail. The observed proton exchange very likely occurs via unsymmetrical methyl bridges, which implies hydrogen–metal interac-

tions. This type of coordination has recently been detected by Shapley *et al.* in a convincing series of experiments (Section IV,A,1).

3. A chemical characterization of the intermediate methyl species proved possible with protic acids derived from anions capable of metal coordination. If the dimetallacyclopropane complex **19** is allowed to react with hydrogen halides such as HBr in diethyl ether, the dinuclear, neutral halo(methyl) compound **108** precipitates. The basic geometry of the bromo(methyl) derivative (Ziegler, Heidelberg) is characterized by a short metal-to-metal bond comparable in length with other rhodium–rhodium bonds bridged by two well-behaved carbonyl ligands (*51, 52*). The most intriguing feature of this molecule, with respect to the metal centers, is its stereochemistry. Although the starting methylene compound **19** displays kinetically stable *S,S*- or *R,R*-configurations with regard to the chiral rhodium centers arising from trans-oriented sets of terminal ligands, its ring-opening product **108** exhibits both the cyclopentadienyl groups and the other terminal functions (CH_3 and Br) in stereochemically rigid positions cis to each other! Nothing other than tautomeric hydrogen exchange can be responsible for this drastic switch of stereochemistry because there is no chance for the trans-bridging hydride to initiate H/D exchange, unless the steric prerequisites of the organometallic fragments around the metal-to-metal bond for such a process are present. If the protonation reaction of **19** employing HBr as described before is conducted at $-80°C$, a light-yellow precipitate is formed which has been characterized unequivocally as the cationic hydrido(methylene) species of type **106.** Upon warming to room temperature, this primary product rapidly undergoes irreversible ring opening, finally yielding the thermally stable bromo(methyl) derivative **108.** This reaction sequence not only provides the first authentic example of a ring-opening process for dimetallacyclopropane-type methylene complexes, but also demonstrates that a nonsynchronous two-step mechanism which requires proton-induced metal-to-metal bond activation operates in this kind of reaction.

The product-determining importance of the nucleophilic counterions is evident from comparison with the chemistry of hydrido (methylene) complexes containing anions that lack coordination capabilities. Moreover, ring opening as discussed here holds promise for a straightforward synthetic entry to the hitherto little investigated class of organometals having alkyl and halide groups attached to adjacent metal centers.

These results prove that direct protonation of the pronounced electron-rich metal-to-metal bond is greatly preferred over proton attack at the methylene bridge, a process that induces a spectacular sequence of reactions proceeding via both intra- and intermolecular pathways. By contrast, protonation of the methylene carbon comes to the fore if an electron-rich metal-to-metal bond is either not present or sterically not available, as is evident from comparison with the results obtained by Wilkinson (95–97) (Scheme 30).

Evidence for a primary attack of the electron-rich rhodium–rhodium bond by a proton receives independent support from experiments dealing with the parent carbonyl complex **20** shown in Scheme 31. This molecule differs from its methylene counterpart **19** solely by replacement of a CO group for the CH_2 ligand. The two compounds are geometrically and electronically equivalent. It is not surprising then that the action of tetrafluoroboric acid upon solutions of **20** in diethyl ether at temperatures below 0°C results in immediate and quantitative formation of a temperature-sensitive precipitate which is reversibly deprotonated by sodium methoxide. Again, we encounter a cationic hydrido complex having a bridging hydrogen functionality. In addition, the chemistry of this 1 : 1 electrolyte very much parallels the fate of the primary protonation product derived from the methylene complex (cf. Schemes 29 and 31). The initially dark-red nitromethane solutions, which are stable at −20°C for many hours, gradually turn grass-green at room temperature. Again, partial deprotonation occurs with concomitant formation of $(\eta^5\text{-}C_5H_5)$Rh-(CO)$_2$ and specifically the C_{3v} isomer of the trinuclear cluster $(\eta^5\text{-}C_5H_5)_3$-Rh$_3$(CO)$_3$, the latter not only being isoelectronic with the methylidyne cluster cation **105** but also sharing its fundamental structural features. This neutral cluster is also formed in more than 90% yield under exactly the same conditions that convert the dinuclear methylene complex to the trinuclear methylidyne cation (50). Once again, activation of the rhodium–rhodium bond by direct protonation constitutes the product-determining prerequisite for the spontaneous acid-induced fragmentation of dinuclear compounds discussed here, and this method has in fact been successfully used as a straightforward synthetic alternative for the well-known photoactivation of metal carbonyls (233).

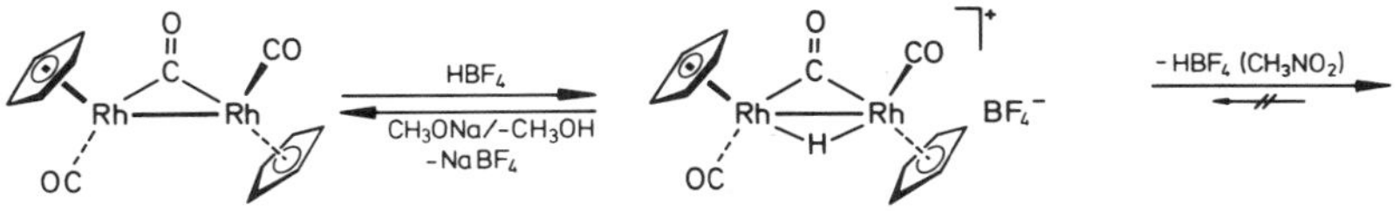

20

$$\tfrac{1}{3} \ (\eta^5\text{-}C_5H_5)_3 Rh_3(\mu\text{-}CO)_3 \ + \ (\eta^5\text{-}C_5H_5)Rh(CO)_2$$

SCHEME 31

IX

CATALYTIC IMPLICATIONS
OF THE METHYLENE BRIDGE

We have learned from the foregoing discussion that methylene bridges are formed under mild conditions from a variety of starting materials and that they exhibit pronounced reactivity to hydrogen and unsaturated hydrocarbons. Beyond that, it is generally observed that carbenes prefer a bridging position to a terminal coordination site. These phenomena established for molecular systems have led to the assessment that surface methylene species must be very significant in certain catalytic reactions of carbon monoxide and hydrogen (Fischer–Tropsch process) (38, 238). For a very simple reason, however, great care has to be exercised in drawing firm mechanistic parallels between metal cluster chemistry and catalytically active metal surfaces. The molecular regimes in discrete di- and polynuclear complexes are mostly coordinatively saturated, whereas metal surfaces, especially if they are relatively flat and close-packed, have no coordinatively saturated surface atoms, even in the presence of chemisorbed species. Furthermore, the stereochemistry of chemisorbed methylene might be quite different from what is established for their molecular models (109) in which the planes defined by the CH_2 group and the dimetallacyclopropane backbone are always approximately ($\pm 4°$) perpendicular to each other (see Section VI). This geometry seems to be less stable for methylene units chemisorbed to relatively bare metal surfaces. EHMO calculations performed by Muetterties et al. (239) for the Ni(111) surface indicate that the other stereochemical extreme (110) prevails, mainly because of optimal hydrogen–metal interaction.

109   110   111   112

113   114

Similar conclusions emphasizing the importance of M—H bonding have been drawn for the related methylidyne (CH) species from ultrahigh-vacuum spectroscopic studies (241) and from theoretical work (239), according to which structure 111, found in a plethora of well-characterized clusters, is less stable for chemisorbed CH units. Once again, hydrogen bonding yields a coordination mode that one might call a "side on" attached carbyne (type **112**). The iron cluster $HFe_4(CH)(CO)_{12}$ obtained by protonation of the anionic carbide precursor represents the first example of this kind of three-center C—H—Fe bonding array (namely, μ_4, η^2-CH) in organocluster chemistry (*36, 37*). Since we have earlier pointed out the stereochemical relationship between ethylene bridging over a metal–metal bond with the structures of dimetallacyclopropanes, one final comment should be added concerning the cluster–surface analogy of these latter systems. The gross geometrical features of **114** are related to the stereochemistry of conventional μ-methylene complexes **109** in that the planes defined by the respective ligands are normal to the M—M vector. This geometry has long been invoked and is, in fact, encountered in most dinuclear metal alkyne complexes. Nevertheless, a steadily increasing number of the planar stereoisomers **113** can be found in the current literature [see *242–244* for leading references]. Configuration **114** was noted to be more stable than **113** for a bare Pt_2 acetylene complex (*38*).

Two significant studies in support of the proposed importance of μ-methylene intermediates in nonstoichiometric CO/H_2 reactions must be mentioned here. First, coordinated carbon monoxide has been reduced with $H_3Al(OR_2)_x$ to give high yields of ethylene. The intermediacy of dinuclear μ-methylene complexes was postulated, and they are thought to account for clean carbene dimerization, as shown in Scheme 32 (*245*). Note, however, that the isomeric ethanediyl alternative $(CO)_xM$—CH_2—CH_2—$M(CO)_x$ has not been excluded as a possible key intermediate, which, alternatively, could equally well account for ethylene formation.

The other example to be discussed in this context comes from Pettit's group. Simultaneous treatment of the iron complex $(\mu\text{-}CH_2)[Fe(CO)_4]_2$ (**35**) with hydrogen and ethylene gives both methane (66%) and propylene (6%), the expected products from the two separate reactions. In addition, ethane ($\sim$600%) is formed, with the actual hydrogenation catalyst still to be determined (*72*). Because simple diazoalkanes provide the cleanest method to metal-attached alkylidenes, and with the expectation that dissociative chemisorption of diazomethane to absorbed CH_2 and free N_2 would occur, the reactions of CH_2N_2 with and without H_2 over various transition metals were examined in a careful study with regard to the product ratio (*73*). It was found, that gas-phase decomposition of the parent diazoalkane upon passage over active Ni, Pd, Fe, Co, Ru, or Cu-

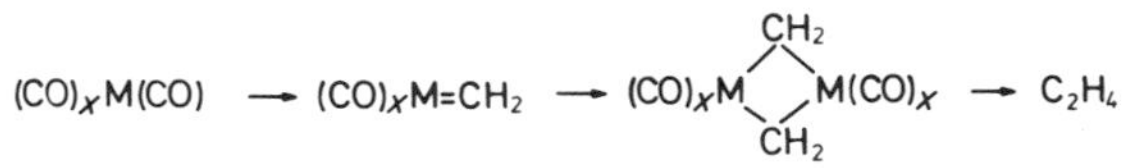

$$(CO)_xM(CO) \longrightarrow (CO)_xM=CH_2 \longrightarrow (CO)_xM \overset{\displaystyle CH_2}{\underset{\displaystyle CH_2}{\diagup\diagdown}} M(CO)_x \longrightarrow C_2H_4$$

SCHEME 32

surfaces at 1 atm and in the temperature range 25–250°C proceeds both rapidly and quantitatively. Ethylene is the predominant product, formed along with minor amounts of xylene-soluble polymers. The nature of the products changes markedly if diazomethane is diluted with hydrogen. Over Co, Fe, or Ru, a mixture of hydrocarbons ranging from 1 to 18 in chain length (and longer) is produced. The hydrocarbons are mainly linear alkanes and monolefins characteristic of those produced in the Fischer–Tropsch reaction (246). With increasing H_2 partial pressure, both the olefin content and the chain length decrease. These results contrast sharply with the gas-phase thermal decomposition of diazomethane (290–400°C) in the absence of any catalyst, where the products consist of merely a mixture of low-molecular weight compounds such as CH_4, C_2H_6, C_2H_4, C_2H_2, and HCN. When excess hydrogen is present, the products are primarily methane and ethane (247). Incidentally, the liquid-phase decomposition of CH_2N_2 under various conditions is well-known to give polymethylene (248).

On Ni and Pd surfaces, a CH_2N_2–N_2 mixture is not only converted to methane but to higher hydrocarbons as well (73). Moreover, CO and H_2 over the same catalysts (1 atm) give CH_4 as the principal product. Most significantly, however, the latter system yields higher linear hydrocarbons when passed over a Ni catalyst at greater pressures (68 atm), with the molecular weight distribution closely resembling that observed for the CH_2N_2–H_2 mixture at 1 atm! It appears reasonable to assume that the principal gas-phase reaction of CH_2 fragments alone on metal surfaces is not polymerization but dimerization and elimination of ethylene from the surface. Discrete di- and polynuclear examples of the postulated key intermediates **115** and **116** have been described in this chapter, although the conversion **115** → **116** has not yet been observed in organometallic model systems. In addition, the very close similarities in the relative distribution of hydrocarbon products resulting from metal-mediated reactions of CO–H_2 on the one hand and CH_2N_2 on the other suggest a common mechanism for these two processes. Methylene groups clearly do polymerize on the aforementioned surfaces, but the polymerization is initiated by metal hydride functionalities, with the growth pattern including the initial steps shown in Scheme 33. The same kind of chemisorbed

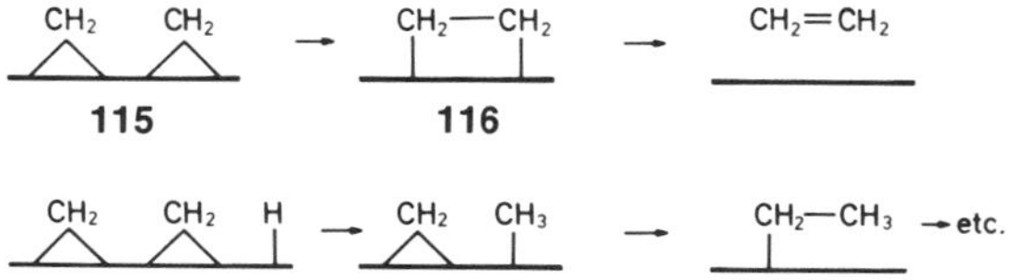

SCHEME 33

CH_2 groups may be produced in the course of metal carbide reduction following dissociative chemisorption of CO. This crude mechanism was originally proposed by Fischer and Tropsch as early as 1926 (249). The reactions of CH_2N_2–H_2 mixtures and CH_2N_2 alone on copper surfaces give further support to this mechanistic proposal. In both cases, only ethylene is produced. This behavior very likely originating from the low tendency of copper to chemisorb hydrogen dissociatively (73).

X
THE GROUP IV μ-METHYLENE CONGENERS

A. Synthesis

There is an array of well-established analogues of the μ-methylene complexes in which the heavier congeners of carbon adopt bridging positions (250–275). Strikingly, the μ-silylene bridge is far less common than germylene, stannylene, or plumbylene derivatives. The preparation of the μ-methylsilylene iron complex $[\mu\text{-Si(H)CH}_3][(\eta^5\text{-C}_5H_5)Fe(CO)_2]_2$ from the reaction of its mononuclear precursor $(\eta^5\text{-C}_5H_5)Fe(CO)_2[Si(H)(CH_3)Cl]$ with the highly nucleophilic metal carbonyl anion $[(\eta^5\text{-C}_5H_5)Fe(CO)]_2^-$ (Na salt) may be mentioned as a good example of a clear-cut synthetic method (254) frequently used for the construction of the analogous systems. Thermally and/or photochemically induced CO elimination is a process typical of such compounds and yields metal-to-metal bonds in addition to carbonyl bridges, e.g., $[\mu\text{-Si(H)CH}_3](\mu\text{-CO})[(\eta^5\text{-C}_5H_5)Fe(CO)]_2$ (254). In some cases, bis-μ-germylene derivatives have been obtained upon photolysis of mononuclear compounds belonging to the above-mentioned type. Thus $(CO)_5Mn[Ge(CH_3)_2Cl]$ gives $[\mu\text{-Ge}(CH_3)_2]_2\{Mn(CO)_4\}_2$ in surprisingly high yields (266).

Similar reactions have been reported for the corresponding iron compound $(\eta^5\text{-C}_5H_5)Fe(CO)_2[Ge(CH_3)_2Cl]$ which mainly gives $[\mu\text{-Ge}(CH_3)_2](\mu\text{-CO})[(\eta^5\text{-C}_5H_5)Fe(CO)]_2$ and minor amounts of its isoelectronic derivative $[\mu\text{-Ge}(CH_3)_2]_2[(\eta^5\text{-C}_5H_5)Fe(CO)]_2$ (266c). Photolysis of bis(metal-carbonyl)germanes is a convenient route to complexes containing bridging dialkylgermylene and bridging carbonyl groups [e.g., $(\mu\text{-R}_2Ge)\{M(CO)_n\}_2 \rightarrow (\mu\text{-R}_2Ge)(\mu\text{-CO})\{M(CO)_{n-1}\}_2 + CO$ (260, 266, 267)]. In several cases the method of choice, especially for the preparation of μ-germylene and μ-stannylene complexes, involves dehydrogenation of organogermanes and -stannanes, respectively (e.g., H_2MR_2; M = Ge, Sn), in the presence of metal carbonyls under photochemical conditions

(*251, 259*). A method that seems to promise general applicability was recently demonstrated by Jutzi *et al.*, who have prepared the dihalogermylene-bridged compound (μ-GeBr$_2$)[W(CO)$_5$]$_2$ by addition of the mononuclear "carbenoidal" dibromogermylene complex (CO)$_5$W=GeBr$_2$ (*269*) to coordinatively unsaturated W(CO)$_5$ fragments (*270, 271*), this route being reminiscent of the synthesis of dimetallacyclopropanes from Fischer-type carbene precursors (see Section IV,B,1). Diazoalkane homologs of composition N$_2$=MR$_2$ (M = Si, Ge, Sn, Pb) are, of course, not available for the introduction of MR$_2$ groups into organometallic molecules.

B. *Structures*

For comparison, several representative dinuclear metal carbonyl derivatives bridged by group IV elements are listed in Table VI. It becomes evident that two major classes must be considered. First, MR$_2$-bridged molecules lacking metal–metal interactions can easily be recognized by (a) wide "internal" M,E,M angles α which are generally greater than ~100° and (b) by large metal–metal separations (cf. [μ-Sn(CH$_3$)$_2$]$_2$\{Fe-(CO)$_4$\}$_2$). No structurally established carbon counterpart of this class is yet available. As a matter of fact, the overwhelming majority of methylene complexes belong to the second subgroup of molecules that do have metal–metal bonds in addition to bridging MR$_2$ functionalities. These members are characterized by far smaller M,E,M angles, the latter ranging between 68 and 86° regardless of the nature of the MR$_2$ bridge (E = C, Si, Ge, Sn, Pb). As expected from the increasing atomic radii within a given group of elements of the periodic table, the heavier homologues of carbon form larger M,E,M angles than the parent dimetallacyclopropanes (Tables I and V). On the other hand, the acuteness of these angles is not only determined by the length of the M—E bond but also by the magnitude of the M—M separation. Thus, for a given M—M bond, larger bridging atoms give more acute angles on simple geometric grounds. It is to be noted, however, that considerable variations in Fe—Ge distances have been found within a series of μ-germylene iron complexes, but no trends are apparent (*268*). A careful survey of all structural data available clearly demonstrates that a bridging R—E—R group is able to adjust its geometry (e.g., the M,E,M angle) to the boundary conditions offered by the organometallic framework, with the possible exception of the methylene bridge which greatly prefers an acute M,C,M angle and, consequently, stabilizes an additional metal–metal bond to a very great extent ("dimetallacyclopropanes," see Section VI).

TABLE VI

Comparative Structural Data of ER$_2$-Bridged Metalcarbonyl Derivatives (E = C, Si, Ge, Sn)

Compound	d(M—M) [pm]	d(M—E) [pm]	∢(M,E,M) [deg]	∢(R,E,R) [deg]	Reference
[μ-C(H)CH$_3$]$_2$[(μ-CO)[cpFe(CO)]$_2$ (cis)	249.9(9)	192(6) 200(5)	77.2	—^{a}	64
[μ-C(H)(CO)$_2$t-C$_4$H$_9$)][(μ-CO)[cpFe(CO)]$_2$ (cis)	—	179.4–192.6	79.4–85.3	—	48
μ-CH$_2$[(η^5-C$_5$H$_5$)Rh(CO)]$_2$	266.49(4)^c	202.9(4)^c 205.5(4)^c	81.7(1)^c 81.0(1)^d	115.9(4)^c 110.4(1)^d	40, 41 56
[μ-Ge(CH$_3$)$_2$]$_2$(μ-CO)[cpFe(CO)]$_2$ (cis)	262.8(1)	234.5(1) 234.7(1)	68.15(3)	103.0(4)	263
[μ-Ge(CH$_3$)$_2$]$_3$[Ru(CO)$_3$]$_3$	293	247	73.4(1)	—^{a}	258
[μ-Ge(C$_2$H$_5$)$_2$]$_2$[Fe(CO)$_4$]$_2$	—^{a}	249(2)	104.5	—^{a}	256
[μ-Sn(CH$_3$)$_2$]$_2$[Fe(CO)$_4$]$_2$	413.1(19)^b 413.9(15)^b	263.1(11) 264.7(8)	102.6(3) 103.7(4)	109(2) 109(2)	261
(μ-GeCl$_2$)[cpFe(CO)$_2$]$_2$	—	235.7(4)	128.4(2)	96.1(2)	255
[μ-Si(C$_6$H$_5$)$_2$]$_2$[Mn(CO)$_4$]$_2$	287.1(2)	240.2(4)	73.39(6)	104.2(3)	252
[μ-Ge(CH$_3$)$_2$](μ-CO)[Mn(CO)$_4$]$_2$	285.4(2)	243.2(2) 247.7(2)	71.13(5)	—^{a}	267
[μ-Ge(C$_6$H$_5$)(Ø$_4$C$_4$H)][Fe(CO)$_4$]$_2$^{e}	278.5(3)	240.8(2) 243.0(2)	70.30(8)	105.5(5)	268
[μ-Ge(CH$_3$)$_2$]$_3$[Fe(CO)$_3$]$_2$	275.0(11)	239.8(4)	70.0(2)	105.3(11)	257
(μ-GeBr$_2$)[W(CO)$_4$]$_2$	337.0(1)^b	258.0(1) 257.2(1)	81.69(4)	99.93(7)	271
(μ-SiH$_2$)$_2$[(η^5-C$_5$H$_5$)Ti]$_2$	336.8(10)^b	215.9(13)	102.8(7)	—^{a}	275

^{a} Not reported.
^{b} Nonbonding distance.
^{c} X-ray diffraction.
^{d} Neutron Diffraction.
^{e} Ø$_4$C$_4$H = —C(C$_6$H$_5$)=C(C$_6$H$_5$)—C(C$_6$H$_5$)=C(H)C$_6$H$_5$.

Curtis *et al.* have stressed the significant contraction of metal–metal bonds when bridged by SiR_2 or GeR_2 groups (*267*). Such shortening is unexpected in view of the extreme crowding in bridged structures of this kind and in view of the geometric strain associated with the acute M,E,M angles. Both effects would tend to lengthen the M—M bond rather than cause the observed contraction. In order to overcome this contradiction, a bonding model in which the bridging ER_2 unit is considered to be a bridging carbenoid ligand has been proposed. The pictorial representation given in Fig. 12 for $[\mu\text{-}M(CH_3)_2]_2\{Mn(CO)_4\}_2$ is derived from Fenske–Hall-type MO calculations performed by Teo *et al.* for the isoelectronic phosphinidene species $[(\mu\text{-}PH_2)_2\{Mn(CO)_4\}_2]^{2+}$ (*276*). A strong metal–bridge interaction of the metal and bridge b_{1u} orbital combinations (Fig. 12a) and a π-type interaction of the carbenoid p-orbitals and the $1b_{3u}$ metal combination (Fig. 12b) is expected, in addition to mixing of the carbenoid p-orbital which alleviates the antibonding character of the metal–metal π^* (b_{2g}) orbital (Fig. 12c). These four-center, two-electron orbitals offer the conceptual advantage of removing the angle strain implied by picturing the M—E bonds as lying along the metal–metal vector. In the carbenoid model, the regions of electron density are directed toward the center of the four-membered cycle and round its periphery in "bent" bonds (Fig. 12b). Hence by delocalizing the M—M bonding orbitals ($a_g + b_{1u}$) and by

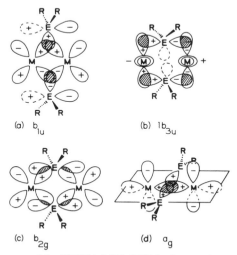

(a) b_{1u} (b) $1b_{3u}$

(c) b_{2g} (d) a_g

FIG. 12. MO diagram of $[\mu\text{-}M(CH_3)_2]_2\{Mn(CO)_4\}_2$ [*e.g.*, M = Ge (*267*)]. Note that metal–d-orbital interactions play a minor role. Reprinted with permission from K. Triplett and M. D. Curtis, *J. Am. Chem. Soc.* **97**, 5747 (1975). Copyright 1975 American Chemical Society.

providing some bonding character to the Mn—Mn antibonding orbitals (b_{2g}, $1b_{3u}$), the net M—M interaction is strengthened and a very short M—M bond (here: Mn—Mn) results (267). This model also accounts for the observed relatively short Mn—Ge bonds without invoking enhanced d-orbital contributions from the bridging group. MO calculations suggest that the d-orbitals (at least on bridging phosphorus ligands) play only a minor role in the bonding anyway (176). Based upon structural data it was also concluded that in a given series of related multiply bridged homodinuclear compounds a bridging carbon monoxide ligand is more effective in strengthening the metal-to-metal bond than a bridging GeR_2 group (267).

C. Fluxionality

In sharp contrast to their μ-methylene counterparts, dialkylgermylene bridges are commonly observed to share the fluxional behavior of the metal carbonyl framework carrying them, although there is clearly no uniform pathway along which these intramolecular ligand-exchange processes proceed. Interestingly, the activation energy of cis–trans interconversion (cis–trans 4:1 at room temperature) of $[\mu\text{-Ge}(CH_3)_2](\mu\text{-CO})[(\eta^5\text{-}C_5H_5)Fe(CO)]_2$ is markedly higher [estimated ≥ 19 kcal/mol from ^1H NMR (266a)] than that of the parent carbonyl $(\mu\text{-CO})_2[(\eta^5\text{-}C_5H_5)Fe(CO)]_2$ (13 ± 2 kcal/mol; 264, 265). This result presumably reflects the lower stability of the nonbridged isomer $(\eta^5\text{-}C_5H_5)_2Fe_2(CO)_3[Ge(CH_3)_2]$, which contains a terminal dimethylgermylene group. In most compounds of the latter type, Lewis bases (e.g., THF, tertiary amines, and phosphines) are coordinated to germanium forming the corresponding ylides, but a few well-established exceptions are known (269). Although the aforementioned example very likely involves cis–trans isomerization according to the Adams–Cotton mechanism originally elaborated in detail for CO scrambling in metal carbonyls (265), there is still some controversy regarding the rearrangement pathways of other nonrigid derivatives with bridging R_2Si, R_2Ge, or R_2Sn groups in place of carbon monoxide. The cobalt compound $[\mu\text{-Ge}(CH_3)_2]_2\{Co(CO)_3\}_2$ (117) may serve as an example illustrating this kind of problem. Due to the folded Co_2Ge_2 framework, methyl groups in two different environments are present in 117. Consequently, the low temperature ^1H-NMR spectrum displays two distinct methyl signals which collapse at $-42.5°C$ to give a single resonance (265). A reverse Berry pseudorotation has been proposed to account for the observed site exchange (Fig. 13). The distorted square-pyramidal coordination at either cobalt center rearranges to trigonal

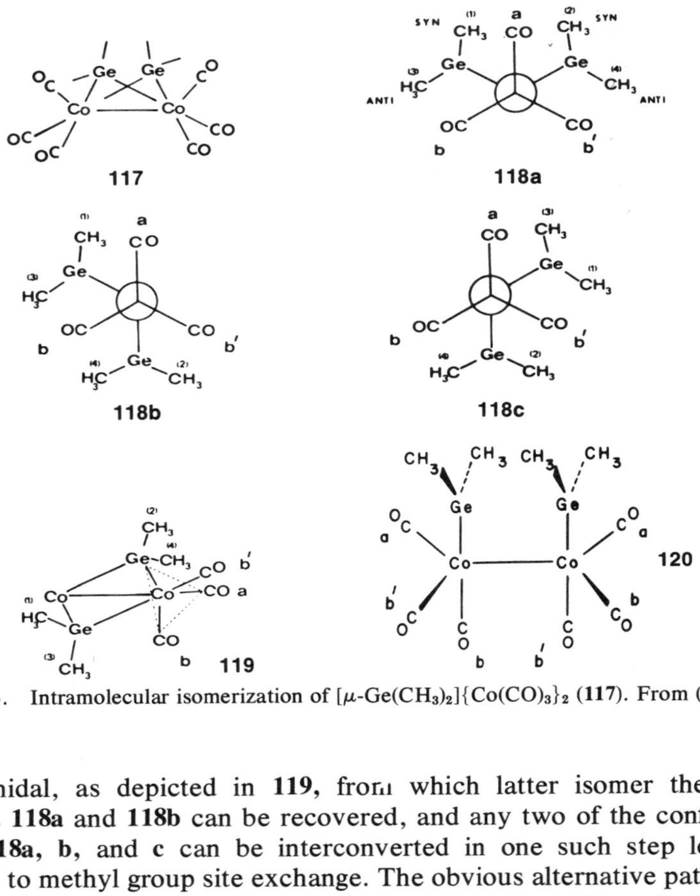

FIG. 13.   Intramolecular isomerization of [μ-Ge(CH$_3$)$_2$]{Co(CO)$_3$}$_2$ (**117**). From (*265*).

bipyramidal, as depicted in **119**, from which latter isomer the conformers **118a** and **118b** can be recovered, and any two of the configurations **118a, b,** and **c** can be interconverted in one such step leading directly to methyl group site exchange. The obvious alternative pathway, which traverses structure **120** by bridge opening, was believed to be less likely (*265*) because of the low driving force evident for carbenoid-like R$_2$Ge=M groups at a time when stable compounds of this type were not yet discovered [cf. (CO)$_5$Cr=Ge(SR)$_2$ (R = mesityl, methyl), (CO)$_5$Cr=GeX$_2$ (X = Cl, Br), or (CO)$_5$Cr=Ge(Mes)$_2$ (Mes = mesityl); (*269*)]. Since steric effects undoubtedly govern the stability of such nonbridged species [(CO)$_5$Cr=Ge(C$_6$H$_5$)$_2$ is *not* known, for example (*269*)], the arguments against a bridge-opening process as the crucial steps of isomerization may still be acceptable although they are no longer convincing ones. In addition to the above-mentioned polytopal rearrangement, the closely related derivative [μ-Ge(CH$_3$)$_2$](μ-CO){Co(CO)$_3$}$_2$ having just *one* germylene bridge may average the environments of the methyl groups via a terminal-bridge CO exchange in which the new Co—CO—Co bridge is formed on the opposite side of the Co—Ge—Co triangle (*266a*). A ^{13}C-NMR study to discriminate between these two possibilities is yet to be done.

XI

ONE STEP BEYOND THE METHYLENE BRIDGE: $\mu(\omega,\omega')$-ALKANEDIYL COMPLEXES

As a kind of extension of the methylene bridge, some longer chained alkanediyl metal complexes are known. They formally derive from type-C compounds (see Sections II–VIII) by adding one or a few more methylene units to the bridging CH_2 function and hence belong to class G as represented by the general formula L_xM—$(CH_2)_x$—ML_x. Typical members of this series are reported for $2 < x < 10$, but there is no reason to assume that the chain length has an upper limit. Despite an earlier prediction (*277*) that the ethylene bridge ($x = 2$), as a matter of principle, suffers from low stability, the rather robust zirconium compound **121** and molecules with similar architecture were obtained by Sinn *et al.* and Kaminsky *et al.* (*277a–281*). The most notable structural features of $(\mu$-$C_2H_4)[(\eta^5$-$C_5H_5)_2Zr$-Cl—$AlEt_3]_2$ are found to be the acute Zr,C,C angles (75.9°), the rather large carbon–carbon distance within the bridging system (155 pm), which is exactly what one would expect for a typical single bond, and finally the missing metal-to-metal bond (*278*). The acuteness of the Zr,C,C angles, though striking in this case, does not represent characteristics inherent in "ethylene"-bridged binuclear metal complexes. For example, the closely related rhenium compound (μ-C_2H_4) $[Re(CO)_5]_2$ (**122**) recently reported by the Beck group has a large Re,C,C angle pushing the metal centers even further from each other. In contrast, the C—C distance is much shorter than in **121** (*219, 282*). Those compounds consequently can no longer be viewed as true ethylene-bridged systems but more or less as ethane-derived organometals in spite of the wide structural variations found even in just these two molecules. In

$(CO)_5Re-CH_2-CH_2-Re(CO)_5$

122

other words, they already rank midway between CH_2-bridged com-
pounds and the ionic species $[(\mu\text{-}C_3H_5)\{(\eta^5\text{-}C_5H_5)Fe(CO)_2\}_2]^+$ (**124**) which
is accessible by hydride abstraction from King's neutral propanediyl
precursor of composition $[\mu\text{-}(CH_2)_3][(\eta^5\text{-}C_5H_5)Fe(CO)_2]_2$ (**123**) (*223*).
Again, no metal-to-metal bond is present in either compound. The close
contact between the iron centers and the β-carbon in **124** is undoubtedly
responsible for the unexpected stability of this nonclassically bonded
carbonium ion. If one accepts any general statement from the hitherto
limited number of compounds, severe metal–β-carbon interaction seems
to be characteristic of these cationic species if the metals are not bonded
to each other (Fig. 14).

Oligo- and polymethylene-bridged complexes can be prepared by reaction of metal carbonyl anions with ω,ω'-dihaloalkanes. This method
proved to be a simple entry to the iron series $[\mu\text{-}(CH_2)_x][(\eta^5\text{-}C_5H_5)$
$Fe(CO)_2]_2$ ($x \geq 3$) (*283*), but yielded quite different products in the case
of $Na[(\eta^5\text{-}C_5H_5)Mo(CO)_3]$ (*283*) and $Na[Mn(CO)_5]$ (*284*). An alternative two-step synthesis of the iron compounds involves the preparation
of the mononuclear ω-haloalkyls $L_xM{-}(CH_2)_x{-}X$, and a subsequent

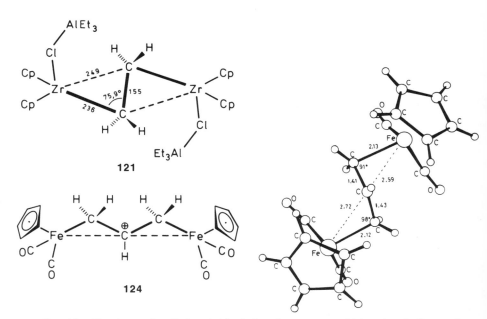

FIG. 14.   Structures of a ethylene-bridged zirconium compound (above) and of a metal-
stabilized cationic 1,3-propanediyl-bridged iron complex (below) including an ORTEP representation of the latter (*223*).

coupling reaction of the latter with the anionic substrates L_xM^-, to yield the bridged systems L_xM—$(CH_2)_x$—ML_x [e.g., $[\mu\text{-}(CH_2)_4][(\eta^5\text{-}C_5H_5)$ $Fe(CO)_2]_2$ (*218, 283*) or $[\mu\text{-}(CH_2)_{10}][(\eta^5\text{-}C_5H_5)Mo(CO)_3]_2$ (*165*)]. All structural data available support the formulation as Class G compounds (see Section I) discussed here as ω,ω'-alkanediyl complexes. Two typical examples may suffice to amplify this point. Constant values (154–156 pm) were found for the carbon–carbon distances of the bridging hydrocarbons in $[\mu\text{-}(CH_2)_x][(\eta^5\text{-}C_5H_5)Fe(CO)_2]_2$ ($x = 3, 4$), and the corresponding C,C,C angles range between 111 and 115° (*218*). It should be noted that some structural relationship is found between $(CH_2)_x$ and the even less common $(CH)_x$ bridged systems. In this context, the iron compound $(\mu\text{-}CH)_4[(\eta^5\text{-}C_5H_5)Fe(CO)_2]_2$, having a 1,3-butadiene-1,4-diyl chain, may be mentioned (*285, 286*). In accord with their alkane character, a fact reliably supported by the ^{1}H- and ^{13}C-NMR data (see Section VII), the chemistry of ω,ω'-alkanediyl complexes approaches the chemistry of ordinary alkyl complexes the longer the hydrocarbon chain grows. Typically, the compound L_xM—$(CH_2)_4$—ML_x undergoes photoinduced alkyl migration to yield the μ-diacyl derivative $L_{x-1}(PR_3)M[C(\!=\!O)$—$(CH_2)_4$—$C(\!=\!O)]M(PR_3)L_{x-1}$ $[L_xM = (\eta^5\text{-}C_5H_5)Fe(CO)_2$, $PR_3 =$ $P(C_6H_5)_3$, $L_{x-1} = (\eta^5\text{-}C_5H_5)Fe(CO)]$ (*153*).

A related compound of proposed structure $(CO)_3Co$—$\overline{(CCF_3)_6}$— $\overline{Co}(CO)_3$ having a six-membered olefin-type hydrocarbon chain spanning a metal–metal bond was isolated, along with some hexakis(trifluoromethyl)benzene, from the reaction of $Co_2(CO)_8$ with hexafluoro-2-butyne at 155°C (*287a*). Interestingly, no alkyne complex $Co_2(CO)_6(\mu\text{-}CF_3C\!\equiv\!CCF_3)$ shows up, although this compound is known to be the major product if the same reaction is carried out at somewhat lower temperatures (110°C) (*288*). A complex of composition $Co_2(CO)_4(CF_3C_2H)_3$ [from $Co_2(CO)_8$ and $CF_3C\!\equiv\!CH$] is thought, based upon ^{19}F-NMR results (*287b*), to have an analogous structure.

We have mentioned that the structural parameters of C_2H_4 bridged compounds can vary over a wide range. Whereas most examples reported do not have metal–metal bonds, there is one conspicuous exception. Theopold and Bergman succeeded in synthesizing the propane-1,3-diyl cobalt derivative **125** from the radical anion $[(\eta^5\text{-}C_5H_5)Co(\mu\text{-}CO)]_2^{\overline{\cdot}}$ and 1,3-dibromopropane (*98, 295*) in 40% yield. This compound is best described as a dimetallacyclopentane, and its chemistry (thermolysis and reaction with CO and phosphines; Scheme 34) supports this view. Formation of cyclopropane (100°C or $I_2/25°C$) is probably the most remarkable feature of this cyclic system. Simple C—C bond formation has never been observed before in ligand-induced or thermal reactions of either mono- or binuclear cyclopentadienylcobalt complexes. The architectural details of

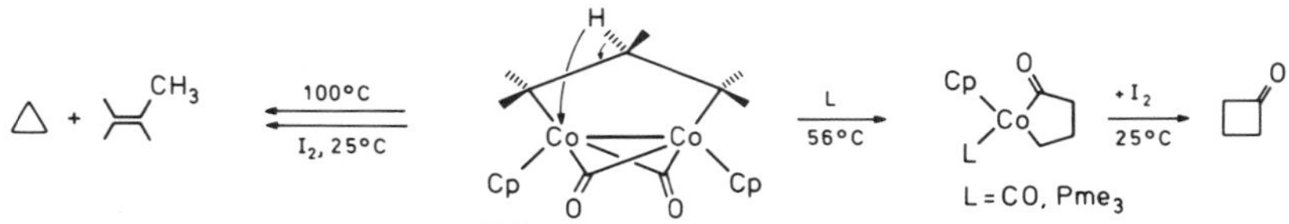

125 dec. 170°C

SCHEME 34

125 are not known, but the five-membered cycle very likely is not planar, although it could be if we compare it with the structure of the trinuclear ruthenium compound **126**, the Ru_2C_3 core of which is strictly coplanar (*289*). This analogy, however, is admittedly overly simplistic since **126**

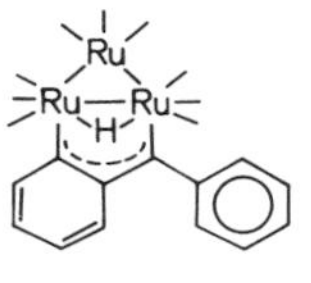

126

reveals a more or less biscarbene-type framework. Bergman's compound nicely demonstrates that a hydrocarbon chain is able to adjust its geometry to a metal–metal bond when winding around a dinuclear frame, thus yielding dimetallacycles; a comparison of compounds **123** and **125** substantiates the impression that the formation of a metal–metal bond will be governed merely by the steric availability of additional coordination sites and not so much by the chain length of the alkane bridge. This view is also justified by the comparison of compounds **125** and **126** with dimetallacyclobutenes of type **126a** reported by Dickson *et al.* (*242–244*) in which the alkyne bridge [e.g., $(\mu\text{-CF}_3C{\equiv}CCF_3)[(\eta^5\text{-C}_5H_5)Rh(CO)]_2$] is σ-bonded to two rhodium atoms and part of an exactly coplanar four-membered cycle (Fig. 15). Removal of one CO group with concomitant σ–π isomerization of the alkyne bridge can be achieved by treatment with trimethylamine oxide (*294*). Similarly, the Ir_2C_2 core of the structurally established dinuclear *o*-phenylene iridium complex $(\mu\text{-C}_6H_4)[(\eta^5\text{-C}_5H_5)Ir(CO)]_2$ barely departs from strict planarity (*290*). From simple structural considerations, ethylene could likewise act as a σ-bridge between two metals, with an extra metal-to-metal bond possibly, but not necessarily, being present at the same time. The tetrafluoroethylene-bridged iron complex $(\mu\text{-}C_2F_4)(\mu\text{-SCH}_3)_2[Fe(CO)_3]_2$ is a perfect example of this idea. Again, the M_2C_2 backbone is planar, more or less representing a 1,2-dimetallated ethane derivative [$d(C{-}C)$ 153.4(7) pm] exhibiting a cisoid configuration

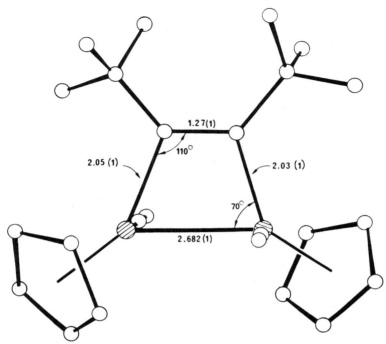

FIG. 15. Structure of the four-membered dimetallacycle [μ-bis(trifluoromethyl)acetylene]bis[carbonyl(η^5-cyclopentadienyl)rhodium](Rh—Rh) (**126a**) (*244*), a typical example of a strictly planar M_2C_2 framework. Bond lengths in picometers.

(*203*). In this particular example, both the nature and the number of bridging methyl-mercapto ligands are responsible for the long metal–metal separation [331.1(1) pm], whereas, among others, the transoid structure of the rhenium compound **122** (*282*) is responsible for the nonbonding distance between the two metal centers.

The methodological feature of another promising synthetic approach, specific to ethanediyl complexes, is nucleophilic attack of metal carbonyl anions on coordinated ethylene. The C_2H_4-bridged compound $(\mu\text{-}C_2H_4)[(\eta^5\text{-}C_5H_5)W(CO)_3]_2$ forms in quantitative yield from the cationic species $[(\eta^5\text{-}C_5H_5)W(CO)_3(\pi\text{-}C_2H_4)]^+$ (BF$_4$ salt) and the anion $[(\eta^5\text{-}C_5H_5)W(CO)_3]^-$. The corresponding molybdenum derivative is far less stable and decomposes, even in the solid state, above about $-20°C$. The same trend was recognized for the mixed systems $(\eta^5\text{-}C_5H_5)(CO)_3M$—$CH_2$—$CH_2$—$Re(CO)_5$ (M = Mo : **127**; M = W : **128**). Decomposition of those species generally yields stoichiometric amounts of free ethylene.

Interestingly enough, both **127** and **128** react with excess [Re(CO)$_5$]$^-$ to form the stable (mp 138°C) crystalline, homodinuclear rhenium compound (μ-C$_2$H$_4$)[Re(CO)$_5$]$_2$ (**122**) in 84% yield (*219, 282*). The structure of **122** [X-ray (*282*)] does not justify a π-ethylene description of the molecule since (a) the dicarbon ligand is not oriented perpendicular to the metal–metal vector, but rather forms a chainlike array with the metals, (b) the C—C distance is quite long [143(2) pm], and (c) each carbon atom of the bridging ligand is attached to a different metal center [*d*(Re—C) 230(1) pm], which is also reflected by the large Re—C—C angle [121(1)°]. These data are fairly consistent with the bonding expected for a ω,ω'-bismetallated ethane, although the C—C distance is admittedly shorter than that of the free ligand (153 pm).

XII
CHALLENGING PROBLEMS AND FUTURE TRENDS

Methylene-bridged organometallic compounds are no longer curiosities. The basic questions concerning their structural behavior, spectroscopic features, and bonding patterns have been settled at least to the extent that we can use the data obtained so far as safe guidelines for further work. The synthetic strategy is well elaborated these days and will be used for the systematic construction of new members in the dimetallacyclopropane series. One of the great preparative challenges is the extension of the known methods for the preparation of multimetal carbido species of type (μ-C)$_x$[ML]$_y$. More sophisticated diazo precursors, such as Ag$_2$C=N$_2$ and other heavy metal derivatives of diazomethane, clearly constitute the starting materials of choice to achieve this goal. Planar coordination around carbon as calculated for certain lithium compounds could be possible even in transition metal systems if steric constraints succeed. Furthermore, dimetallacyclopropenes are not yet known as stable species, and the question has to be asked as to why and under what conditions the metal–metal frames escape unsaturation, and, consequently, how they can be prepared by synthetic chemists. As far as the reactivity of μ-methylene complexes is concerned, much effort will be focused on catalysis-related problems. For example, chain-growth homologation within the coordination sphere of molecular bimetallic units has not been accomplished, although such processes work readily on metal surfaces (Section IX). Finally, methylene mobility in high nuclearity clusters should be studied in order to define the extent to which a CH$_2$ unit prefers a certain metal and the metals which are most active in C—H activation.

Metal-centered combination of methylene with other small molecules such as a CO or another CH_2 group are also expected to suggest boundary conditions under which such reactions occur. Work in this area will undoubtedly be directed to technically relevant processes, but it will take a major amount of straightforward organometallic benchwork to clarify the basic problems.

ACKNOWLEDGMENTS

I wish to thank my resourceful co-workers Christine Bauer, Helmut Biersack, José Gimeno, John Huggins, Willibald Kalcher, Gangolf Kriechbaum, Johann Plank, Barbara Reiter, Doris Riedel, Martina Smischek, Ilona Schweizer, Isolde Steffl, and Josef Weichmann, all of whom have participated in our research on μ-methylene complexes since 1974. My colleagues Ivan Bernal, Dore Clemente, Bill Geiger, Thomas F. Koetzle, Carl Krüger, B. Rees, and Manfred L. Ziegler have assisted through X-ray diffraction, neutron diffraction, and electrochemical studies. I also thank Bob Bergman, David Curtis, Dennis Lichtenberger, Peter Hofmann, Roald Hoffmann, Selby Knox, Fritz Kreißl, Earl Muetterties, John Shapley, Gordon Stone, and Geoffrey Wilkinson for providing preprints of their results prior to publication, and to Karin Kilgert for her assistance in preparation of the manuscript. Generous support for our studies by the Deutsche Forschungsgemeinschaft, the Fonds der Chemischen Industrie, Degussa Hanau, Badische Anilin- & Sodafabrik AG, Chemische Werke Hüls, and Hoechst AG is gratefully acknowledged as well.

ADDENDUM

Since this article was completed, several papers concerning methylene-bridged metal compounds have either appeared in the literature or have been submitted for publication. A brief discussion follows:

1. The reactions of several noncyclic α-diazoketones [*e.g.*, benzoyl(phenyl)diazomethane (**129a**)] with the rhodium dimer **21** have been studied and shown to exceed the stage of μ-methylene complexation discussed in Section IV,A,1 for other diazoalkanes. The novel metallacycles **130**, formed in near quantitative yields, arise from attack on a rhodium carbonyl function by the pronounced nucleophilic ketocarbene oxygen atoms (Scheme 35). An X-ray structure determination of compound **130a** revealed the presence of a folded five-membered ring system Rh—CR=CR—O—C=O that acts as a π-olefin-type ligand to the other metal center (Fig. 16). The geometry around the "methylene" carbon C-4 violates the "orthogonality criterion" for true μ-methylene ligands in the sense that the dihedral angle (113.2°) between the two planes defined by C-3, C-4, C-41 and Rh-1, C-4, Rh-2 deviates considerably from 90°. The syntheses of **130a–d** represent the first successful cycloaddition

	a	b	c	d
R	C_6H_5	$C_6H_4OCH_3$-*p*	CH_3	CH_3
R'	C_6H_5	$C_6H_4OCH_3$-*p*	C_6H_5	CH_3

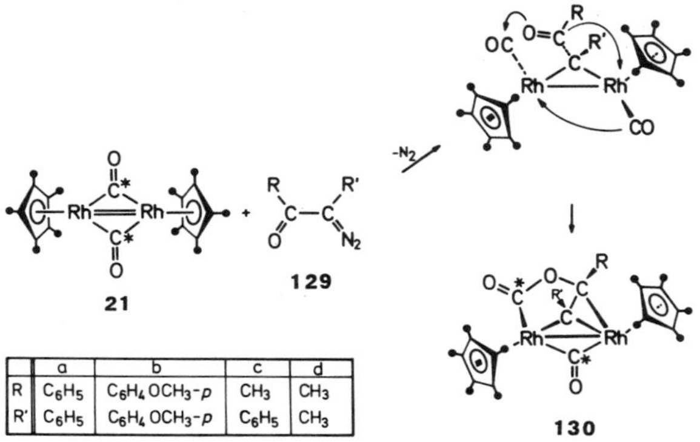

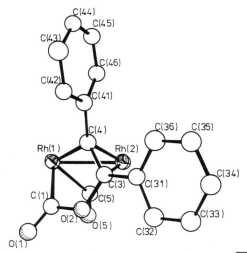

SCHEME 35

reactions of ketocarbenes to dinuclear metal carbonyl frameworks (*296*). A discussion of the "orthogonality criterion" is given in Ref. *296*. μ-Methylene ligands have been classified according to their π-acceptor–σ-donor behavior by means of IR spectroscopy (*55*).

FIG. 16. ORTEP representation of the cycloadducts $(\eta^5\text{-}C_5Me_5)_2\overline{Rh[C(C_6H_5)=C(C_6H_5)}$ $\underline{-O-C}(=O)]Rh(\mu\text{-}CO)(\eta^5\text{-}C_5Me_5)$ (**130a**). The pentamethylcyclopentadienyl ligands are omitted for clarity. Important bond lengths [pm] and angles [deg]: Rh(1)–Rh(2) 269.7(1), Rh(1)–C(4) 202.8(7), Rh(2)–C(4) 205.6(6), Rh(1)–C(1) 201.4(6), Rh(2)–C(3) 219.8(6), Rh(1)–C(5) 206.3(8), C(3)–C(4) 142.5(9), C(3)–O(2) 141.0(7), C(1)–O(1) 120.4(8), C(1)–O(2) 138.5(9); Rh(1)–C(4)–Rh(2) 82.6(2), Rh(1)–C(5)–Rh(2) 84.7(3), Rh(1)–C(4)–C(3) 108.1(4), Rh(1)–C(4)–C(41) 121.1(5), Rh(2)–C(4)–C(41) 127.3(5).

2. The preparation of the first stable examples of metallated μ-methylene complexes has been achieved by the diazo method (see Section IV,A,1). Reaction of Buchner's reagent **131** with either $[(\eta^5\text{-}C_5Me_5)Co(\mu\text{-}CO)]_2$ (**132**) or $[(\eta^5\text{-}C_5Me_5)Rh(\mu\text{-}CO)]_2$ (**21**) in boiling tetrahydrofuran cleanly yields the tetranuclear mercuriomethylene derivatives **133** (M = Co) and **134** (M = Rh), respectively, compounds in which two dimetallacyclopropane units are linked together by mercury (Scheme 36) (*297*). This particular subdivision of dimetallacyclopropanes is likely to expand very soon since numerous heterodiazoalkanes are now accessible as "easy-to-make" compounds. It appears as if one of the "challenging problems" (see Section XII) will be solved without difficulty, not least because the silver diazoalkane $AgC(=N_2)[P(=O)(C_6H_5)_2]$ has been trapped in the stable rhodium complex $(\eta^5\text{-}C_5Me_5)_2Rh_2(\mu\text{-}CO)[C(Ag)(=N_2)\{P(=O)(C_6H_5)_2\}]$ (*153*).

3. μ-Methylene iron compounds of composition $(\mu\text{-}CRR')[(\eta^5\text{-}C_5H_5)\text{-}Fe(NO)]_2$ (**135**) having NO ligands in terminal or bridging positions are accessible by carbene transfer from appropriate diazoalkanes to the metal-to-metal double bond of $[(\eta^5\text{-}C_5H_5)Fe(\mu\text{-}NO)]_2$ (**26**). It was shown that a nitrosyl function is more resistant to bridge opening than a carbonyl bridge in structurally analogous systems (*298*).

4. A series of μ-methylene cobalt complexes has been synthesized through carbene addition to the metal–metal "double" bond of $[(\eta^5\text{-}C_5Me_5)Co(\mu\text{-}CO)]_2$ (**132**). These reactions require higher temperatures than the analogous synthesis of the isostructural rhodium complexes, but once again proceed in practically quantitative yields. The structural dependence of these and similar μ-methylene complexes upon the nature of the metals, the methylene bridges, and the peripheral ligands has been discussed (*296, 299*).

5. A significant contribution to the catalysis-related problem of stepwise reduction of coordinated carbon monoxide to a metal-attached methylene unit has been reported by Steinmetz and Geoffroy (*300*).

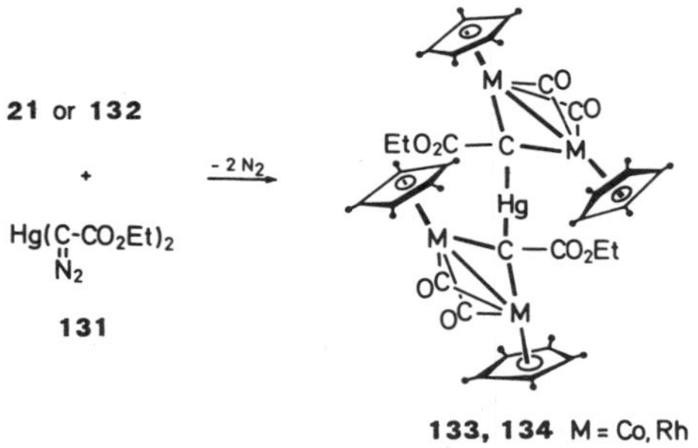

21 or 132

+

$Hg(\overset{\text{O}}{\underset{N_2}{C}}\text{-}CO_2Et)_2$

131

$\xrightarrow{-2N_2}$

133, 134 M = Co, Rh

Scheme 36

$Os_3(CO)_{12}$ reacts with the well-known hydride transfer reagent K[BH(O-i-$C_3H_7)_3$] in tetrahydrofuran at 0°C to give first the expected formyl anion $[Os_3(CO)_{11}(CHO)]^-$ (characterization in solution by IR and ^{1}H NMR). Subsequent treatment of this species with Brønsted acids (H_3PO_4 or CF_3COOH) gives the neutral μ-methylene cluster $Os_3(CO)_{11}(\mu$-$CH_2)$ in 20–30% yield. Deuterium-labeling experiments using Li[BDEt$_3$] and CF_3COOD support a mechanism that involves both a hydroxycarbene and a hydroxymethyl intermediate acting as the immediate precursors of the final CH_2 complex. Note that according to this mechanism, both methylene hydrogens derive from the borohydride reducing agent, and that the function of the acid is to remove the carbonyl oxygen, presumably as H_2O (the idealized net reaction is shown below the rule):

$$Os_3(CO)_{12} + [BHR_3]^- \rightarrow [Os_3(CO)_{11}(CHO)]^- + \text{``}BR_3\text{''}$$
$$[Os_3(CO)_{11}(CHO)]^- + H^+ \rightarrow Os_3(CO)_{11}(=CHOH)$$
$$Os_3(CO)_{11}(=CHOH) + [Os_3(CO)_{11}(CHO)]^- \rightarrow [Os_3(CO)_{11}(CH_2OH)]^- + Os_3(CO)_{12}$$
$$[Os_3(CO)_{11}(CH_2OH)]^- + H^+ \rightarrow Os_3(CO)_{11}(\mu\text{-}CH_2) + H_2O$$

$$\overline{Os_3(CO)_{12} + 2\,[BHR_3]^- + 2\,H^+ \rightarrow Os_3(CO)_{11}(\mu\text{-}CH_2) + H_2O + 2\,\text{``}BR_3\text{''}}$$

6. The results of an independent investigation of $(\mu$-$CH_2)[(\eta^5$-$C_5H_5)Mn(CO)_2]_2$ by means of UV vapor-phase photoelectron spectroscopy and CNDO quantum chemical calculations have been reported by Granozzi *et al.* (*301*).

7. The first paramagnetic metal methylene complex of the proposed structure $(\eta^5$-$C_5H_5)Ti(\mu$-$CH_2)(\mu$-$Cl)Ti(\eta^5$-$C_5H_5)$ has been detected by ESR spectroscopy (*302*). This compound reacts with terminal olefins of the type CH_2=CR_2 with exchange of the metal-bound CH_2 fragment with the terminal CH_2 group of the olefin.

8. The "lactone" complexes $(C_4O_2R^1,R^2)Co_2(CO)_7$ react with alkynes $R^3C{\equiv}CR^4$ to yield the known complexes $(\mu,\eta^2$—$R^3C{\equiv}CR^4)Co_2(CO)_6$ and two isomeric compounds 135 and 136 of composition $(C_4O_2R^1,R^2)$-

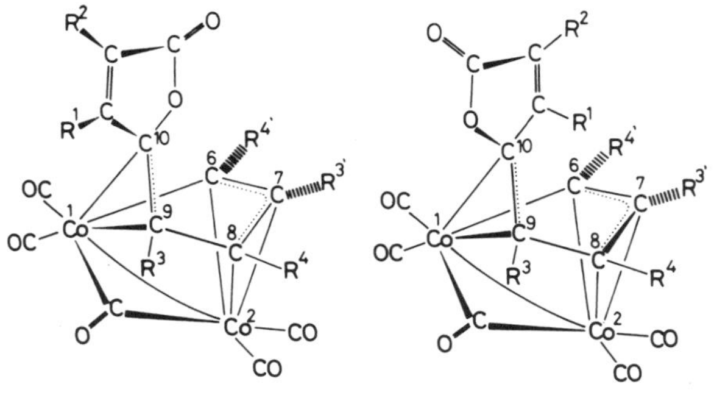

135 136

$(R^3C{\equiv}CR^4)Co_2(CO)_5$. Spectroscopic results and an X-ray structural study show that both isomers contain metallacyclic frameworks (303).

9. Stone and co-workers (304) have synthesized the first μ-CH_2 complexes of a tetranuclear heterometallic cluster. Treatment of $[Os_3Pt(\mu$-$H)_2(CO)_{10}\{P(C_6H_{11})_3\}]$ with CH_2N_2 under kinetic control affords a single orange isomer $[Os_3Pt(\mu$-$H)_2(\mu$-$CH_2)(CO)_{10}\{P(C_6H_{11})_3\}]$ which in solution isomerizes to a second red isomer. The structures in the solid state of both isomers have been determined by X-ray crystallography, and in solution by ^1H- and ^2H-NMR spectroscopy. The spectroscopic studies suggest that the orange **137** and red **138** isomers are formed from $[Os_3Pt(\mu$-$H)_2(CO)_{10}\{P(C_6H_{11})_3\}]$ (**139**) via the steps shown in Scheme 37. The transient μ-CH_3 species are postulated to account for $^2H/^1H$ scrambling between μ-CH_2 and μ-H sites.

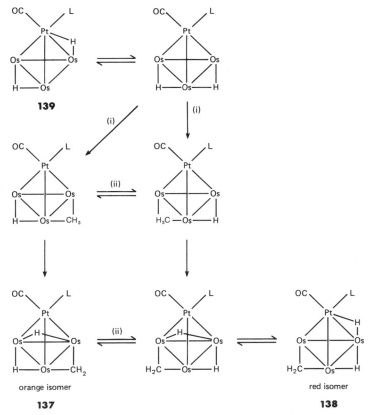

SCHEME 37. L = $P(C_6H_{11})_3$, CO ligands omitted for clarity. (i) CH_2N_2, (ii) rotation of $Pt(CO)\{P(C_6H_{11})_3\}$ fragment about an axis perpendicular to the Os_3 plane.

10. The trimethylaluminium method (see Section IV,B,4) was successfully employed for the synthesis of the mixed methyl/μ-methylene rhodium complex $[(\eta^5\text{-}C_5Me_5)Rh(\mu\text{-}CH_2)CH_3]_2$ which exists in both the cis and trans isomers (yields 15 and 28%, respectively). The magnitude of the ^{1}H- and ^{13}C-NMR chemical shifts of the methylene bridges are once again in excellent agreement with the general observations reported in Section VII. Notably, the cis isomer was easier to pyrolyse than the trans complex, but the gaseous products resulting from thermolysis exhibit about the same composition of individual hydrocarbons (48% CH_4, 20% C_2H_4, 2% C_2H_6, 30% C_3H_6); no cyclopropane was detected in any decomposition, with propane and C_4 hydrocarbons being produced in only negligible amounts (305). The relevance of the decomposition pathways to mechanistic issues of the *Fischer–Tropsch* process has been emphasized.

11. Balch *et al.* (306) have given a detailed report on novel A-frame-type methylene bridged palladium complexes of composition $Pd_2(dpm)_2(\mu\text{-}CHR)X_2$ (X = I, Br, Cl; R = H, CH_3) and $[Pd_2(dpm)_2(\mu\text{-}CH_2)L_2]^{2+}$ (L = pyridine, methylisocyanide). It was demonstrated that the Pd—C (methylene) bond resists insertion of carbon monoxide, isocyanides, or sulfur dioxide. Protonation of $Pd_2(dpm)_2(\mu\text{-}CH_2)I_2$ with fluoroboric acid yields the compound $[Pd_2(dpm)_2(\mu\text{-}I)(CH_3)I]BF_4$ in which the bridging methylene group of the precursor has been converted into a terminal methyl group, a process that has also been encountered in a previous example (52).

12. The well-established diazoalkane route (see Section IV,A,1) as well as the phosphorous ylid $(C_6H_5)_3P{=}CH_2$ have been used to prepare some new dinuclear μ-methylene ruthenium complexes (cf. CRR' = C(H)CO$_2$Et, C(C$_6$H$_5$)$_2$, CH$_2$) from the dimetallacycle $(\eta^5\text{-}C_5H_5)_2Ru_2(CO)(\mu\text{-}CO)\{\mu\text{-}C(O)C_2(C_6H_5)_2\}$ in good yields (307).

13. A rather striking example of a μ-dicyanomethylene gold complex, $[\mu\text{-}C(CN)_2][Au\{P(C_6H_5)_3\}]_2$, has been synthesized by Russian workers who, however, did not realize this compound to represent a methylene-bridged organometal (308). The geometrical data of the dimetallacyclopropane skeleton are typical of many other compounds as discussed in Section VI of this review [d(Au—Au) = 291.2(1) pm; d(Au—C) = 209.5(9) and 210.7(9), respectively; $\sphericalangle$(Au,C,Au') = 87.7(3), $\sphericalangle$(CN,C,CN) = 113.4(8)].

14. The μ-propylidene(2) complex $[\mu\text{-}C(CH_3)_2](\mu\text{-}CO)[(\eta^5\text{-}C_5H_5)\text{-}Ru(CO)]_2$ has been established by X-ray diffraction methods to exist as a *cis* isomer in the crystalline state. However, both *cis* and *trans* isomers were detected in solution by NMR techniques. The interconversion of these two compounds is concluded to involve bridge-terminal carbene

exchange (309), a process that has not been observed in any previous μ-methylene compound (see Section VIII,A).

15. The first example of irreversible methylene bridge-opening induced by a nonprotic reagent has been observed with the μ-diphenylmethylene rhodium compound $[\mu\text{-}C(C_6H_5)_2][\mu\text{-}CO][(\eta^5\text{-}C_5Me_5)_2Rh_2]$. Addition of sulfur dioxide instantaneously gives the derivative $(\mu\text{-}SO_2)[(\eta^5\text{-}C_5Me_5)_2\text{-}Rh_2(CO)\{C(C_6H_5)_2\}]$ having terminal diphenylcarbene and carbon monoxide ligands. A possible explanation of this unusual addition/rearrangement process is the rather high π-acceptor capability of sulfur dioxide that retains this ligand in the bridging position. The formation of the above-mentioned compound has been shown by low-temperature IR measurements ($-80°C$) to proceed through initial addition of SO_2 to the Rh≡Rh double bond of the precursor compound, with the bridge-opening process occuring subsequently at temperatures above approximately $-40°C$. Consistent with this observation is the high-yield formation of the stable isomeric cobalt complex of composition $(\mu\text{-}CO)(\mu\text{-}CH_2)(\mu\text{-}SO_2)[(\eta^5\text{-}C_5Me_5)_2Co_2]$ from the Co≡Co double-bonded methylene compound $(\mu\text{-}CO)(\mu\text{-}CH_2)[(\eta^5\text{-}C_5Me_5)_2Co_2]$ and sulfur dioxide; as expected from the higher stability of triply-bridged cobalt complexes compared to their rhodium congeners, no ring opening was observed with this particular derivative (310, 325).

16. An extended series of methylene cobalt complexes of type $(\mu\text{-}CRR')[(\eta^5\text{-}C_5Me_5)Co(CO)]_2$ has been disclosed by our group (311). Thus the parent compound with R,R' = H can be made from $[(\eta^5\text{-}C_5Me_5)Co(\mu\text{-}CO)]_2$ and diazomethane; clean decarbonylation yielding $(\mu\text{-}CH_2)(\mu\text{-}CO)[(\eta^5\text{-}C_5Me_5)_2Co_2]$ is achieved by heating the dicarbonyl precursor in boiling THF (yield 90%) (310, 311, 325).

17. The bridging carbonyl group of the propylidene(2) ruthenium complex $[\mu\text{-}C(CH_3)_2](\mu\text{-}CO)[(\eta^5\text{-}C_5H_5)Ru(CO)]_2$ has been converted, via a μ-ethylidyne species, to an ethylidene bridge by successive treatment with methyl lithium and tetrafluoroboric acid. The structure of the resulting bis(μ-alkylidene) derivative $[\mu\text{-}C(H)CH_3][\mu\text{-}C(CH_3)_2][(\eta^5\text{-}C_5H_5)\text{-}Ru(CO)]_2$ has been determined by X-ray diffraction techniques (trans configuration, symmetrical bridges, $d(Ru\text{—}Ru') = 270.1(1)$ pm, $d(Ru\text{—}C)$ bridges = 207.3(3)–210.8(3) pm; "internal angles α" [see Section VI] = 79.7(1)° and 81.0(1)°, respectively; mutually coplanar dimetallacyclopropane frames, with the interplanar angle amounting to 177°). There is no evidence of fluxional trans ⇌ cis isomerism in solution (NMR, ≤135°C) (312). Decomposition at 200°C yields ethylene (ex μ-C(H)CH_3), propene (ex μ-C(CH_3)_2), but the major hydrocarbon product is 1-methyl-2-butene (~70%) which results from coupling of the two different carbenes present in the complex. The absence of butenes(2) or tet-

ramethylethylene indicates that the predominant pyrolysis product $(CH_3)_2C=C(H)CH_3$ is formed *intra*molecularly and suggests the viability of metal-assisted methylene/methylene combination in catalytic processes such as the Fischer–Tropsch synthesis (*312, 326*).

18. The French group of Rudler has reported several insertion reactions of alkynes into a μ-alkylidene carbon tungsten bond (*313, 314*).

19. There is a conference report on the preparation of the methylene-bridged heterodinuclear species $(\eta^5\text{-}C_5H_5)_2Zr(\mu\text{-}Cl)(\mu\text{-}CH_2)RuCl[P(C_6H_5)_3]_2$. This compound results in yields of about 10% from the photoassisted reaction of dimethyl zirconocene with $RuCl_2[P(C_6H_5)_3]_3$ (toluene, $+25°C$) under elimination of methane. Although the mechanism of this conversion remains open to further debate, the final compound represents the first example of a μ-methylene complex having both an early transition metal and a late transition metal attached to the CH_2 bridge (*315*). Hydrogen induces irreversible rupture of the Ru—Cl— and the Ru—CH_2 bonds even at atmospheric pressure, with $(\eta^5\text{-}C_5H_5)_2Zr(CH_3)Cl$ being one of the mononuclear products formed in this degradation reaction.

20. The spectroscopic data of the bis(μ-methylene) rhodium complex $[(1.5\text{-}COD)Rh(\mu\text{-}CH_2)]_2$ have now been reported. The ^{1}H-NMR chemical shift of the methylene bridge ($\delta CH_2 = 7.3$; $^2J_{Rh,H} = 1.5$ Hz) once again is in good agreement with the general rules outlined in Section VII. The same is true for the analogous iridium derivative of composition $[(1.5\text{-}COD)Ir(\mu\text{-}CH_2)]_2$ ($\delta \underline{CH_2} = 7.75$ [toluene-d_8; $-5°C$]; δCH_2 145.0 [benzene-d_6; $+20°C$]). The synthesis of both these compounds is interesting in that they derive from the corresponding μ-methyl species by destructive α-hydrogen abstraction followed by hydrogen, *not* methane, elimination (*316*).

21. The "γ-lactone-type" cobalt complexes described in Section III,A have further been studied with regard to spectroscopy (*317*), reactivity (*318*), and their catalytic activity for carbon monoxide/alkyne reactions (*319*).

22. The Bergman group reports on several interesting observations in the field of methylenecobalt chemistry. Thus a number of terminal and internal μ-alkylidene complexes in the cobalt series has been prepared. These complexes have the empirical formula $(\mu\text{-}CR^1R^2)[(\eta^5\text{-}C_5H_5)Co(CO)]_2$. Isomers having doubly bridging carbonyl groups have been detected spectroscopically, the relative amounts of each isomer depending in a sensitive way upon the alkylidene substituents. Complexes in which both R^1 and R^2 are alkyl groups lose one molecule of CO, forming the corresponding unsaturated dinuclear complexes. Upon thermolysis, the above-mentioned internal alkylidene complexes (R^1 and $R^2 = n$-alkyl)

lead to alkenes. However, they also undergo a surprising rearrangement to terminal alkylidene complexes (R^1 = H, R^2 = n-alkyl′), according to the general scheme

$$
\begin{array}{ccc}
R^1\diagdown\quad\diagup CH_2-R^2 & & R^1-CH_2\diagdown\quad\diagup R^2 \\
C & \rightleftharpoons & C \\
M\!\!-\!\!\!-\!\!\!-\!\!M & & M\!\!-\!\!\!-\!\!\!-\!\!M
\end{array}
$$

Isotope labeling studies have provided evidence that both metal/hydrogen migration and π/π-allyl interconversion are involved as primary steps in this rearrangements (*320*). Another paper reports the synthesis and reactions of the first benzodimetallacyclohexene, a rare example of a six-membered metallacycle containing two metals in the ring. Alkylation of $Na^+[Cp_2Co_2(CO)_2]^-$ (Cp = η^5-C_5H_5) with α,α'-dibromo-o-xylene gives the desired metallacycle, benzo-2,3-bis(η^5-cyclopentadienyl-μ-carbonyl-

$$
(\eta^5\text{-}C_5H_5)Co\underset{\underset{O}{\overset{\displaystyle\|}{C}}}{\overset{CH_2}{\diagup}}\!\!-\!\!\underset{\underset{O}{\overset{\displaystyle\|}{C}}}{\overset{CH_2}{\diagdown}}Co(\eta^5\text{-}C_5H_5)
\qquad\qquad
(\eta^5\text{-}C_5H_5)Co\!\!\begin{array}{c}\diagup CH_2\\ \diagdown CH_2\end{array}
$$

cobalta)-cyclohexene. This compound undergoes slow decomposition in solution to give $CpCo(CO)_2$ and the previously unknown mononuclear complex (η^5-cyclopentadienyl)(o-xylylene)cobalt (*321*). In addition, principal chemical features of $(\mu\text{-}CH_2)[(\eta^5\text{-}C_5H_5)Co(CO)]_2$ were studied (*322*).

23. Finally, mention should be made of the synthesis of the trinuclear μ-methylene anion of composition $[\{\mu\text{-}C(H)CO_2C_2H_5\}\{Fe_3(CO)_{10}(\mu\text{-}H)\}]^-$. Once again, the diazoalkane has been used to add the carbene to the corresponding metalcarbonyl framework of the $[HFe_3(CO)_{11}]^-$ precursor. The structure of the product was inferred from spectroscopic data (*323*).

The voluminous Addendum Section covers only a relatively short period of time (six months), and hence convincingly demonstrates how incredibly fast the methylene bridge is extending these days. A few more papers had appeared when this review approached the final page proof stage (*327–333*). Another review article covering this field undoubtedly seems desirable in due course.

REFERENCES

1. E. O. Fischer and A. Maasböl, *Angew. Chem.* **76**, 745 (1964); *Angew. Chem., Int. Ed. Engl.* **3**, 580 (1964).
2. E. O. Fischer, G. Kreis, C. G. Kreiter, J. Müller, G. Huttner, and H. Lorenz, *Angew. Chem.* **85**, 618 (1973); *Angew. Chem., Int. Ed. Engl.* **12**, 564 (1973).
3. E. O. Fischer, *Pure Appl. Chem.* **24**, 407 (1970); **30**, 353 (1972).
4. F. A. Cotton and C. M. Lukehardt, *Prog. Inorg. Chem.* **16**, 487 (1972).
5. For review, see D. J. Cardin, B. Cetinkaya, and M. F. Lappert, *Chem. Rev.* **72**, 545 (1972); see also M. F. Lappert, *J. Organomet. Chem.* **100**, 139 (1975).
6. For review, see E. O. Fischer, *Adv. Organom. Chem.* **14**, 1 (1976); see also E. O. Fischer, *Angew. Chem.* **86**, 651 (1974) (Nobel Lect., Stockholm, 1973).
7. For review, see E. O. Fischer and U. Schubert, *J. Organomet. Chem.* **100**, 59 (1975).
8. For review, see W. A. Herrmann, *Angew. Chem.* **90**, 855 (1978); *Angew. Chem., Int. Ed. Engl.* **17**, 800 (1978).
9. For review, see R. R. Schrock, *Acc. Chem. Res.* **12**, 98 (1979); see also R. R. Schr ck and P. R. Sharp, *J. Am. Chem. Soc.* **100**, 2389 (1978).
10. F. Huq, W. Mowat, A. C. Skapski, and G. Wilkinson, *J. Chem. Soc. Chem. Comn un.* p. 1477 (1971).
11. W. Mowat and G. Wilkinson, *J. Chem. Soc. Trans. Dalton* p. 1120 (1973).
12. W. A. Herrmann, *Angew. Chem.* **86**, 895 (1974); *Angew. Chem. Int. Ed. Engl.* **13**, 812 (1974).
13. W. A. Herrmann, *J. Organomet. Chem.* **97**, 1 (1975).
14. R. Andersen, A. L. Galyer, and G. Wilkinson, *Angew. Chem.* **88**, 692 (1976); *Angew. Chem. Int. Ed. Engl.* **15**, 609 (1976).
15. E. O. Fischer, G. Huttner, T. L. Lindner, A. Frank, and F. R. Kreißl, *Angew. Chem.* **88**, 228 (1976); *Angew. Chem., Int. Ed. Engl.* **15**, 231 (1976).
16. M. H. Chisholm, F. A. Cotton, M. Extine, and B. R. Stults, *Inorg. Chem.* **15**, 2252 (1976).
17. H. Schmidbauer, W. Scharf, and H. J. Füller, *Z. Naturforsch. Teil B* **32**, 858 (1977).
18. M. H. Chisholm, F. A. Cotton, M. Extine, and C. A. Murillo, *Inorg. Chem.* **17**, 696 (1978).
19. M. Nitay, W. Priester, and M. Rosenblum, *J. Am. Chem. Soc.* **100**, 3620 (1978).
20. L. N. Lewis, J. C. Huffman, and K. G. Caulton, *J. Am. Chem. Soc.* **102**, 403 (1980).
21. B. A. Dolgoplosk, E. I. Tinyakova, I. S. Guzman, E. L. Vollerstein, N. N. Chigir, G. N. Bondarenko, O. K. Sharaev, and V. A. Yakovlev, *J. Organomet. Chem.* **201**, 249 (1980).
22. M. Bochmann, G. Wilkinson, A. M. R. Gales, M. B. Hursthouse, and G. Wilkinson, *J. Chem. Soc. Dalton Trans.* p. 1799 (1980).
23. E. O. Fischer, W. Kellerer, B. Zimmer-Gasser, and U. Schubert, *J. Organomet. Chem.* **199**, C24 (1980).
24. T. V. Ashworth, J. A. K. Howard, and F. G. A. Stone, *J. Chem. Soc. Dalton Trans.* p. 1609 (1980).
25. M. J. Chetcuti, M. Green, J. C. Jeffery, F. G. A. Stone, and A. A. Wilson, *J. Chem. Soc. Chem. Commun.* p. 948 (1980).
26. T. V. Ashworth, M. J. Chetcuti, J. A. K. Howard, F. G. A. Stone, S. J. Wisbey, and P. Woodward, *J. Chem. Soc. Dalton Trans.*, in press (1982).
27. J. A. K. Howard, J. C. Jeffery, M. Laguna, R. Navarro, and F. G. A. Stone, *J. Chem. Soc. Dalton Trans.*, in press (1982).
28. D. F. Shriver, *J. Organomet. Chem.* **94**, 259 (1975), and references cited therein.
29. F. A. Hodali and D. F. Shriver, *Inorg. Chem.* **18**, 1236 (1979).

30. H. A. Hodali, D. F. Shriver, and C. A. Ammlung, *J. Am. Chem. Soc.* **100**, 5239 (1978).
31. D. N. Clark and R. R. Schrock, *J. Am. Chem. Soc.* **100**, 6774 (1978).
32. J. B. Keister, *J. Organomet. Chem.* **190**, C36 (1980).
33. For review, see B. R. Penfold and B. H. Robinson, *Acc. Chem. Res.* **6**, 73 (1973).
34. For review, see D. Seyferth, *Adv. Organomet. Chem.* **14**, 97 (1976).
35. For recent examples, see J. B. Keister and T. L. Horling, *Inorg. Chem.* **19**, 2304 (1980); J. R. Fritch and K. P. C. Vollhardt, *Angew. Chem.* **92**, 570 (1980); *Angew. Chem., Int. Ed. Engl.* **19**, 559 (1980); M. J. Chetcuti, M. Green, J. A. K. Howard, J. C. Jeffery, G. N. Pain, S. J. Porter, F. G. A. Stone, and A. A. Wilson, *J. Chem. Soc. Chem. Commun.* p. 1057 (1980); W. I. Bailey, Jr., F. A. Cotton, and J. D. Jamerson, *J. Organomet. Chem.* **173**, 317 (1977); H. Beurich and H. Vahrenkamp, *Angew. Chem.* **90**, 915 (1978); *Angew. Chem. Int. Ed. Engl.* **17**, 863 (1978); see also P. Mitrprachachon, Ph.D. Thesis, Bristol Univ. (1980); A. J. Canty, B. F. G. Johnson, J. Lewis, and J. R. Norton, *J. Chem. Soc. Chem. Commun.* p. 1331 (1972); T. I. Voyevodskaya, I. M. Pribytkova, and Y. A. Ustynyuk, *J. Organomet. Chem.* **37**, 187 (1972); G. Fachinetti, *Angew. Chem.* **91**, 657 (1979); *Angew. Chem. Int. Ed. Engl.* **18**, 619 (1979); R. B. King and C. A. Harmon, *Inorg. Chem.* **15**, 879 (1976); G. Fachinetti, *J. Chem. Soc. Chem. Commun.* p. 397 (1979).
36. M. Tachikawa and E. L. Muetterties, *J. Am. Chem. Soc.* **102**, 4541 (1980).
37. M. A. Beno, J. M. Williams, M. Tachikawa, and E. L. Muetterties, *J. Am. Chem. Soc.* **102**, 4542 (1980).
38. For review, see E. L. Muetterties, *J. Organomet. Chem.* **200**, 177 (1980).
39. W. A. Herrmann, B. Reiter, and H. Biersack, *J. Organomet. Chem.* **97**, 245 (1975).
40. W. A. Herrmann, C. Krüger, R. Goddard, and I. Bernal, *Angew. Chem.* **89**, 342 (1977); *Angew. Chem., Int. Ed. Engl.* **16**, 334 (1977).
41. W. A. Herrmann, C. Krüger, R. Goddard, and I. Bernal, *J. Organomet. Chem.* **140**, 73 (1977).
42. W. A. Herrmann, *Chem. Ber.* **111**, 1077 (1978).
43. W. A. Herrmann, I. Schweizer, M. Creswick, and I. Bernal, *J. Organomet. Chem.* **165**, C17 (1979).
44. W. A. Herrmann and I. Schweizer, *Z. Naturforsch. Teil B* **33**, 1128 (1978).
45. M. Creswick, I. Bernal, W. A. Herrmann, and I. Steffl, *Chem. Ber.* **113**, 1377 (1980).
46. K. K. Mayer and W. A. Herrmann, *J. Organomet. Chem.* **182**, 361 (1979).
47. M. Creswick, I. Bernal, and W. A. Herrmann, *J. Organomet. Chem.* **172**, C39 (1979).
48. W. A. Herrmann, J. Plank, I. Bernal, and M. Creswick, *Z. Naturforsch., Teil B* **35**, 680 (1980).
49. W. A. Herrmann, J. Plank, E. Guggolz, and M. L. Ziegler, *Angew. Chem.* **92**, 660 (1980); *Angew. Chem., Int. Ed. Engl.* **19**, 651 (1980).
50. W. A. Herrmann, J. Plank, and D. Riedel, *J. Organomet. Chem.* **190**, C47 (1980).
51. W. A. Herrmann, J. Plank, M. L. Ziegler, and B. Balbach, *J. Am. Chem. Soc.* **102**, 5906 (1980).
52. W. A. Herrmann, J. Plank, D. Riedel, M. L. Ziegler, K. Weidenhammer, E. Guggolz, and B. Balbach, *J. Am. Chem. Soc.* **103**, 63 (1981).
53. W. A. Herrmann, J. Plank, M. L. Ziegler, and P. Wülknitz, *Chem. Ber.* **114**, 716 (1981).
54. W. A. Herrmann, C. Bauer, J. Plank, W. Kalcher, D. Speth, and M. L. Ziegler, *Angew. Chem.* **93**, 272 (1981); *Angew. Chem., Int. Ed. Engl.* **20**, 193 (1981).
54a. W. A. Herrmann, J. Plank, C. Bauer, M. L. Ziegler, E. Guggolz, and R. Alt, *Z. Anorg. Allgem. Chem.*, in press (1982).
55. W. A. Herrmann, C. Bauer, G. Kriechbaum, H. Kunkely, M. L. Ziegler, D. Speth, and E. Guggolz, *Chem. Ber.*, in press (1982).

56. F. Takusagawa, A. Fumagalli, T. F. Koetzle, and W. A. Herrmann, *Inorg. Chem.* **20**, 3060 (1981).

57. M. Creswick, I. Bernal, B. Reiter, and W. A. Herrmann, *Inorg. Chem.*, in press.

58. W. A. Herrmann, *Jahrb. Akad. Wiss. (Göttingen)* p. 15 (1979).

59. W. A. Herrmann and C. Bauer, *J. Organomet. Chem.* **204**, C21 (1981).

59a. W. A. Herrmann and C. Bauer, *Chem. Ber.* **115**, 14 (1982).

60. W. A. Herrmann, G. Kriechbaum, C. Bauer, M. L. Ziegler, and E. Guggolz, *Angew. Chem.*, in press.

61. W. A. Herrmann, in "Transition Metal Chemistry, Current Problems and the Biological As Well As the Catalytical Relevance," (A. Müller and E. Diemann, eds.), pp. 127–156. Verlag Chemie, Weinheim, 1981.

61a. W. A. Herrmann, *Pure Appl. Chem.*, in press (1982).

62. D. L. Davies, A. F. Dyke, A. Endesfelder, S. A. R. Knox, P. J. Naish, A. G. Orpen, D. Plaas, and G. E. Taylor, *J. Organomet. Chem.* **198**, C43 (1980).

63. A. F. Dyke, S. A. R. Knox, and P. J. Naish, *J. Organomet. Chem.* **199**, C47 (1980).

64. A. F. Dyke, S. A. R. Knox, P. J. Naish, and A. G. Orpen, *J. Chem. Soc. Chem. Commun.* p. 441 (1980).

65. A. F. Dyke, S. A. R. Knox, P. J. Naish, and G. E. Taylor, *J. Chem. Soc. Chem. Commun.* p. 803 (1980).

66. J. A. Beck, S. A. R. Knox, G. H. Riding, G. E. Taylor, and M. J. Winter, *J. Organomet. Chem.* **202**, C49 (1980).

67. S. A. R. Knox, personal communication (1980); see also A. F. Dyke *et al.*, *J. Chem. Soc. Chem. Commun.* p. 537 (1981).

68. J. Levisalles, H. Rudler, Y. Jeannin, and F. Dahan, *J. Organomet. Chem.* **178**, C8 (1979).

69. J. Levisalles, H. Rudler, Y. Jeannin, and F. Dahan, *J. Organomet. Chem.* **187**, 233 (1980).

70. J. Levisalles, H. Rudler, F. Dahan, and Y. Jeannin, *J. Organomet. Chem.* **188**, 193 (1980).

71. J. Levisalles, F. Rose-Munch, H. Rudler, J. C. Daran, Y. Dromzee, and Y. Jeannin, *J. Chem. Soc. Chem. Commun.* p. 685 (1980).

72. C. E. Sumner, Jr., P. E. Riley, R. E. Davis, and R. Pettit, *J. Am. Chem. Soc.* **102**, 1752 (1980).

73. R. C. Brady, III and R. Pettit, *J. Am. Chem. Soc.* **102**, 6181 (1980).

74. R. Pettit, *Int. Symp. Homogeneous Catal., 2nd, Düsseldorf, 1980*.

75. R. B. Calvert and J. R. Shapley, *J. Am. Chem. Soc.* **99**, 5225 (1977).

76. R. B. Calvert, J. R. Shapley, A. J. Schultz, J. M. Williams, S. L. Suib, and G. D. Stucky, *J. Am. Chem. Soc.* **100**, 6240 (1978).

77. R. B. Calvert and J. R. Shapley, *J. Am. Chem. Soc.* **100**, 6544 (1978).

78. R. B. Calvert and J. R. Shapley, *J. Am. Chem. Soc.* **100**, 7726 (1978).

79. R. J. Lawson and J. R. Shapley, *Inorg. Chem.* **17**, 2963 (1978).

80. A. J. Schultz, J. M. Williams, R. B. Calvert, J. R. Shapley, and G. D. Stucky, *Inorg. Chem.* **18**, 319 (1979).

81. A. D. Clauss, P. A. Dimas, and J. R. Shapley, *J. Organomet. Chem.* **201**, C31 (1980).

82. P. A. Dimas, E. N. Duesler, R. J. Lawson, and J. R. Shapley, *J. Am. Chem. Soc.* **102**, 7787 (1980).

83. J. R. Shapley *et al.*, unpublished observations (1980).

84. J. Cooke, W. R. Cullen, M. Green, and F. G. A. Stone, *J. Chem. Soc. Chem. Commun.* p. 170 (1968); *J. Chem. Soc. A* p. 1872 (1969).

85. T. V. Ashworth, M. Berry, J. A. K. Howard, M. Laguna, and F. G. A. Stone, *J. Chem. Soc. Chem. Commun.* p. 43 (1979).

86. T. V. Ashworth, J. A. K. Howard, M. Laguna, and F. G. A. Stone, *J. Chem. Soc. Dalton Trans.* p. 1593 (1980).
87. M. Berry, J. A. K. Howard, and F. G. A. Stone, *J. Chem. Soc. Dalton Trans.* p. 1601 (1980).
88. T. V. Ashworth, M. Berry, J. A. K. Howard, M. Laguna, and F. G. A. Stone, *J. Chem. Soc. Dalton Trans.* p. 1615 (1980).
89. M. Berry, J. Martin-Gil, J. A. K. Howard, and F. G. A. Stone, *J. Chem. Soc. Dalton Trans.* p. 1625 (1980).
90. J. A. K. Howard, K. A. Mead, J. R. Moss, R. Navarro, F. G. A. Stone, and P. Woodward, *J. Chem. Soc. Dalton Trans.* p. 743 (1981).
91. G. M. Dawkins, M. Green, J. C. Jeffery, and F. G. A. Stone, *J. Chem. Soc. Chem. Commun.* p. 1120 (1980).
92. For a recent review, see T. V. Ashworth, M. J. Chetcuti, L. J. Farrugia, J. A. K. Howard, J. C. Jeffery, R. Mills, G. N. Pain, F. G. A. Stone, and P. Woodward, *ACS Symp. Ser.* No. 155, pp. 299–313 (1981).
93. J. Martin-Gil, J. A. K. Howard, R. Navarro, and F. G. A. Stone, *J. Chem. Soc. Chem. Commun.* p. 1168 (1979).
94. N. M. Boag, M. Green, R. M. Mills, G. N. Pain, F. G. A. Stone, and P. Woodward, *J. Chem. Soc. Chem. Commun.* p. 1171 (1980).
95. R. A. Anderson, R. A. Jones, G. Wilkinson, M. B. Hursthouse, and K. M. Abdul Malik, *J. Chem. Soc. Chem. Commun.* p. 865 (1977).
96. M. B. Hursthouse, R. A. Jones, K. M. Abdul Malik, and G. Wilkinson, *J. Am. Chem. Soc.* **101**, 4128 (1979).
97. R. A. Jones, G. Wilkinson, A. M. R. Galas, M. B. Hursthouse, and K. M. Abdul Malik, *J. Chem. Soc. Dalton Trans.* p. 1771 (1980).
98. K. H. Theopold and R. G. Bergman, *J. Am. Chem. Soc.* **102**, 5694 (1980).
99. F. N. Tebbe, *J. Am. Chem. Soc.* **100**, 3611 (1978).
100. F. N. Tebbe, G. W. Parshall, and G. S. Reddy, *J. Am. Chem. Soc.* **100**, 3611 (1978).
101. F. N. Tebbe and R. L. Harlow, *J. Am. Chem. Soc.* **102**, 6149 (1980).
102. M. P. Brown, J. R. Fisher, J. S. Franklin, R. J. Puddephatt, and K. R. Seddon, *J. Chem. Soc. Chem. Commun.* p. 749 (1978).
103. M. P. Brown, J. R. Fisher, R. J. Puddephatt, and K. R. Seddon, *Inorg. Chem.* **18**, 2808 (1979).
104. R. Korswagen, R. Alt, D. Speth, and M. L. Ziegler, *Angew. Chem.* **93**, 1073 (1981); *Angew. Chem., Int. Ed. Engl.* **20**, 1049 (1981).
105. "International Union of Pure and Applied Chemistry, Nomenclature of Inorganic Chemistry," 2nd ed. Butterworth, London, 1971.
106. "International Union of Pure and Applied Chemistry, Nomenclature of Organic Chemistry," 2nd ed. Butterworth, London, 1971.
107. H. W. Sternberg, J. G. Shukys, C. D. Donne, R. Markby, R. A. Friedel, and J. Wender, *J. Am. Chem. Soc.* **81**, 2339 (1959).
108. G. Albanesi and M. Tovagliere, *Chim. Ind. (Milan)* **41**, 189 (1959).
109. G. Albanesi, *Chim. Ind. (Milan)* **46**, 1169 (1964).
110. G. Albanesi and E. Gavezzotti, *Chim. Ind. (Milan)* **47**, 1322 (1965).
111. G. Albanesi and E. Gavezzotti, *Atti Accad. Naz. Lincei, Cl. Sci. Fis., Mat. Nat., Rend.* **41**, 497 (1966).
112. H. Greenfield, H. W. Sternberg, R. A. Friedel, J. Wotiz, R. Markby, and H. Wender, *J. Am. Chem. Soc.* **78**, 120 (1956).
113. For review, see R. S. Dickson and P. J. Fraser, *Adv. Organomet. Chem.* **12**, 812 (1974).
114. O. S. Mills and G. Robinson, *Inorg. Chim. Acta* **1**, 61 (1967).

115. J. C. Sauer, R. D. Cramer, V. A. Engelhardt, T. A. Ford, H. E. Holmquist, and B. W. Howk, *J. Am. Chem. Soc.* **81**, 3677 (1959).
116. D. J. S. Guthrie, J. U. Khand, G. R. Knox, J. Kollmeier, P. L. Pauson, and W. E. Watts, *J. Organomet. Chem.* **90**, 93 (1975).
117. G. Pályi, G. Váradi, A. Vizi-Orosz, and L. Markó, *J. Organomet. Chem.* **90**, 85 (1975).
118. G. Váradi, I. Vecsei, I. Ötvös, G. Pályi, and L. Markó, *J. Organomet. Chem.*, in press (1982).
119. G. Váradi and G. Pályi, *Inorg. Chim. Acta* **20**, L33 (1976).
120. I. T. Horváth, G. Pályi, L. Markó, and G. Andretti, *J. Chem. Soc. Chem. Commun.* p. 1054 (1979).
121. P. A. Elder, D. J. S. Guthrie, J. A. D. Jeffreys, G. R. Knox, J. Kollmeier, P. L. Pauson, D. A. Symon, and W. E. Watts, *J. Organomet. Chem.* **120**, C13 (1976).
122. J. A. D. Jeffreys, *J. Chem. Soc. Dalton Trans.* p. 435 (1980).
123. G. Pályi and G. Váradi, *J. Organomet. Chem.* **86**, 119 (1975).
124. W. A. Herrmann, *Angew. Chem.* **86**, 556 (1974); *Angew. Chem., Int. Ed. Engl.* **13**, 599 (1974).
125. W. A. Herrmann, *Chem. Ber.* **108**, 486 (1975).
126. W. A. Herrmann and J. Plank, unpublished observations (1977–1980); J. Plank, Ph.D. Thesis, Univ. Regensburg (1980).
127. D. A. Clemente, B. Rees, G. Bandoli, M. C. Biagini, B. Reiter, and W. A. Herrmann, *Angew. Chem.* **93**, 920 (1981); *Angew. Chem., Int. Ed. Engl.* **20**, 887 (1981).
128. For a recent monograph, see M. Regitz, "Diazoalkane—Eigenschaften und Synthesen." Thieme, Stuttgart, 1977; see also references cited in Ref. 8.
129. (a) R. R. Schrock, *J. Am. Chem. Soc.* **97**, 6577 (1975); (b) R. R. Schrock and P. R. Sharp, *J. Am. Chem. Soc.* **100**, 2389 (1978).
130. M. D. Rausch, W. H. Boon, and H. G. Alt, *J. Organomet. Chem.* **141**, 299 (1977).
131. M. Green, J. A. K. Howard, A. Laguna, M. Murray, J. L. Spencer, and F. G. A. Stone, *J. Chem. Soc. Chem. Commun.* p. 451 (1975).
132. F. G. A. Stone, *J. Organomet. Chem.* **100**, 257 (1975).
133. B. L. Booth, R. N. Haszeldine, P. R. Mitchell, and J. J. Cox, *J. Chem. Soc. Chem. Commun.* p. 529 (1967).
134. W. A. Herrmann and I. Schweizer, *Z. Naturforsch. Teil B*, **33**, 911 (1978).
135. M. L. Ziegler, K. Weidenhammer, and W. A. Herrmann, *Angew. Chem.* **89**, 577 (1977); *Angew. Chem., Int. Ed. Engl.* **16**, 555 (1977).
136. W. A. Herrmann, I. Steffl, M. L. Ziegler, and K. Weidenhammer, *Chem. Ber.* **112**, 1731 (1979).
137. J. B. Keister and J. R. Shapley, *J. Am. Chem. Soc.* **98**, 1056 (1976).
138. (a) A. J. Deeming and M. Underhill, *J. Chem. Soc. Chem. Commun.* p. 277 (1973); (b) J. P. Yesinowski and D. Bailey, *J. Organomet. Chem.* **65**, C27 (1974).
139. For review, see E. L. Muetterties, *Science* **196**, 839 (1977).
140. E. L. Muetterties, *Angew. Chem.* **90**, 577 (1978); *Angew. Chem., Int. Ed. Engl.* **17**, 545 (1978).
141. For review, see E. L. Muetterties, *J. Organomet. Chem.* **200**, 177 (1980).
142. W. A. Herrmann and I. Bernal, *Angew. Chem.* **82**, 186 (1977); *Angew. Chem., Int. Ed. Engl.* **16**, 172 (1977).
143. I. Bernal, J. D. Korp, G. M. Reisner, and W. A. Herrmann, *J. Organomet. Chem.* **139**, 321 (1977).
144. W. P. Fehlhammer, W. A. Herrmann, and K. Öfele, *in* "Handbuch der Präparativen Anorganischen Chemie" (G. Brauer, ed.), 3rd ed., Vol. III, pp. 1799–2033. Enke, Stuttgart, 1981.
145. L. S. Pu and A. Yamamoto, *J. Chem. Soc. Chem. Commun.* p. 9 (1974).

146. For a comprehensive review of carbene olefin addition, see R. A. Moss, *in* "Carbenes" (M. Jones, Jr. and R. A. Moss, eds.), Vol. 1, p. 153. Wiley (Interscience), New York, 1973.
147. Concerning the problems connected with the assignment of metal–metal bond orders, see Ref. 54, ftn. 10; see also I. Bernal, M. Creswick, and W. A. Herrmann, *Z. Naturforsch, Teil B* **34**, 1345 (1979).
148. A. Nutton and P. M. Maitlis, *J. Organomet. Chem.* **166**, C21 (1979).
149. *Nachr. Chem. Tech. Lab.* **28**, 788 (1980).
150. P. Hofmann, *Angew. Chem.* **91**, 591 (1979); *Angew. Chem., Int. Ed. Engl.* **18**, 554 (1979).
150a. P. Hofmann, *Fres. Z. Analyt. Chem.* **304**, 262 (1980).
151. A. R. Pinhas, T. A. Albright, P. Hofmann, and R. Hoffmann, *Helv. Chim. Acta* **63**, 29 (1980).
152. J. Plank, Ph.D. Thesis, Univ. Regensburg, 1980.
153. W. A. Herrmann and C. Bauer, unpublished observations (1980); cf. C. Bauer, Dipl. Thesis, Univ. Regensburg (1981).
154. F. A. Cotton, personal communication (1979).
155. M. L. Ziegler and W. A. Herrmann *et al.*, unpublished observations (1979–1981).
156. W. A. Herrmann, *ACS Symp. Binucl. Met. Compd., New Orleans, La. 1980*.
157. M. L. Ziegler and W. A. Herrmann *et al.*, unpublished observations (1980) (X-ray structure of [(η^5-C$_5$Me$_5$)Rh(μ-CO)]$_2$); J. L. Calderon, S. Fontana, E. Frauendorfer, V. W. Day, and D. A. Iske, *J. Organomet. Chem.* **64**, C16 (1974). (X-Ray structure of [(η^5-C$_5$H$_5$)Fe(μ-NO)]$_2$.
158. R. Colton, M. J. McCormick, and C. D. Pannan, *Aust. J. Chem.* **31**, 1425 (1978); R. Colton, M. J. McCormick, and C. D. Pannan, *J. Chem. Soc. Chem. Commun.* p. 823 (1977); S. D. Robinson, *Inorg. Chim. Acta* **27**, L108 (1978).
159. Note, however, that neither Pd$_2$Cl$_2$ (μ-dpam)$_2$ nor Pt$_2$Cl$_2$ (μ-dpam)$_2$ contain a direct metal–metal bond, cf. M. P. Brown, A. N. Keith, L. Manojlović-Muir, K. W. Muir, R. J. Puddephatt, and K. R. Seddon, *Inorg. Chim. Acta* **34**, 1223 (1979).
160. M. P. Brown, R. J. Puddephatt, M. Rashidi, and K. R. Seddon, *J. Chem. Soc. Dalton Trans.* p. 516 (1978).
161. M. P. Brown, R. J. Puddephatt, M. Rashidi, and K. R. Seddon, *J. Chem. Soc. Dalton Trans.* p. 1540 (1978).
162. M. Cowie and T. G. Southern, *J. Organomet. Chem.* **193**, C46 (1980).
163. R. J. Klingler, W. M. Butler, and M. D. Curtis, *J. Am. Chem. Soc.* **100**, 5034 (1978).
164. M. D. Curtis, *Gordon Conf. Organomet. Chem., Andover, N.H., 1980*.
164a. L. Messerle and M. D. Curtis, *J. Am. Chem. Soc.* **102**, 7789 (1980).
165. W. A. Herrmann and G. Kriechbaum, unpublished observations (1980).
166. R. B. King and M. S. Saran, *J. Am. Chem. Soc.* **94**, 1784 (1972).
167. R. B. King and M. S. Saran, *J. Am. Chem. Soc.* **95**, 1811 (1973).
168. For the structure of the related μ-dicyanovinylidene complex, see R. M. Kirchner and J. A. Ibers, *J. Organomet. Chem.* **82**, 243 (1974).
169. F. Seel and G. V. Röschenthaler, *Angew. Chem.* **82**, 182 (1970); *Angew. Chem., Int. Ed. Engl.* **9**, 166 (1970).
170. F. Seel and G.-V. Röschenthaler, *Z. Anorg. Allg. Chem.* **386**, 297 (1971).
170a. F. Seel and R. D. Flaccus, *J. Fluorine Chem.* **12**, 81 (1978).
171. M. Green, J. A. K. Howard, A. Laguna, M. Murray, J. L. Spencer, and F. G. A. Stone, *J. Chem. Soc. Chem. Commun.* p. 451 (1975); M. Green, A. Laguna, J. L. Spencer, and F. G. A. Stone, *J. Chem. Soc. Dalton Trans.* p. 1010 (1977).
172. E. O. Fischer and H.-J. Beck, *Angew. Chem.* **82**, 44 (1970); *Angew. Chem., Int. Ed. Engl.* **9**, 72 (1970).

173. C. P. Casey and T. J. Burkhardt, *J. Am. Chem. Soc.* **95**, 5833 (1975); C. P. Casey, T. J. Burkhardt, C. A. Bunnell, and J. C. Calabrese, *J. Am. Chem. Soc.* **99**, 2127 (1977).
174. E. O. Fischer, E. Offhaus, J. Müller, and D. Nöthe, *Chem. Ber.* **105**, 3027 (1972).
175. E. O. Fischer, T. L. Lindner, H. Fischer, G. Huttner, P. Friedrich, and F. R. Kreißl, *Z. Naturforsch., Teil B* **32**, 648 (1977).
176. A. F. Dyke, S. A. R. Knox, P. J. Naish, and G. E. Taylor, *J. Chem. Soc. Chem. Commun.* p. 409 (1980).
177. K. Folting, J. C. Huffman, L. N. Lewis, and K. G. Caulton, *Inorg. Chem.* **18**, 3483 (1979).
178. A. N. Nesmeyanov, A. B. Antonova, N. E. Kolobova, and K. N. Anisimov, *Izv. Akad. Nauk SSSR, Ser. Khim.* p. 2873 (1974).
179. A. N. Nesmeyanov, G. G. Aleksandrov, A. B. Antonova, K. N. Anisimov, N. E. Kolobova, and Y. T. Struchkov, *J. Organomet. Chem.* **110**, C36 (1976).
180. A. B. Antonova, N. E. Kolobova, P. V. Petrovsky, B. V. Lokshin, and N. S. Obezyuk, *J. Organomet. Chem.* **137**, 55 (1977).
181. E. O. Fischer, V. Kiener, D. St. P. Bunbury, E. Frank, P. F. Lindley, and O. S. Mills, *J. Chem. Soc. Chem. Commun.* p. 1378 (1968).
182. P. F. Lindley and O. S. Mills, *J. Chem. Soc. A* p. 1279 (1969).
183. E. O. Fischer, V. Kiener, and R. D. Fischer, *J. Organomet. Chem.* **16**, P60 (1969).
184. G. Huttner and D. Regler, *Chem. Ber.* **105**, 2726 (1972).
185. E. L. Muetterties, *ACS Symp. Binucl. Met. Compd., New Orleans, La. 1980.*
186. T. R. Halbert, M. E. Leonowicz, and D. J. Maydonovitch, *J. Am. Chem. Soc.* **102**, 5101 (1980).
187. R. B. Bates, L. M. Kroposki, and D. E. Potter, *J. Org. Chem.* **37**, 560 (1972); M. E. Jung and R. B. Blum, *Tetrahedron Lett.* **43**, 3791 (1977).
188. The first CCl_2 complex, $Fe(TPP)(CCl_2)(H_2O)$ [TPP = *meso*-tetraphenylporphyrinato], was reported in 1977: D. Mansuy, M. Lange, J.-C. Chottard, P. Guerin, P. Morliere, D. Brault, and M. Rougé, *J. Chem. Soc. Chem. Commun.* p. 648 (1977); D. Mansuy, M. Lange, J.-C. Chottard, J.-F. Bartoli, B. Chevrier, and R. Weiss, *Angew. Chem.* **90**, 828 (1978); *Angew. Chem., Int. Ed Engl.* **17**, 781 (1978). For a recent example, see G. R. Clark, K. Marsden, W. R. Roper, and L. J. Wright, *J. Am. Chem. Soc.* **102**, 1206 (1980).
189. For chlorocarbene complexes, see A. J. Hartsthorn, M. F. Lappert, and K. J. Turner, *J. Chem. Soc. Chem. Commun.* p. 929 (1975); E. O. Fischer, W. Kleine, and F. R. Kreißl, *J. Organomet. Chem.* **107**, C23 (1976).
190. See D. Seyferth, *Acc. Chem. Res.* **5**, 65 (1972), and references therein.
191. B. L. Booth, G. C. Casey, and R. N. Haszeldine, *J. Chem. Soc. Dalton Trans.* p. 403 (1980).
192. P. Hong, N. Nishii, K. Sonogashira, and N. Hagihara, *J. Chem. Soc. Chem. Commun.* p. 993 (1972).
193. P. Hong and K. Sonogashira, *Symp. Organomet. Chem., 20th, Kyoto* Abstr. No. 207 (1972).
194. H. Ueda, Y. Kai, N. Yasuoka, and N. Kasai, *Symp. Organomet. Chem., 21st, Sendai, Jpn.* Abstr. No. 214 (1973).
195. T. Yamamoto, A. R. Garber, J. R. Wilkinson, C. B. Boss, W. E. Streib, and L. J. Todd, *J. Chem. Soc. Chem. Commun.* p. 354 (1974).
196. H. Ueda, Y. Kai, N. Yasuoka, and N. Kasai, *Bull. Chem. Soc. Jpn.* **50**, 2250 (1977).
197. K. Schorpp and W. Beck, *Z. Naturforsch., Teil B* **28**, 738 (1973); W. A. Herrmann, *Angew. Chem.* **86**, 345 (1975); *Angew. Chem., Int. Ed. Engl.* **13**, 335 (1975); A. D. Redhouse and W. A. Herrmann, *Angew. Chem.* **88**, 652 (1976); *Angew. Chem., Int. Ed.*

Engl. **15**, 615 (1976); W. A. Herrmann, J. Plank, M. L. Ziegler, and K. Weidenhammer, *J. Am. Chem. Soc.* **101**, 3133 (1979).

198. P. Hong, K. Sonogashira, and N. Hagihara, *Tetrahedron Lett.* **16**, 1105 (1971); D. A. Young, *Inorg. Chem.* **12**, 482 (1973).
199. W. A. Herrmann, J. Gimeno, J. Weichmann, M. L. Ziegler, and B. Balbach, *J. Organometal. Chem.* **213**, C26 (1981); W. A. Herrmann and J. Weichmann, unpublished results (1981).
200. For review, see N. Calderon, J. P. Lawrence, and E. A. Ofstead, *Adv. Organomet. Chem.* **17**, 449 (1979).
201. R. Aumann, H. Wormann, and C. Krüger, *Angew. Chem.* **88**, 640 (1976); *Angew. Chem., Int. Ed. Engl.* **15**, 609 (1976).
202. B. F. G. Johnson, J. W. Kelland, J. Lewis, A. L. Mann, and P. R. Raithby, *J. Chem. Soc. Chem. Commun.* p. 547 (1980).
203. J. J. Bonnet, R. Mathieu, R. Poilblanc, and J. A. Ibers, *J. Am. Chem. Soc.* **101**, 7487 (1979).
204. M. R. Churchill, B. G. DeBoer, J. R. Shapley, and J. B. Keister, *J. Am. Chem. Soc.* **98**, 2357 (1976).
205. F. N. Tebbe, G. W. Parshall, and D. W. Ovenall, *J. Am. Chem. Soc.* **101**, 5074 (1979).
206. E. L. Muetterties, *Inorg. Chem.* **14**, 951 (1975); for a recent review, see Ref. 200.
207. R. R. Schrock, *J. Am. Chem. Soc.* **98**, 5399 (1976).
208. H.-J. Sinn, W. Kaminsky, H.-J. Vollmer, and R. Woldt, *Angew. Chem.* **92**, 396 (1980); *Angew. Chem., Int. Ed. Engl.* **19**, 390 (1980).
209. D. C. Calabro, D. L. Lichtenberger, and W. A. Herrmann, *J. Am. Chem. Soc.* **103**, 6852 (1981).
210. S. Shaik, R. Hoffmann, C. R. Fisel, and R. H. Summerville, *J. Am. Chem. Soc.* **102**, 4555 (1980).
211. F. R. Kreißl, P. Friedrich, T. L. Lindner, and G. Huttner, *Angew. Chem.* **89**, 325 (1977); *Angew. Chem., Int. Ed. Engl.* **16**, 314 (1977).
211a. F. R. Kreißl, personal communication (1981).
212. O. Orama, U. Schubert, F. R. Kreißl, and E. O. Fischer, *Z. Naturforsch., Teil B* **35**, 82 (1980).
213. J. Carty, N. Mott, J. Taylor, and J. E. Yule, *J. Am. Chem. Soc.* **100**, 3051 (1978).
214. A. J. Carty, N. J. Taylor, N. H. Paile, W. Smith, and J. G. Yule, *J. Chem. Soc. Chem. Commun.* p. 41 (1976).
215. J. R. Shapley, M. Tachikawa, M. R. Churchill, and R. A. Lashewycz, *J. Organomet. Chem.* **162**, C39 (1978).
216. M. R. Churchill and R. A. Lashewycz, *Inorg. Chem.* **18**, 848 (1979).
217. M. R. Churchill and B. G. DeBoer, *Inorg. Chem.* **16**, 1141 (1977).
218. L. Pope, P. Sommerville, M. Laing, K. J. Hudson, and J. R. Moss, *J. Organomet. Chem.* **112**, 309 (1976).
219. B. Olgemüller and W. Beck, *Chem. Ber.* **114**, 867 (1981).
220. O. Ermer, J. D. Dunitz, and I. Bernal, *Acta Crystallogr., Sect. B* **29**, 2278 (1973).
221. B. E. Geiger *et al.*, unpublished observations (1979–1980).
222. E. O. Fischer, R. L. Clough, G. Besl, and F. R. Kreißl, *Angew. Chem.* **88**, 584 (1976); *Angew Chem. Int. Edit. Engl.* **15**, 543 (1976); cf. $(\eta^5\text{-}C_5H_5)Mn(CO)_2[C(CH_3)_2]$ ($\delta C(Carbene) = 372.75$, acetone-$d_6$) with $(\mu\text{-}CH_2)[(\eta^5\text{-}C_5H_5)Mn(CO)_2]_2$ ($\delta C(carbene) = 150.03$, CD_2Cl_2; see Table 4).
223. M. Laing, J. R. Moss, and J. Johnson, *J. Chem. Soc. Chem. Commun.* p. 656 (1977).
224. R. F. Bryan, P. T. Greene, M. J. Newlands, and D. S. Field, *J. Chem. Soc. A* p. 3068 (1970).

225. R. Ginsburg, L. M. Cirjak, and L. F. Dahl, *J. Chem. Soc. Chem. Commun.* p. 468 (1979).
226. W. I. Bailey, Jr., D. M. Collins, F. A. Cotton, J. C. Baldwin, and W. C. Kaska, *J. Organomet. Chem.* **165**, 373 (1979).
227. O. S. Mills and J. P. Nice, *J. Organomet. Chem.* **10**, 337 (1967).
228. W. A. Herrmann, J. Plank, M. L. Ziegler, and E. Guggolz, unpublished observations (1980).
229. J. C. T. R. Burckett-St. Laurent, M. R. Caira, R. B. English, R. J. Haines, and L. R. Nassimbeni, *J. Chem. Soc. Dalton Trans.* p. 1077 (1977).
230. H. Vahrenkamp. *Chem. Ber.* **107**, 3867 (1974).
231. O. S. Mills and A. D. Redhouse, *J. Chem. Soc. Chem. Commun.* p. 444 (1966).
232. O. S. Mills and A. D. Redhouse, *J. Chem. Soc. A* p. 1282 (1968).
233. W. A. Herrmann, J. Plank, and D. Riedel, *Angew. Chem.* **92**, 961 (1980); *Angew. Chem., Int. Ed. Engl.* **19**, 937 (1980).
234. R. G. Bergman, *ACS Symp. Binucl. Met. Compd., New Orleans, La. 1980*.
235. V. G. Andrianov, Y. T. Struchkov, and E. R. Rossinskaja, *J. Chem. Soc. Chem. Commun.* p. 338 (1973).
235a. W. A. Herrmann and J. Huggins, unpublished observations (1980).
236. S. H. Pine, R. Zahler, D. A. Evans, and R. H. Grubbs, *J. Am. Chem. Soc.* **102**, 3270 (1980).
237. T. R. Howard, J. B. Lee, and R. H. Grubbs, *J. Am. Chem. Soc.* **102**, 6878 (1980).
238. For reviews, see E. L. Muetterties and J. Stein, *Chem. Rev.* **79**, 479 (1979); W. A. Herrmann, *Angew. Chem., Int. Ed. Engl.* No. 2, in press (1982).
239. R. Gravin and E. L. Muetterties, unpublished observations, cited in Ref. 38.
240. G. Wilkinson *et al.*, unpublished observations.
241. J. E. Demuth and H. Ibach, *Surf. Sci.* **78**, L238 (1978).
242. R. Dickson, S. H. Johnson, H. P. Kirsch, and D. J. Lloyd, *Acta Crystallogr., Sect. B* **33**, 2057 (1977).
243. R. S. Dickson, G. N. Pain, and M. F. Mackay, *Acta Crystallogr., Sect. B* **35**, 2321 (1979).
244. R. S. Dickson, C. Mok, and G. Pain, *J. Organomet. Chem.* **166**, 385 (1979), and references cited therein.
245. C. Masters, C. van der Woude, and J. A. van Doorn, *J. Am. Chem. Soc.* **101**, 1633 (1979).
246. For reviews, see G. Henrici-Olivé and G. Olivé, *Angew. Chem.* **88**, 144 (1976); *Angew. Chem., Int. Ed. Engl.* **15**, 136 (1976); C. Masters, *Adv. Organomet. Chem.* **17**, 61 (1979); W. A. Herrmann, *Angew. Chem.*, in press (1982).
247. W. J. Dunning and C. C. McCain, *J. Chem. Soc. B* p. 68 (1966).
248. See, e.g., A. G. Nasini, L. Trosserelli, and G. Saini, *Makromol. Chem.* **44/46**, 550 (1961).
249. F. Fischer and H. Tropsch, *Brennst.-Chem.* **7**, 97 (1926).
250. For reviews, see E. H. Brooks and R. J. Cross, *Organomet. Chem. Rev., Sect. A* **6**, 227 (1970); A. Bonny, *Coord. Chem. Rev.* **25**, 229 (1978); C. S. Cundy, B. M. Kingston, and M. F. Lappert, *Adv. Organomet. Chem.* **11**, 253 (1973).
251. For bridging ER_2-ligands (E = Ge, Sn, Pb), see F. G. A. Stone, *in* "New Pathways in Inorganic Chemistry" (Ebsworth, Maddock, and Sharpe, eds.), Chap. 12, pp. 283–302. Cambridge Univ. Press, London and New York, 1968; "Comprehensive Inorganic Chemistry," Vol. 2. Pergamon, Oxford, 1973.
252. G. L. Simon and L. F. Dahl, *J. Am. Chem. Soc.* **95**, 783 (1973).
253. B. J. Aylett and H. M. Colquhoun, *J. Chem. Res., Synop.* No. 1, p. 148 (1977).

254. W. Malisch and W. Ries, *Angew. Chem.* **90**, 140 (1978); *Angew. Chem., Int. Ed. Engl.* **17**, 120 (1978).

255. M. A. Bush and P. Woodward, *J. Chem. Soc. A* p. 1883 (1967).

256. J.-C. Zimmer and H. Huber, *C. R. Hebd. Seances Acad. Sci., Ser. C* **267**, 1685 (1968).

257. M. J. Bennett, W. Brooks, M. Elder, W. A. G. Graham, D. Hall, and R. Kummer, *J. Am. Chem. Soc.* **92**, 209 (1970).

258. J. Howard, S. A. R. Knox, F. G. A. Stone, and P. Woodward, *J. Chem. Soc. Chem. Commun.* p. 1477 (1970).

259. S. A. R. Knox and F. G. A. Stone, *J. Chem. Soc. A* p. 2874 (1971), and references cited therein.

260. A. J. Cleland, S. A. Fieldhouse, B. H. Freeland, and R. J. O'Brien, *J. Organomet. Chem.* **32**, C15 (1971).

261. M. Elder, *Inorg. Chem.* **8**, 2703 (1969); J. Howard and P. Woodward, *J. Chem. Soc. A* p. 3648 (1971).

262. M. Elder and D. Hall, *Inorg. Chem.* **8**, 1424 (1969).

263. R. D. Adams, M. D. Brice, and F. A. Cotton, *Inorg. Chem.* **13**, 1080 (1974).

264. R. D. Adams and F. A. Cotton, *J. Am. Chem. Soc.* **92**, 5003 (1970).

265. F. A. Cotton, *in* "Dynamic Nuclear Magnetic Resonance Spectroscopy" (L. M. Jackman and F. A. Cotton, eds.), pp. 496–499. Academic Press, New York, 1975.

266. (a) R. C. Job and M. D. Curtis, *Inorg. Chem.* **12**, 2514 (1973); (b) R. C. Job and M. D. Curtis, *Inorg. Chem.* **12**, 2510 (1973); (c) M. D. Curtis and R. C. Job, *J. Am. Chem. Soc.* **94**, 2153 (1972).

267. K. Triplett and M. D. Curtis, *J. Am. Chem. Soc.* **97**, 5747 (1975), and literature cited therein.

268. M. D. Curtis, W. M. Butler, and J. Scibelli, *J. Organomet. Chem.* **191**, 209 (1980).

269. P. Jutzi and W. Steiner, *Angew. Chem.* **88**, 720 (1976); *Angew. Chem., Int. Ed. Engl.* **15**, 684 (1976); P. Jutzi and W. Steiner, *Angew. Chem.* **89**, 675 (1977); *Angew. Chem., Int. Ed. Engl.* **16**, 639 (1977); P. Jutzi, W. Steiner, and K. Stroppel, *Chem. Ber.* **113**, 3357 (1980).

270. P. Jutzi and K. Stroppel, *Chem. Ber.* **113**, 3366 (1980).

271. C. Burschka, K. Stroppel, and P. Jutzi, *Acta Crystallogr.*, in press (1982).

272. R. M. Sweet, C. J. Fritchie, Jr., and R. A. Schunn, *Inorg. Chem.* **6**, 749 (1967).

273. E. W. Abel and S. Moorhouse, *Inorg. Nucl. Chem. Lett.* **7**, 905 (1971).

274. C. D. Garner and R. G. Senior, *Inorg. Nucl. Chem. Lett.* **10**, 609 (1974).

275. G. Hencken and E. Weiss, *Chem. Ber.* **106**, 1747 (1973).

276. B. K. Teo, M. B. Hall, R. F. Fenske, and L. F. Dahl, *J. Organomet. Chem.* **70**, 413 (1974).

277. J. E. Ellis, *J. Organomet. Chem.* **86**, 22 (1975).

277a. For review, see H. J. Sinn and W. Kaminsky, *Adv. Organomet. Chem.* **18**, 99 (1980).

278. W. Kaminsky, J. Kopf, H. Sinn, and H.-J. Vollmer, *Angew. Chem.* **88**, 688 (1976); *Angew. Chem., Int. Ed. Engl.* **15**, 629 (1976).

279. Preparation of CH_2–CH_2–bridged zirconium compounds; W. Kaminsky and H. Sinn, *Liebigs Ann. Chem.* p. 424 (1975).

280. NMR study of CH_2–CH_2–bridged zirconium compounds: W. Kaminsky and H.-J. Vollmer, *Liebigs Ann. Chem.* p. 438 (1975).

281. Structure of $(\eta^5$-$C_5H_5)_2(Cl)Zr$–CH_2–$CH(AlEt_2)_2$: W. Kaminsky, J. Kopf, and G. Thirase, *Liebigs Ann. Chem.* p. 1531 (1974).

282. W. Beck and B. Olgemüller, *J. Organomet. Chem.* **127**, C45 (1977).

283. R. B. King, *Inorg. Synth.* **2**, 531 (1963).
284. R. B. King, *J. Am. Chem. Soc.* **85**, 1922 (1963); C. P. Casey and R. L. Anderson, *J. Am. Chem. Soc.* **93**, 3554 (1971).
285. M. R. Churchill and J. Wormald, *Inorg. Chem.* **8**, 1936 (1969); M. R. Churchill, J. Wormald, W. P. Giering, and G. F. Emerson, *J. Chem. Soc. Chem. Commun.* p. 1217 (1968).
286. R. E. Davis, *J. Chem. Soc. Chem. Commun.* p. 1218 (1968).
287. (a) R. S. Dickson and D. B. W. Yawney, *Aust. J. Chem.* **22**, 533 (1969); (b) *Aust. J. Chem.* **20**, 77 (1967).
288. J. L. Boston, D. W. A. Sharp, and G. Wilkinson, *J. Chem. Soc. A* p. 3488 (1962).
289. A. W. Parkins, E. O. Fischer, G. Huttner, and D. Regler, *Angew. Chem.* **82**, 635 (1970); *Angew. Chem., Int. Ed. Engl.* **9**, 633 (1970).
290. M. D. Rausch, R. G. Gastinger, S. A. Gardner, R. K. Brown, and J. S. Wood, *J. Am. Chem. Soc.* **99**, 7870 (1977), and literature cited therein.
291. For reviews and leading references on vinylidene complexes, see P. J. Stang, *Chem. Rev.* **78**, 383 (1978); R. B. King, *N.Y. Ann. Acad. Sci.* **239**, 171 (1974); For recent papers, see H. Berke, *J. Organomet. Chem.* **185**, 75 (1980); H. Berke, *Chem. Ber.* **113**, 1370 (1980); H. Berke, *Z. Naturforsch., Teil B* **35**, 86 (1980); L. N. Lewis, J. C. Huffmann, and K. G. Caulton, *Inorg. Chem.* **19**, 1246 (1980).
292. E. O. Fischer, T. L. Lindner, G. Huttner, P. Friedrich, F. R. Kreißl, and J. O. Besenhard, *Chem. Ber.* **110**, 3397 (1977).
293. E. O. Fischer, E. W. Meineke, and F. R. Kreißl, *Chem. Ber.* **110**, 1140 (1977).
294. R. S. Dickson and G. N. Pain, *J. Chem. Soc. Chem. Commun.* p. 277 (1979).
295. R. G. Bergman, *Pure Appl. Chem.* **53**, 161 (1981).
296. C. Bauer, E. Guggolz, W. A. Herrmann, G. Kriechbaum, and M. L. Ziegler, *Angew. Chem., Int. Ed. Engl.* **21**, in press (1982); *Angew. Chem. Suppl.*, in press.
297. W. A. Herrmann and J. M. Huggins, *Chem. Ber. 115*, 396 (1982).
298. W. A. Herrmann and C. Bauer, *Chem. Ber.*, in press.
299. W. A. Herrmann, J. M. Huggins, B. Reiter, and C. Bauer, *J. Organomet. Chem. 214*, C19 (1981).
300. G. R. Steinmetz and G. L. Geoffroy, *J. Am. Chem. Soc.* **103**, 1278 (1981).
301. G. Granozzi, E. Tondello, M. Casarin, and D. Ajò, *Inorg. Chim. Acta* **48**, 73 (1981).
302. P. J. Krusic and F. N. Tebbe, *J. Am. Chem. Soc.*, in press.
303. G. Váradi, I. T. Horváth, G. Pályi, L. Markó, Y. L. Slovokhotov, and Y. T. Struchkov, *J. Organomet. Chem.* **206**, 119 (1981).
304. M. Green, D. R. Hankey, M. Murray, A. G. Orpen, and F. G. A. Stone, *J. Chem. Soc. Commun.* p. 689 (1981).
304a. I. Bernal, M. Creswick, and R. A. Lalancette, unpublished results (1981) (low temperature x-ray diffraction of $(\mu\text{-}CH_2)[(\eta^5\text{-}C_5H_4CH_3)Mn(CO)_2]_2$.
304b. Yu. T. Struchkov, *Pure Appl. Chem.* **52**, 741 (1980).
305. K. Isobe, D. G. Andrews, B. E. Mann, and P. M. Maitlis, *J. Chem. Soc. Chem. Commun.* p. 809 (1981).
306. A. L. Balch, C. T. Hunt, Ch.-Li Lee, M. M. Olmstead, and J. P. Farr, *J. Am. Chem. Soc.* **103**, 3764 (1981).
307. D. L. Davies, A. F. Dyke, S. A. R. Knox, and M. J. Morris, *J. Organomet. Chem.* **215**, C30 (1981).
308. E. I. Symslova, E. G. Perevalova, V. P. Dyadchenko, K. I. Grandberg, Yu. L. Slovokhotov, and Yu. T. Struchkov, *J. Organomet. Chem.* **215**, 269 (1981).
309. A. F. Dyke, S. A. R. Knox, K. A. Mead, and P. Woodward, *J. Chem. Soc. Chem. Commun.* p. 861 (1981).

310. W. A. Herrmann, M. Smischek, and Ch. Bauer, unpublished results (1981).
311. W. A. Herrmann, *Session Lect. Int. Conf. Organomet. Chem., Xth, Toronto, Canada, 1981,* Abstr. S9.
312. M. Cooke, D. L. Davies, J. E. Guerchais, S. A. R. Knox, K. A. Mead, J. Roué, and P. Woodward, *J. Chem. Soc. Chem. Commun.* p. 862 (1981).
313. J. Levisalles, F. Rose-Munch, H. Rudler, J.-C. Daran, Y. Dromzée, and Y. Jeannin, *J. Chem. Soc. Chem. Commun.* p. 152 (1981).
314. H. Rudler, F. Rose, J. Levisalles, J. C. Daran, Y. Dromzée, and Y. Jeannin, *Int. Conf. Organomet. Chem., Xth, Toronto, Canada, 1981,* Abstr. 5A01.
315. S. Sabo, B. Chaudret, D. Gervais, and R. Poilblanc, *Int. Conf. Organomet. Chem., Xth, Toronto, Canada, 1981, Abstr.* 5A04.
316. G. F. Schmidt, E. L. Muetterties, M. A. Beno, and J. M. Williams, *Proc. Natl. Acad. Sci. U.S.A.* **78,** 1318 (1981).
317. G. Váradi, S. Vastag, and G. Pályi, *Atti Accad. Sci. Ist.* Bologna, *Cl. Sci.* Ser. XIII 223, (1979).
318. L. Bencze, V. Galamb, A. Guttmann, and G. Pályi, *Accad. Naz. Lincei Roma* **68,** 437 (1980).
319. G. Pályi, G. Váradi, I. T. Horváth, *J. Mol. Catal.* **13,** in press (1981).
320. K. H. Theopold and R. G. Bergman, *J. Am. Chem. Soc.,* in press.
321. W. H. Hersh and R. G. Bergman, *J. Am. Chem. Soc.,* in press.
322. K. H. Theopold and R. G. Bergman, *J. Am. Chem. Soc.* **103,** 2489 (1981).
323. F. Mathieu, *J. Organomet. Chem.* **215,** C57 (1981).
324. I. R. McKeer and M. Cowie, *J. Organomet. Chem.,* in press (1982).
325. W. A. Herrmann, J. M. Huggins, C. Bauer, M. Smischek, H. Pfisterer, and M. L. Ziegler, *J. Organomet. Chem.,* in press (1982).
326. W. A. Herrmann, *Angew. Chem.* **21** No. 2, in press (1982). (Review on "Organometallic Aspects of the Fischer–Tropsch Synthesis.")
327. P. Jandik, U. Schubert, and H. Schmidbauer, *Angew. Chem.* **94,** 74 (1982); *Angew. Chem. Suppl.* **1982,** 1.
328. J. Boso Lee, G. J. Gajda, W. P. Schaefer, T. R. Howard, T. Ikariya, D. A. Straus, and R. H. Grubbs, *J. Am. Chem. Soc.* **103,** 7358 (1981).
329. A. D. Clauss, J. R. Shapley, and S. R. Wilson, *J. Am. Chem. Soc.* **103,** 7387 (1981).
330. F. G. A. Stone, *Inorg. Chim. Acta* **50,** 33 (1981).
331. M. Röper, H. Strutz, and W. Keim, *J. Organomet. Chem.* **219,** C5 (1981).
332. J. C. Jeffery, I. Moore, H. Razay, and F. G. A. Stone, *J. Chem. Soc. Chem. Commun.* p. 1255 (1981).
333. F. W. Hartner, Jr., and J. Schwartz, *J. Am. Chem. Soc.* **103,** 4979 (1981).

ADVANCES IN ORGANOMETALLIC CHEMISTRY, VOL. 20

Nucleophilic Displacement at Silicon: Recent Developments and Mechanistic Implications

R. J. P. CORRIU and C. GUERIN

Laboratoire des Organométalliques
Equipe de Recherche Associée au C.N.R.S. N° 554
Université des Sciences et Techniques du Languedoc,
Montpellier, France

I

INTRODUCTION

In recent years, mainly from the 1960s on, a great number of stereochemical studies of substitution at silicon by nucleophiles have been carried out. The earlier work, up to 1964, is the subject of an important monograph by Sommer (*1*); more recent studies have also been summarized by him (*2*) and others [see Prince (*3*) and Fleming (*4*)] in comprehensive surveys. Recently, we collected the main known features in this area (*5*). The intent of this chapter is to focus upon the latest developments concerning nucleophilic displacement at silicon and to survey the following topics:

1. The factors controlling the stereochemistry of substitution, i.e., the nature of the leaving group and the nucleophile.

2. Frontier-orbital approximations. A consistently useful frontier-orbital approximation for nucleophilic displacement at silicon was ex-

pounded by Anh and Minot (6). This approach will be discussed and used to give a qualitatively rational explanation for numerous data.

II
NUCLEOPHILIC DISPLACEMENT AT SILICON:
CONTROLLING FACTORS OF THE STEREOCHEMISTRY

Displacement reactions must proceed either with retention or inversion of configuration at silicon. The change in stereochemistry is mainly a function of the nature of the leaving group and of the electronic character of the nucleophile.

As outlined in our previous review (5), coupling reactions of a nucleophilic reagent Nu^- with an organosilane $R_3Si—X$ occur only by a one-stage nucleophilic mechanism [Eq. (1)]. The halosilanes are not prone to one-electron transfer which could lead to a two-stage mechanism [Eq. (2)].

$$Nu^- + R_3Si—X \rightarrow R_3Si—Nu + X^- \tag{1}$$

$$Nu^- + R_3Si—X \xrightarrow[\text{transfer}]{e^-} Nu^. + \overline{R_3Si—X^.} \rightarrow R_3SiNu + X^- \tag{2}$$

Furthermore, a quasicyclic S_Ni–Si mechanism (1, 2) (Scheme 1) cannot be invoked to explain the retention of configuration. The use of a basic solvent or of a cryptand leads to a shift of the stereochemistry to retention (7, 8) and to a parallel increase of the substitution rate (9). These data are in complete opposition to the S_Ni–Si process. Because the controlling force of such a mechanism is electrophilic assistance to the cleavage of the Si—X bond, the use of a more basic solvent or of a cryptand would shift the stereochemistry toward inversion.

In the following sections we shall examine the factors controlling the stereochemistry of substitution at silicon.

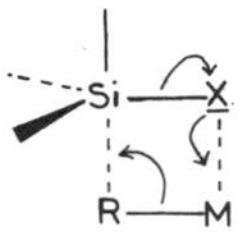

SCHEME 1. S_Ni–Si mechanism.

A. Influence of the Leaving Group

1. Stereochemistry and the Nature of the Leaving Group

As shown in Table I, for nucleophiles of diverse nature, the stereochemical outcome depends strongly upon the nature of the leaving group. The following comments can be made:

1. The Si—H bond is displaced with retention of configuration. Inversion occurs only in two cases, i.e., coupling reactions with Ph_2CHLi (10)

TABLE I

INFLUENCE OF LEAVING GROUP ON STEREOCHEMISTRY

1-NpPhRSi—X + Nu⁻ → 1-NpPhRSi—Nu + X⁻

Substrate	No.	Nu	Predominant stereo-chemistry
1-NpPh(R)Si—H	1	R'Li (R' = alkyl,p-$CH_3OC_6H_4CH_2$, allyl,	RN 10, 11
(R = Me, Et,		$PhCH_2$)	
i-Pr, Vi)	2	Ph_2CHLi	IN 10
	3	t-BuOH/t-BuOK	RN 12
	4	KOH	RN 13, 14
	5	$LiNH_2$	RN 15
	6	$LiAlD_4$	RN 16
1-NpPh(R)Si—OMe	7	R'Li (R' = alkyl, p-$CH_3OC_6H_4CH_2$, allyl	RN 10, 11
(R = Me, Et,	8	R'Li (R' = $PhCH_2$, Ph_2CH)	IN 10, 11
i-Pr, Vi)	9	KOH	RN 13, 17, 18
	10	$LiAlH_4$	RN 13, 17, 19
	11	AlH_3/Et_2O	RN 13
1-NpPh(R)Si—F	12	R'Li (R' = alkyl)	RN 11, 20
(R = Me, Et,	13	R'Li (R' = allyl, $PhCH_2$,	IN 11, 20
i-Pr, Vi)		p-$MeOC_6H_4CH_2$, Ph_2CH)	
	14	$LiAlH_4$	IN 19, 21
1-NpPh(R)Si—X	15	R'Li (R' = alkyl)	RN 22, 23
(R = Me;	16	R'Li (R' = allyl, $PhCH_2$)	IN 22, 23
X = SMe, SPh)	17	$LiAlH_4$	IN 22, 23
	18	ROH	IN 22
1-NpPh(R)Si—X	19	R'Li	IN 11, 20, 23
(X = Cl, Br;	20	KOH	IN 13, 24
R = Me, Et,	21	MeOH	IN 13, 17, 24
i-Pr, Vi,	22	$LiAlH_4$	IN 17, 19
neo-C_5H_{11})	23	AlH_3/Et_2O	IN 13
	24	CH_3COO^-	IN 13, 17, 24

or with alcohols in presence of Raney nickel (25). The latter reaction certainly occurs by a different mechanism. Activation of the Si—H bond proceeds with adsorption on the Ni surface and formation of a pentacoordinate silicon intermediate. To illustrate this assumption, we can note that

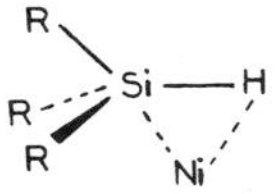

SCHEME 2

the metal–silicon bond of the $\pi Cp(CO)_2Mn(H)$–SiR_3 complexes is cleaved by nucleophiles with inversion of configuration. Structural investigations indicate a partial single bond between Si and hydrogen and thus a pentacoordinate geometry around silicon (26).

2. Chloro- and bromosilanes mainly undergo inversion.

3. The Si—F and Si—SR bonds show parallel behavior, leading either to retention or inversion according to the nature of the nucleophile.

4. The Si—OR bond is displaced mainly with retention by organolithiums, except for some charge-delocalized reagents such as $PhCH_2Li$ and Ph_2CHLi (Table I, No. 8). Similar behavior is observed with Grignard reagents (5).

Bifunctional organosilanes [1-NpRSi(X)Y; R = phenyl or ferrocenyl] are good models to study the relative ability of X and Y leaving groups to be displaced. Typical reactions are shown in Table II; the data lead to the following conclusions:

1. The (H) substituent appears to be the worst leaving group. With 1-NpPhSi(H)OR and 1-NpFcSi(H)F derivatives, only displacement of F and OR occurs, so both of these are better leaving groups than H.

2. At the opposite extreme, the chlorine substituent is the best leaving group. With 1-NpPhSi(Cl)OMen and 1-NpFcSi(F)Cl, displacement of Cl only occurs with inversion, whatever the nucleophile.

3. Comparison of F and OR substituents is more complex (32). However, when the reaction takes place with inversion, only highly stereoselective displacement of the fluorine atom occurs.

These experimental facts show mainly parallel chemical and stereochemical data for mono- and bifunctional organosilanes and allow us to conclude that the relative ease of displacement is

$$Cl > F > OR > H$$

TABLE II
INFLUENCE OF THE LEAVING GROUP: BEHAVIOR OF SOME
BIFUNCTIONAL ORGANOSILANES[a]

Substrate	Nucleophile	Product (stereochemistry)	Reference
1-NpPhSi(Cl)OMen	RLi or RMgX	1-NpPhSi(R)OMen	(IN) 27, 28
1-NpPhSi(H)OMen	RLi	1-NpPhSi(R)H	(RN) 27, 28
	R^1MgX (R^1 = Me, Et, n-Bu)	1-NpPhSi(R^1)H	(RN) 27
	R^2MgX (R^2 = allyl, PhCH$_2$)	1-NpPhSi(R^2)H	(IN) 28
1-NpFcSi(F)Cl	RLi or RMgX	1-NpFcSi(F)R	(IN) 29
1-NpFcSi(H)F	RLi	1-NpFcSi(H)R	(RN) 29
	RMgX (R = Me, Et, allyl)	1-NpFcSi(H)R	(IN) 29
1-NpFcSi(F)OEt	RLi (R = Ph, allyl, PhC≡C)	1-NpFcSi(R)OEt	(IN) 30, 31
	allyl MgBr	1-NpFcSi(allyl)OEt	(IN) 30, 31

[a] Fc = ferrocenyl.

This order is closely parallel to the reactivity order observed in coupling reactions between monofunctional organosilanes R_3Si—X and Grignard reagents or organolithiums (33). In general, rate constants vary by a factor of between 1 and 50 for reactions giving either retention or inversion. In the case of the F and OMe leaving groups, for reactions leading to inversion, rate constants are related by a factor of 10^3 ($k_F/k_{OMe} > 10^3$) (33). Furthermore, this empirical order parallels a general change of the stereochemistry from inversion to retention.

2. Explanation of the Dominant Influence

The above data suggest that a simple relationship between the nature of the leaving group and some of its physical properties may exist. For instance, Sommer (1, 2) classified leaving groups X (R_3Si—X) according to their respective basicity. For good leaving groups such as Cl or Br, whose conjugate acids have a pK_a smaller than about 5, the favored stereochemistry is inversion of configuration. For poor leaving groups such as H or OR, whose conjugate acids have pK_a larger than about 10, the predominant stereochemistry is retention of configuration.

However, examination of Tables I and II shows clearly that this explanation is not sufficiently general. For instance, the Si—SR and Si—F bonds show very similar stereochemical behavior, leading either to retention or inversion. Following the Sommer's rule, the SR group (for instance MeSH, $pK_a \sim 10$) would be a poor leaving group and be displaced mainly with retention. At the opposite extreme, the fluoro substituent (HF, $pK_a \sim$

3.5–4) would be a good leaving group and undergo inversion. Moreover, it does not explain why Si—SR, Si—F, and Si—OR bonds are substituted with inversion or retention according to the nature of the nucleophile.

A satisfactory explanation also cannot be made in terms of polarizability of the leaving group. The highly polarizable Si—Cl and Si—SR bonds are always displaced with inversion. However, the Si—F bond, which is far less polarizable than Si—OR and Si—H bonds, is more easily cleaved by nucleophiles with inversion of configuration. It is also impossible to explain the closely similar behavior of F and SR groups. However, the relative tendency of Si—X bonds to undergo inversion and the polarizability order run parallel in the same group of the periodic table. The order of polarizability and ability for inversion are

$$Br > Cl \gg F, SR \gg OR$$

Thus, it is quite difficult to derive a relationship between substitution stereochemistry and any physical property of the leaving group (basicity, electronegativity, or polarizability). This failure prompted us to consider an empirical relationship between the observed stereochemistry (inversion or retention) and the ability of the leaving group to be displaced (23, 33).

Ability to be cleaved   OTs, OCOR, Br, Cl > SR, F > OR > H
Stereochemistry   IN ——————————————————————————→ RN

We can also introduce the OTs and OCOR leaving groups in the above classification near chloro and bromo groups (2). They show high reactivity and a stereochemical behavior similar to that of chlorosilanes, leading only to inversion (34).

Finally, the following points are pertinent:

1. Stereochemical data show that the above empirical order can generally be extended to various acyclic organosilanes (2, 5). Moreover, cyclic

```
        Ph                      Ph                      C₆F₅
        |                       |                       |
1-Np — Si — X           Me — Si — X             Me — Si — X
        |                       |                       |
        R                       R                       R
        1                       2                       3

R = Et, i-Pr, Vi        R = t-BuCH₂               R = 1-Np
                            i-Pr                       Ph
                            Et
                            Ph₃Si
                            Ph₃Ge
```

TABLE III

INFLUENCE OF LEAVING GROUP ON STEREOCHEMISTRY

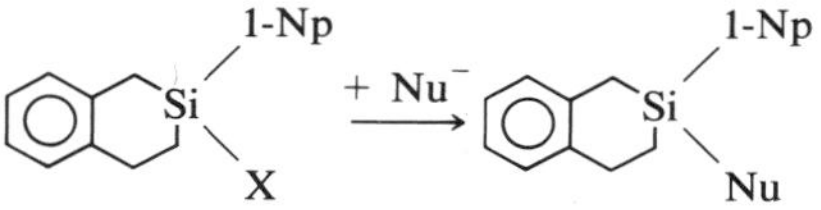

Nucleophile	Predominant stereochemistry			
	Si—Cl	Si—F	Si—OMe	Si—H
RLi (R = aryl and alkyl)	RN	RN	RN	RN
RLi (R = benzyl or allyl)	IN	RN	RN	RN
RMgX (R = alkyl)	IN	RN	RN	—
RMgX (R = allyl or benzyl)	IN	IN	RN	—
LiAlH₄	IN	Racemic	RN	—

geometry does not change the dominant influence of the leaving group on the stereochemistry, as shown in Table III for 1-naphthyl-2-sila-2-tetrahydro-1,2,3,4-naphthalene (*35, 36*).

Furthermore, optically active systems containing both a silicon atom and a leaving group (SR or OR) in the ring have been investigated (*37, 38*). The SR group favors inversion, compared to the analogous OR group. Stereochemical data are summarized in Table IV.

2. Bifunctional organosilanes (Table II) lead to the same stereochemical changes when varying the nature of the leaving group.

3. The tendency of halosilanes to be racemized by nucleophiles (HMPA, etc.) is parallel to the tendency of the Si-X bonds to be displaced

TABLE IV

STEREOCHEMICAL BEHAVIOR OF ENDOCYCLIC LEAVING GROUPS

Nucleophile	Predominant stereochemistry	
	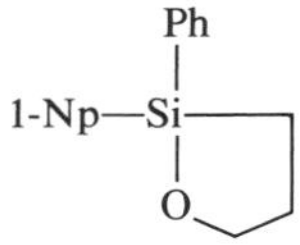 1-Np—Si (Ph, O)	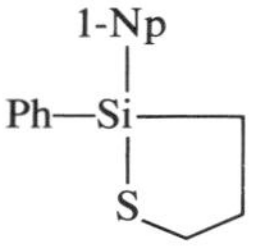 Ph—Si (1-Np, S)
Dibal/hexane	RN	RN
LiAlH₄	RN	IN
RLi (R = alkyl)	RN	RN
RMgX (R = alkyl)	RN	IN

with inversion (39):

$$\equiv Si\!-\!Br > \equiv Si\!-\!Cl > \equiv Si\!-\!F$$

4. Inversion is observed mainly with Si—X bonds which are able to be stretched under the influence of an attacking nucleophile. For instance, the X-ray crystal structure analysis of the five-membered chloro-(N-chlorodimethylsilylacetamido)methyldimethylsilane (4) shows the chlor-

4

ine atom in an apical position trans to the oxygen (40): the Si—Cl bond is considerably longer than in ordinary tetravalent chlorosilanes [2.348 Å instead of 2.01 Å (41)].

This observation, that the aptitude of X to be displaced with inversion, is parallel to the tendency of the Si—X bond to be stretched under the influence of an attacking nucleophile, does not agree with the apicophilicity order stated from dynamic NMR studies on stable pentacoordinate phosphoranes (42), which is as follows:

$$F > H > Cl > Br > Ph > SPh > OR > R$$

An NMR study of (5) at various temperatures allows us to assess this ability to form a pentacoordinate structure (6) (43), and to state the apicophilicity of various substituents at silicon.

5   **6**

3. *Nuclear Magnetic Resonance Studies on Pentacoordinate Silicon Derivatives: Relative Apicophilicity of X Groups at Silicon*

Detailed ¹H NMR spectroscopy and X-ray investigation by Van Koten and Noltes have shown that the tin atom in $(2\text{-}Me_2NCH_2C_6H_4)R_2SnBr$ (7) (R = Ph) is pentacoordinate as a result of intramolecular Sn—N coordina-

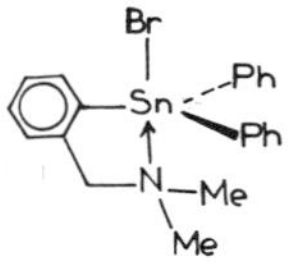

7

tion (*44*). A similar approach was applied to silicon in order to assess its ability to form a pentacoordinate structure and to compare the relative apicophilicity of groups bonded to a pentacoordinate silicon center (*43*). The most significant data will be summarized in the following sections.

a. ¹H-NMR Data. The equilibrium between **5** and **6** can be studied at low temperatures by ¹H-NMR spectroscopy [Eq. (3)]. The formation of an intramolecular Si—N bond induces diastereotopy of the two Me groups. For each product, the temperature of coalescence is given in Table V. The above data agree with intramolecular Si—N coordination and the formation of a pentacoordinate species. The following points are noteworthy:

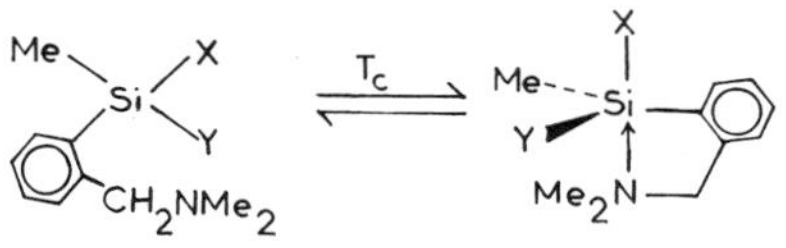

$$(3)$$

5   **6**

T_c = Temperature of coalescence

1. In the case of X = H or OR, even at $-100°C$, no intramolecular Si—N coordination is detected.

TABLE V

TEMPERATURE OF COALESCENCE FOR $(2\text{-Me}_2\text{NCH}_2\text{C}_6\text{H}_4)\text{MeSi(X)Y}$ COMPOUNDS

X = Y		Y = H		Y = 1-Np	
X	$T_c(°C)$	X	$T_c(°C)$	X	$T_c(°C)$
H[a]	—	O-*t*-Bu	—	H	—
OEt[a]	—	SEt	−65	OMe	—
F	−85	F	−52	F	−65
OCOCH₃	+20	OCOCH₃	−46	Cl	−35
		Cl	−40	OCOCH₃	−35
		Br	−10		

[a] No coalescence occurs.

2. The fluorine atom, which is the most electronegative group attached to silicon, shows a low capability for forming pentacoordinate complexes.

3. In contrast, groups such as SR, OCOCH₃, Cl, and Br show a notable tendency toward intramolecular coordination.

Thus the ease of pentacoordination at the silicon atom can be stated as

$$H, OR < F, SR < OAc, Cl, Br$$

Furthermore, the ¹H-NMR spectra of 6 and 8 are very similar: An X-ray investigation of the chlorogermane (8) shows that the germanium atom has a distorted trigonal bipyramidal (TBP) geometry (45). Thus we can assume a TBP geometry for 6 at low temperatures. This assumption is also supported by a recent work reported by Voronkov et al. (46). (Aroyloxymethyl)trifluorosilanes (9) show the presence of an intramolecular coor-

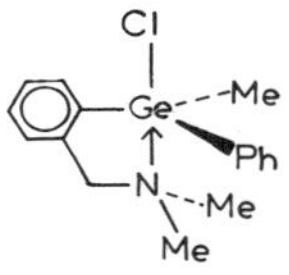

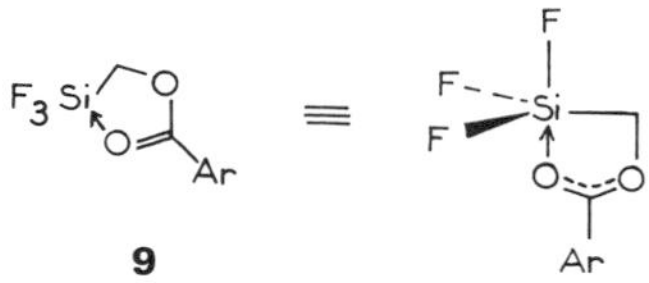

dinate F₃Si←O=C bond, which remains intact even in the gaseous and liquid states. The pentacoordinate character of silicon has been confirmed by X-ray diffraction. Furthermore, the structure shows two fluorines in equatorial positions; the third fluorine atom is apical.

b. ¹⁹F-NMR Data.   The ¹⁹F-NMR spectra of 10 at low temperatures are also consistent with a TBP geometry (45). Moreover, the ¹⁹F-NMR spectroscopic study of the equilibrium shown in Eq. (4) allows a compari-

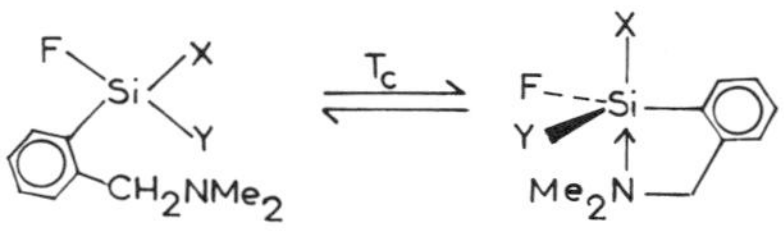

$$X = Y = F \; ; \; X = F, Y = H \; ; \; X = F, Y = Me \; ;$$

$$X = Cl, Y = Me \; ; \; X = O\text{-}t\text{-}Bu, Y = Me$$

(4)

son of the relative apicophilicity of groups bonded to a pentacoordinate silicon center. For instance, the low temperature spectrum of (2-NMe₂-CH₂C₆H₄)MeSiF₂ in nonpolar media displays two singlets at $\delta = -111$

ppm (F_a) and $\delta = -154$ ppm (F_e), each corresponding to one fluorine atom. Similarly, the ^{19}F-NMR spectrum of $(2\text{-}NMe_2CH_2C_6H_4)SiF_3$ consists of a triplet (F_a, $\delta = -128$ ppm) and a doublet (F_e, $\delta = -148$ ppm): the triplet at low field corresponds to one apical fluorine, whereas the doublet at high field accounts for two equatorial fluorines. Thus the fluorine chemical shifts provide valuable information about the structure around silicon and the relative apicophilicity of the groups bonded to silicon. The following orders are found (45):

$$Cl > F, \quad F > OR, \quad \text{and} \quad F > H$$

It is noteworthy to find chlorine more apicophilic than fluorine, although the latter is more electronegative. For instance, it explains the stereochemical behavior of 1-NpFcSiClF (29): only chlorine is displaced by carbon nucleophiles and inversion is the predominant stereochemistry.

In conclusion, all these observations suggest the presence of a five-coordinate silicon intermediate both in nucleophilic displacements at silicon and in the racemization of halosilanes as previously suggested (5, 39). The ability to form a pentacoordinate silicon adduct is not controlled by the electronegativity of the substituents at silicon, but by the tendency of the Si—X bond to be stretched under the influence of a nucleophile:

$$OAc, Cl, Br > F, SR > OR, H$$

This order is parallel to the ability of the Si—X bond to be displaced with inversion

$$OCOR, Br, Cl > F, SR > OR > H$$
$$IN \xrightarrow{\hspace{4cm}} RN$$

and to the tendency of halosilanes to be racemized:

$$F < Cl < Br$$

Apparently, the tendency of an Si—X bond to be stretched under the influence of a nucleophile controls its ability to be cleaved with inversion. Furthermore, it is noteworthy that similar effects are found in nucleophilic displacements at phosphorus (47).

B. Influence of the Nucleophile

The stereochemical outcome is greatly influenced by the nature of the attacking nucleophile (Tables I and II). For the same leaving group, either inversion and retention can be observed, depending on the nature of the entering group. For instance, with carbon organometallics (R^-M^+) the nature of the carbanion and of the metal counterion is of importance:

1. Aryl, phenylethynyl, and alkylorganolithiums (i.e., species with a well-localized negative charge) react mainly with retention. Inversion is observed only with the best Cl or Br leaving groups (Table I, No. 19). On the other hand, allyl or benzyllithiums in which the negative charge is delocalized over an sp^2 carbon system always react with inversion, except with hydrogen, the poorest leaving group (Table I, No. 1).

2. Concerning the influence of the metal counterion M^+, the Grignard reagents show a general shift toward inversion compared to organolithiums. For instance, alkyllithiums lead to retention with the borderline Si—F and Si—SR bonds; the corresponding Grignard reagents react with predominant inversion (5).

It is very clear that these data cannot be explained only in terms of the ability of the leaving group to be cleaved, as discussed in Section II,A.

1. *Pseudorotation and Stereochemical Data*

An explanation could be achieved by the consistent application of the mechanistic concepts used to explain data at tetravalent phosphorus compounds (48). Most of the available data on nucleophilic displacements are usually explained by an apical entry of the nucleophile: displacement of the leaving group occurs from an apical position (Scheme 3). The first invoked TBP intermediate is the most stable one on the basis of the relative apicophilicity of the groups attached to phosphorus; the more electronegative ligands prefer apical positions. Since both inversion and retention can proceed, the retention pathway has been rationalized on the basis of intramolecular ligand exchange by pseudorotation (Scheme 3).

It should be noted that pseudorotation does not explain the retention as stereochemistry in nucleophilic displacements: it only yields a picture of observed stereochemical events. Retention is the consequence of a nucleophilic attack at 90° with respect to the leaving group.

The above question can be answered by examining the following data:

a. Monofunctional Organosilanes. In the case of monofunctional organosilanes, the "preference rules" stated with tetravalent phosphorus

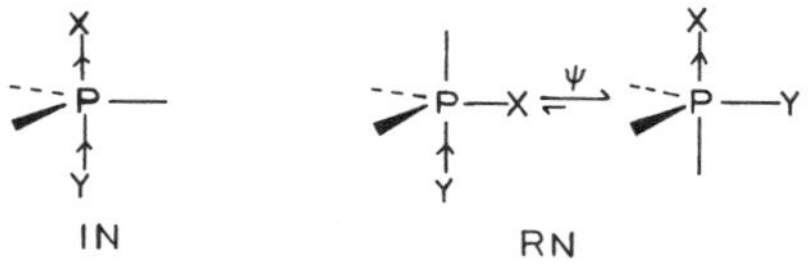

SCHEME 3

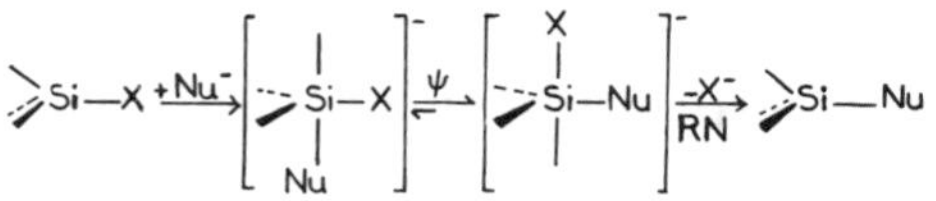

$$\underset{\diagup}{\diagup}Si-X \xrightarrow{+Nu^-} \left[\begin{array}{c} \vdots \\ \diagdown Si-X \\ \vert \\ Nu \end{array} \right]^- \underset{\rightleftarrows}{\overset{\psi}{\rightleftarrows}} \left[\begin{array}{c} X \\ \vert \\ \diagdown Si-Nu \\ \vert \end{array} \right]^- \xrightarrow[RN]{-X^-} \underset{\diagup}{\diagup}Si-Nu$$

SCHEME 4

compounds are not followed for nucleophilic substitutions with retention. Only one electronegative group is attached to silicon; it is also the leaving group. Apical attack of the nucleophile places the only electronegative group in an equatorial position (Scheme 4).

In order to illustrate this point, we report the opposite behavior of the allyl and n-butyl anions in the substitution of an optically active fluorosilane, 1-NpPhMeSi—F. The first leads to inversion (91% IN) and the latter to retention (98% RN) (*49*), when the reactions with allyl- and n-butyllithium are carried out in the presence of a cryptand specific for lithium cation. The use of a cryptand makes it possible to study the difference in behavior of the anion species alone, since it avoids the possibility of electrophilic assistance by Li^+ which could affect the apicophilicity of the fluorine atom.

Kanberg and Muetterties (*50*) have shown by NMR spectroscopy that fluorine substituents always occupy the apical positions in pentacoordinated anions such as $RSiF_4^-$ or $R_3SiF_2^-$. Thus, using the rules for phosphorus chemistry, the most stable intermediates will be **12** and **13**,

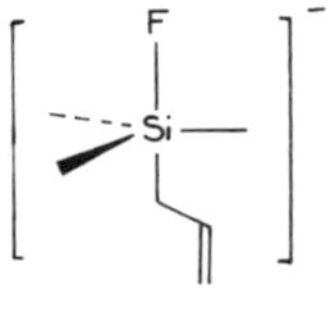

$$\left[\begin{array}{c} F \\ \vert \\ \cdots Si- \\ \end{array} \right]^- \qquad \left[\begin{array}{c} F \\ \vert \\ \cdots Si- \\ \end{array} \right]^-$$

12 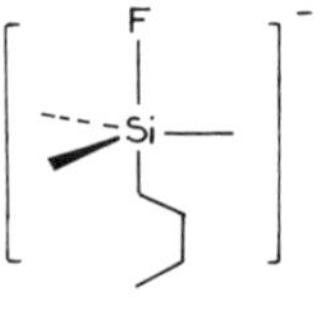   **13**

whatever the entering nucleophiles. Structures **12** and **13** explain only inversion: they cannot lead to the substitution of the fluorine atom with retention. *Thus the "preference rules" stated to explain the stereochemical data reported in phosphorus chemistry fail in describing the changes of stereochemistry observed with the nature of the nucleophile.*

b. Solvent Effects. We have found that small solvent or salt effects can change the stereochemistry. Some pertinent data are summarized in Table VI.

Using the preference rules of phosphorus chemistry, the most stable intermediates would be the same whatever the nature of the solvent and the experimental conditions. So, it is impossible to explain in this way the changes of stereochemistry with medium effects or with the nature of

TABLE VI

X	Nu⁻ (solvent)	Predominant stereochemistry (SS %)	Reference
F	$CH_3MgBr(Et_2O)$	IN (81)	7b
	$CH_3MgBr(THF)$	RN (93)	7b
	$(CH_3)_2Mg(Et_2O)$	RN (98)	7b
OMe	$CH_2{=}CHCH_2MgBr(Et_2O)$	IN (93)	7b
	$CH_2{=}CHCH_2MgBr(DME)$	RN (78)	7b
	$(CH_2{=}CH{-}CH_2)_2Mg(DME)$	RN (75)	7b
F	$LiAlH_4/LiBr(Et_2O)$	RN (95)	51
	$LiAlH_4(Et_2O)$	Racemic	52
	$AlH_4^{-\,a}$	IN (65)	51

^{a} $AlH_4^- = LiAlH_4 +$ a cryptand specific of Li^+.

nucleophile. Thus the answer must be found in other concepts and, in particular, in the intimate structure of the nucleophile.

2. Electronic Character and Stereochemistry

The stereochemistry at silicon is extremely sensitive to the nature of the nucleophiles (Tables I, II, and VI). As a consequence, the stereoselectivity, i.e., either percentage of RN or percentage of IN, is a quite sensitive and reliable measure of the dependence of the mechanism upon small changes in the structure of the anion, the metal, or the solvent. The analysis of the experimental data can take the following form:

1. Hard nucleophiles (alkyl anions) give predominant retention of configuration. At the opposite extreme, softer nucleophiles, such as benzyl or allyl anions, lead mainly to inversion. The following examples are quite significant:

(a) p-MeOC₆H₄CH₂Li cleaves the Si—OMe bond with retention (Table I, No. 7) instead of inversion for C₆H₅CH₂Li (Table I, No. 8). The p-methoxybenzyl anion, in which the negative charge is more localized on the carbon reactive center because of the electron-donor methoxy group, is a harder reagent than the benzyl anion. Consequently, the stereochemistry is shifted toward retention.

TABLE VII

COMPARISON OF STEREOCHEMICAL BEHAVIOR OF RLi AND RMgX

Substrate	Nucleophile	Predominant stereochemistry	Reference
1-NpPh(R)Si—F (R = Me, Et, i-Pr, Vi)	R'Li (R' = Me, Et, n-Pr, n-Bu)	RN	11, 20
	R'MgX (R' = Me, n-Bu)	IN	7a
1-NpFc(H)Si—F	RLi (R = Me, Et, CH₂=CH—CH₂	RN	29
	RMgX (R = Me, Et, CH₂=CH—CH₂	IN	29
1-NpPhMeSi—X (X = SMe or SPh)	RLi (R = Et, n-Bu, neo-C₅H₁₁)	RN	22, 23
	RMgX (R = Et, n-Bu)	IN	23
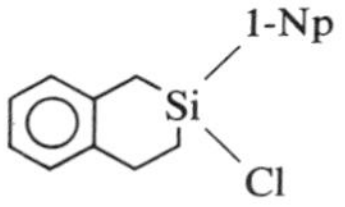	RLi (R = Me, Et, n-Pr, n-Bu)	RN	35
	RMgX (R = Me, Et, n-Pr, n-Bu)	IN	36

(b) The Ph₂CH⁻ anion, in which the negative charge is particularly well delocalized, leads only to inversion, even with hydrogen as the leaving group (Table I, Nos. 2, 8, and 13).

2. Concerning the nature of the metal, the softer (harder) the cation M⁺, the more covalent (ionic) the R—M bond of the organometallic and the softer (harder) the anion R⁻ (53). Thus the organolithiums will be harder reagents than the analogous Grignard reagents. Organolithiums react mainly with retention, compared to Grignard reagents. Some significant data are summarized in Table VII. In addition, organolithium and organosodium compounds behave similarly (54).

3. An increase of the solvating power of the solvent (Et₂O < THF < DME) leads to a shift of the stereochemistry toward retention. Striking data are reported in Table VIII for 1-NpPhViSi—F (7a). The solvating power can modify the electronic character of the nucleophile. A Grignard reagent is softer in ether than in THF or DME (Scheme 5).

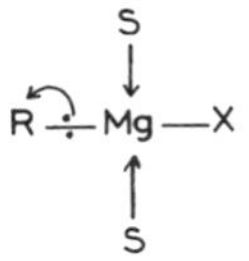

$$R \overset{\displaystyle S}{\underset{\displaystyle S}{-Mg—X}}$$

SCHEME 5

TABLE VIII

INFLUENCE OF SOLVENT ON STEREOCHEMISTRY
(R_3SiX = 1-NpPhViSiF)

Nucleophile	Solvent	Predominant stereochemistry (%)	
EtMgBr	Et_2O	IN	(65)
	THF	IN	(66)
	DME	RN	(59)
Et_2Mg	Et_2O	IN	(59)
	THF	RN	(67)
	DME	RN	(67)
n-BuMgBr	Et_2O	IN	(67)
	THF	RN	(59)
	DME	RN	(71)
(n-Bu)$_2$Mg	Et_2O	IN	(69)
	THF	RN	(75)
	DME	RN	(74)

4. Significant data concerning the influence of the solvation were reported by Sommer and Fujimoto (55) in the case of nucleophilic substitutions by alkoxides at Si—OMe, Si—F, and Si—SMe bonds (Table IX).

The most important fact that emerges from these data is the stereochemical cross-over from retention to inversion when the alcohol content of the medium is increased. Whatever the nature of the leaving group, inversion becomes predominant either with ROLi or RONa. These observations suggest an explanation of the above data in terms of electronic factors. The RO^-M^+ species in benzene have a localized negative charge on the oxygen atom; i.e., they act as hard anions and react with retention. When the alcohol ratio is increased, the charge is dispersed by strong hydrogen-bonding interactions (Scheme 6). Such dispersal of the negative charge produces a softer species which reacts predominantly with inversion.

Thus a close relationship between the stereochemistry of substitution at silicon and the electronic character of the attacking nucleophile exists. For the same leaving group at silicon, the hard reagents tend to lead to retention whereas soft reagents give inversion, as summarized for organometallics in Scheme 7. Significant data are summarized in Table X.

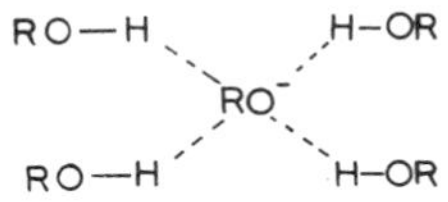

RO—H   H—OR

RO

RO—H   H—OR

SCHEME 6

TABLE IX

NUCLEOPHILIC DISPLACEMENTS BY ALKOXIDES

$R_3Si—X + n\text{-BuOH}/n\text{-BuOM} \rightarrow R_3Si—O\text{-}n\text{-Bu}$

(M = Li or Na)

Organosilane[a]	Metal alkoxide	ROH/ROM (mol/mol)	ROH/R$_3$SiX (mol/mol)	ROH (vol% in solvent)	Predominant stereochemistry (%)
R$_3$Si—OMe	n-BuOLi	10	2.6	2.3	RN (100)
	n-BuOLi	10	15	16.7	RN (89)
	n-BuOLi	72	47	51.2	IN (65)
	n-BuOLi	210	79	100	IN (>81)
	n-BuONa	15	2.6	2.3	RN (95)
	n-BuONa	28	15	16.7	IN (76)
	n-BuONa	180	49	52.0	IN (82)
	n-BuONa	450	83	100	IN (77)
R$_3$Si—F	n-BuOLi	2	1.31	3	RN (92)
	n-BuOLi	14.3	1.11	15.8	RN (62)
	n-BuOLi	32	1.37	50.0	IN (88)
	n-BuONa	1.5	1.16	2.0	RN (94)
	n-BuONa	13.7	1.23	15.8	RN (56)
	n-BuONa	38	1.19	46.8	IN (80)
R$_3$Si—SMe	n-BuOLi	0.7	1.15	0.8	RN (70)
	n-BuOLi	3.0	1.13	3.4	IN (61)
	n-BuOLi	39	1.22	50	IN (85)
	n-BuONa	2.4	1.30	3.2	IN (58)
	n-BuONa	36	1.32	50	IN (72)

[a] R$_3$Si = 1-NpPhMeSi. The reactions were carried out in benzene solvent.

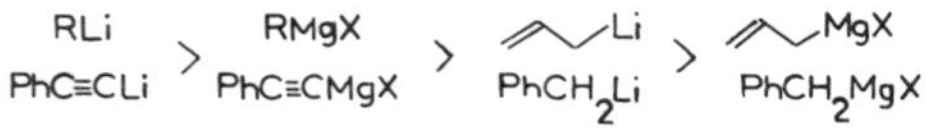

$$\begin{array}{ccccccc}
\text{RLi} & & \text{RMgX} & & \text{\reflectbox{/}\hspace{-2pt}\diagup\text{Li}} & & \text{\reflectbox{/}\hspace{-2pt}\diagup\text{MgX}} \\
\text{PhC≡CLi} & > & \text{PhC≡CMgX} & > & \text{PhCH}_2\text{Li} & > & \text{PhCH}_2\text{MgX}
\end{array}$$

SCHEME 7.   Stereochemical behavior of organometallics (R = alkyl or aryl).

Consistent with these observations are the experiments with phenoxide anions (56) and the relationship between the stereochemistry at silicon and the regioselectivity of attack on α-enones (57).

a. Coupling Reactions with Phenoxides.   In these reactions, the electronic character of the negative charge on the oxygen atom can be changed by varying the substituent Y in the para position. Some data are given in Table XI.

An electron-releasing substituent (OMe group), which increases the negative charge on the nucleophilic center and thus its hardness, favors retention as the stereochemical outcome. On the other hand, a significant decrease of stereoselectivity in retention or predominant inversion occurs with an electron-withdrawing substituent such as —NO₂. In this case, we are faced with a greater delocalization of the negative charge over the aromatic system and thus with a softer nucleophile.

The effect of a variable delocalization of the negative charge is increased when carrying out the reaction with a cryptand specific for the Na⁺ cation. With *p*-methoxyphenoxide, the dissociation of the ion pair

TABLE X

STEREOCHEMISTRY AT SILICON AND ELECTRONIC CHARACTER OF ORGANOMETALLICS

Substrate	Nucleophile	Predominant stereochemistry (%)	Reference
R | 1—NpPhSi—F R = Et or Vi	Ph—C≡C—Li	RN (100)	11
	MeLi	RN (93)	11
	n-PrLi	RN (97)	11
	n-BuLi	RN (93)	11
	Me—⟨O⟩—Li	RN (98)	11
	EtMgBr	IN (65)	7a
	PhCH₂Li	IN (82)	11
	/\/—MgBr	IN (93)	7a
	PhCH₂MgCl	IN (91)	7a
Vi | 1—NpPhSi—F	PhC≡CMgBr (THF)	RN (76)	7a
	n-BuMgBr (THF)	RN (59)	7a
	/\/—MgBr (THF)	IN (62)	7a

TABLE XI

STEREOCHEMISTRY OF REACTIONS OF R_3SiX WITH PHENOXIDES[a]

$$Y-\!\!\!\bigcirc\!\!\!-O^-M^+ \quad R_3Si-O-\!\!\!\bigcirc\!\!\!-Y$$

R_3Si-X	$\underline{Y}$	M^+	$[\alpha]_D$ (deg)	Stereo-chemistry
Ph	MeO	Na$^+$	6	RN
|	MeO	—[b]	8	RN
1—Np—Si—SPh	H	Na$^+$	2	IN
|	H	—[b]	5	IN
(+) Me	NO$_2$	Na$^+$	3	IN
	NO$_2$	—[b]	8	IN
Ph	MeO	Na$^+$	−4.5	RN
|	MeO	—[b]	−6	RN
1—Np—Si—Me	H	Na$^+$	3	RN
|	H	—[b]	−2.5	IN
(+) F	NO$_2$	Na$^+$	0.5	RN
	NO$_2$	—[b]	−4	IN

[a] Solvent = THF.
[b] Reactions carried out with a cryptand specific for Na$^+$.

increases the negative charge on the oxygen atom, and the proportion of retention increases. On the other hand, with a softer anion such as p-nitrophenoxide, complexing the Na$^+$ cation favors the delocalization of the negative charge and thus inversion.

Thus the changes of the stereochemistry always depend on electronic factors: the more delocalized nucleophilic anions react mainly with inversion, the ones with more concentrated charge with retention.

b. Parallel between the Stereochemistry at Silicon and the Regioselectivity of Attack at α-Enones. The regioselectivity of nucleophilic additions to α-enones have been extensively studied (58). The experimental data, especially that of Seyden-Penne et al. (59), and theoretical studies (60–63) showed clearly that the proportion of 1,2 to 1,4 addition is controlled mainly by electronic factors.

In the case of alanes (Table XII), we have shown that reagents that favor 1,2-addition to α-enones react with retention at silicon, whereas those that favor 1,4-addition react with inversion (57).

Similar data are obtained with organometallics. The change from inversion to retention is in the order (57):

$$R_2CuMgX > RMgX > RLi$$

TABLE XII

REDUCTION REACTIONS OF SOME OPTICALLY ACTIVE SILANES AND REPRESENTATIVE ENONES

Reaction number	Reagent (solvent)[a]	1-NpPhMeSiX (55%) X = OMe (%)	1-NpPhMeSiX (55%) X = F (%)	1-Np/Si/F silane (%)	1-Np/Ph–Si/O (THF) silane $[\alpha]_D$ (deg)	Cyclohexenone[c] (%)	Benzalacetone[c] (%)
1	Dibal (hexane)	RN (100)	RN (98)	RN (100)	+19 RN	1,2 (100)	1,2 (100)
2	Dibal (Et$_2$O)	RN (99)	IN (90)	RN (80)	+16	1,2 (92)	1,2 (75)
3	Dibal (TMEDA–hexane)	RN (95)	IN (90)	RN (70)	+15 RN	1,2 (60)	1,2 (14)
4	Dibal (THF)			RN (100)	+10 RN	1,2 (85)	1,2 (10)
5	AlH$_3$ (Et$_2$O)	RN (92)	IN (96)	IN (70)	+16 RN	1,2 (97)	1,2 (80)
6	AlH$_3$ (THF)	RN (90)	IN (100)	IN (75)	+ 8 RN	1,2 (87)	1,2 (25)
7	(*i*-BuO)$_2$AlH (Et$_2$O)	RN (90)	IN (100)	IN (80)	−14 IN	1,2 (87)	1,2 (10)
8	(*i*-BuO)$_2$AlH (THF)	RN (90)	IN (100)	IN (80)	−16 IN	1,2 (80)	1,2 (8)
9	(EtS)$_2$AlH (THF)	RN (85)	IN (100)	IN (80)	−18 IN	1,2 (40)	1,2 (6)
10	H$_2$AlI (THF)	RN (55)				No reaction	1,2 (0)

[a] TMEDA = tetramethylethylenediamine, THF = tetrahydrofuran, Dibal = *i*-Bu$_2$AlH.

[b] Predominant stereochemistry: the $[\alpha]_D$ for optically pure R$_3$Si*H are known. A predominant stereochemistry of 90% inversion indicates a reaction path that is 90% invertive and 10% retentive, giving a product that is 80% optically pure.

[c] % 1,4-addition = 100% − (1,2-addition)%.

Organolithiums give only 1,2-addition (58), organocuprates lead mainly to attack at C-4 (64), and organomagnesium reagents show intermediate behavior. Furthermore, this parallel is strengthened by the inversion at silicon observed in the substitution of the Si—OMe by $(PhCHCN)^-Li^+$ (57). This soft reagent is known to give predominant 1,4-addition with α-enones (65).

In concluding this section on the influence of the nature of the nucleophile, it is important to stress the dominant influence of the nucleophile on the stereochemistry at silicon. This effect cannot be interpreted in terms of the stability of the intermediate on the basis of the "apicophilicity rule" as stated in phosphorus chemistry. It fails to explain the retention of configuration as stereochemical outcome. No better explanation can be extracted from the quasicyclic S_Ni–Si mechanism (1, 2). On the other hand, data obtained with various nucleophiles show clearly that the stereochemistry is controlled by the electronic character of the nucleophile. In other words, this factor at first determines the geometry of attack of the nucleophile at silicon, which leads in a first determinant step to the formation of a pentacoordinate intermediate (33). We proposed the following:

1. Concerning the reactions occurring with inversion, the stereochemistry requires a back-side attack of the nucleophile opposite the leaving group. The skeleton of intermediate **14** corresponds to apical entry and departure.

$$\left[\begin{array}{c} X \\ | \\ \cdots\text{Si} — \\ | \\ Nu \end{array} \right]^-$$

14

2. Concerning substitution at silicon occurring with retention, a nucleophilic attack 90° to the leaving group is required. For the same leaving group, this geometry of attack is only determined by the electronic character of the nucleophile. Two different mechanisms can be advanced: (a) apical attack of the nucleophile, leading to an intermediate such as **15** with the electronegative leaving group in an equatorial position or (b) equatorial attack of the nucleophile, the leaving group being apical in a trigonal bipyramid like intermediate **16**. Apical entry of the nucleophile implies that the leaving group is equatorial. Kanberg and Muetterties (50) have shown that the electronegative fluorine atoms always occupy the apical positions of pentacoordinated anions such as $RSiF_4^-$ or $R_2SiF_3^-$.

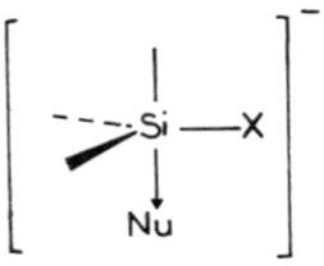

15 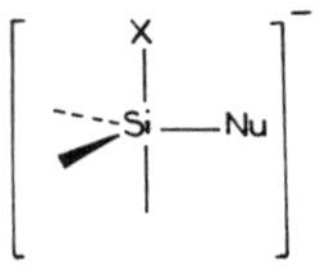   **16**

Thus we can assume that the most stable intermediate resulting from apical attack of the nucleophile is **14** and not **15**. This intermediate **14** explains only inversion at silicon.

On the other hand we can note that equatorial attack is strongly supported by Martin's elegant structural investigations on pentacoordinated silicon species (66). The siliconates **17** and **18** are obtained by nucleophilic attack (Ph⁻ or 4-Me₂N-C₅H₄N) at a tetravalent silicon atom **19**. Their

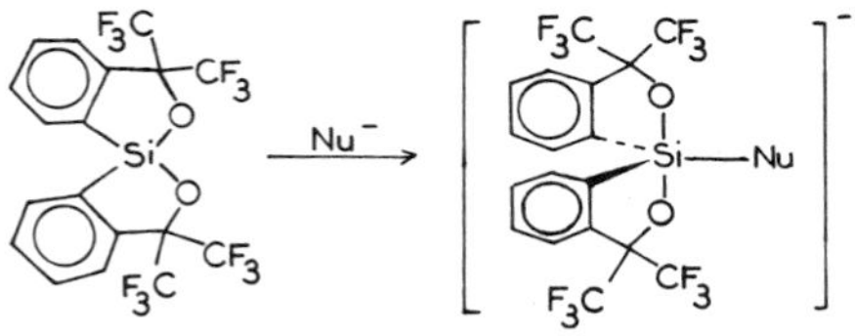

19

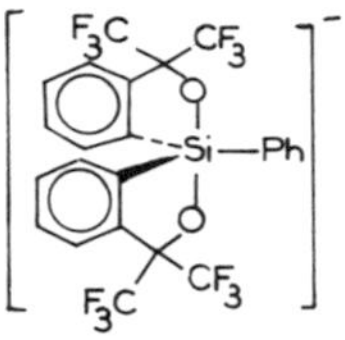

17 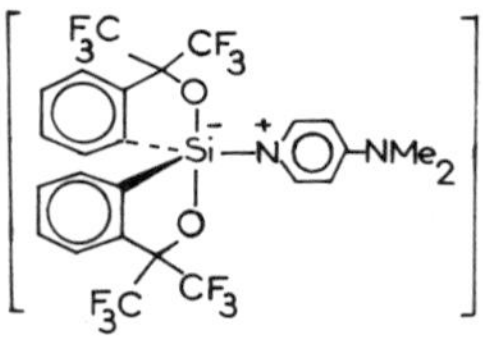   **18**

structures, established by NMR spectroscopy, confirm clearly that the electronegative groups (oxygen atoms) are apical, the two five-membered rings apical–equatorial, and the entering group in an equatorial position.

That equatorial attack can explain the retention at silicon is also well supported by data reported by Minot (6c) for the calculation of the HOMO of a pentacoordinate D_{3h} intermediate (SiX_5^-). Briefly, we can summarize the main conclusions as follows:

1. The geometry of SiH_5^- with nearly equal bond lengths in the equatorial and apical positions suggests the two kinds of stereochemistry, i.e., either retention or inversion of configuration.

2. A substituent of low level (soft reagent) shows a high capacity to stabilize a negative charge in the apical position. It weakens the bond in

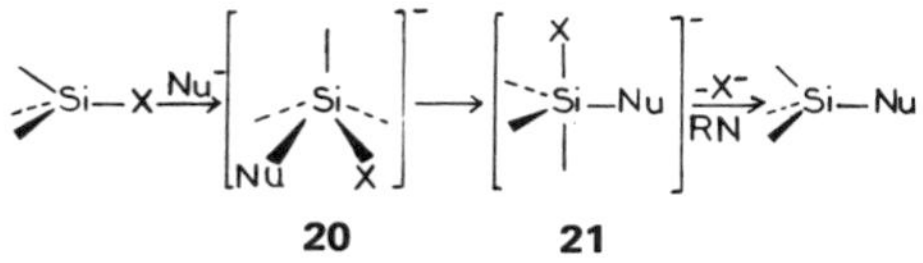

20   **21**

SCHEME 8

the other apical position, whereas it strengthens the equatorial bond. The consequence is approach of a soft nucleophile 180° to the leaving group, and therefore decomposition of the intermediate with inversion.

3. Equatorial groups of high level (hard or small reagents) stabilize a negative charge in a perpendicular apical position. They weaken a perpendicular apical bond; in contrast, they strengthen the trans-equatorial bond. The stereochemical consequence is approach of the nucleophile 90° to the leaving group. Therefore, it implies retention as the stereochemical outcome.

Consequently, equatorial attack seems the most reasonable and simplest way to describe approach of the nucleophile 90° to the leaving group, leading to the most stable pentacoordinated intermediate. However, it is also possible to postulate a square pyramidal structure such as **20** as a transition state that would change into intermediate **21** (Scheme 8). Such a structure was previously proposed in phosphorus chemistry by Hudson *et al.* to explain the alkaline hydrolysis of cyclic phosphonamidates (*67*).

III

NUCLEOPHILIC DISPLACEMENTS AT SILICON: A RATIONALIZATION OF THE STEREOCHEMISTRY BY A FRONTIER-ORBITAL APPROXIMATION

Anh and Minot (*6*) reported a rationalization of the stereochemistry at silicon by an extension of Salem's treatment of the Walden inversion (*68*). The frontier-orbital approximation is assumed, i.e., the major interaction during the reaction occurs between the HOMO of the nucleophile and the LUMO of the substrate σ^*_{Si-X}. The calculated structure of the latter is shown below, with the big lobes of the hydrid AOs pointing toward each other (Scheme 9).

Front-side attack, corresponding to an attack on the big lobe of silicon, leads to retention. When unfavorable, out-of-phase overlap between the nucleophile and the orbitals of the leaving group predominates, nucleophilic attack occurs at the rear of the molecule, opposite X, leading to

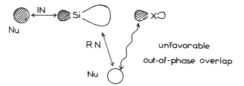

SCHEME 9.   Structure of the substrate LUMO, $\sigma^*_{Si-X}(R_3Si-X = H_3Si-Cl$ or $H_3Si-F)$.

inversion. Therefore, retention and inversion can be considered the result of a fine balance between the in-phase and out-of-phase orbital overlap between the nucleophile and the LUMO of the substrate σ^*_{Si-X}. In other words, it is possible to increase the probability of retention of configuration by increasing the favorable interaction between the nucleophile and the reaction center and/or by decreasing the unfavorable interaction between the nucleophile and the leaving group.

Salem's frontier-orbital treatment is consistent with the fact that retention of configuration is a commonly observed stereochemical outcome at silicon (Tables I and II), whereas there is still no proven example of an S_N2 reaction with retention at carbon (69). Because Si—X bonds are significantly longer than C—X bonds, the unfavorable interaction between X and the nucleophile for front-side RN attack is less for silicon than for carbon (Scheme 9). The valence orbitals also change from 2s and 2p for carbon to 3s and 3p for silicon and therefore become more diffuse and capable of better overlap with the nucleophile at longer distances. Consequently, the probability of attack with retention is enhanced.

The shape of the σ^*_{Si-x} LUMO is quite sensitive to the leaving group X. The calculations show that a group X of high electronegativity, for instance X = F in the case of H_3SiF, increases the s character around silicon, leading to a more dissymetric hybrid atomic orbital at silicon with a bigger lobe between Si and X. Increasing the electronegativity of the leaving group will therefore favor retention of configuration. At the same time, the stereochemical outcome will be influenced by the length of the Si—X bond and by the size of the valence orbitals around X (Scheme 9).

The influence of the nucleophile is analyzed as follows. A hard reagent is usually a small one (70), with contracted valence orbitals. Its long-range, out-of-phase unfavorable overlap with the leaving group will be negligible and front-side attack leading to retention is therefore possible. On the other hand, a soft nucleophile usually has diffuse valence orbitals (70). It has therefore a sizable out-of-phase overlap with the leaving group and the stereochemistry is shifted toward inversion (rear-side attack).

The size of the reagent is not the only controlling factor since a change in the hardness of the nucleophile implies a modification of its valence

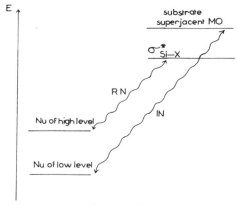

E

substrate
superjacent MO

σ^*_{Si-X}

R N

IN

Nu of high level

Nu of low level

SCHEME 10

orbitals and also of its HOMO level (Scheme 10). When the level is high (i.e., in a hard nucleophile with contracted negative charge), frontier-orbital interaction is predominant. When the level is low (i.e., in a soft nucleophile with voluminous valence orbitals), the relative importance of the nucleophile HOMO–substrate superjacent MO interaction is increased. The structure of the latter is shown below (Scheme 11) for the Si—F bond.

The big lobe of the Si hybrid orbital points to the rear in the superjacent MO, which therefore favors inversion of configuration. As a consequence, the authors are led to the conclusion that a nucleophile with a high-lying HOMO will give more retention of configuration than a nucleophile with a low-lying HOMO. It has been shown that in aprotic solutions, the HOMO of a hard anion lies at a higher energy than that of a soft anion (71). Thus hard nucleophiles lead to retention; at the opposite extreme, soft nucleophiles react with inversion.

Before we discuss the experimental observations, we can summarize briefly the essential features of the above theoretical approach as follows:

1. The major interactions during the reaction between a nucleophile and an organosilane R_3Si—X occur between the HOMO of the nucleophile and the LUMO of the substrate σ^*_{Si-X}.

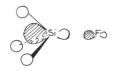

SCHEME 11.   Structure of the superjacent MO of SiH_3F.

2. The shape of σ^*_{Si-X} directs the stereochemistry. A group X of high electronegativity increases the s character at silicon. This implies a bigger lobe between Si and X. Thus increasing the electronegativity of the leaving group will favor the retention (Scheme 9).

3. The stereochemistry is the result of a balance between the in-phase and out-of-phase orbital overlap between the nucleophile and the substrate LUMO.

Our purpose is now to show that the Anh and Minot approach gives a qualitatively rational explanation for most of the stereochemical data at silicon.

A. Stereochemical Changes with the Leaving Group

1. Substitution of the Si—H Bond

Although the Si—H bond is quite short (41) and hydrogen has a low electronegativity, retention of configuration is commonly observed with H as leaving group (Table I and Section II). The hydrogen 1s orbital is small and overlaps little with the orbitals of the nucleophile. Furthermore, hydrogen is the only leaving group with no lone pair; the H—Nu repulsion is therefore drastically reduced, and retention is favored (Scheme 9).

The stereochemical behavior of Ph₂CHLi, i.e., cleavage of the Si—H bond with predominant inversion, can be explained as follows. We are faced with a very soft nucleophile of low level (Scheme 10). The substrate superjacent MO–nucleophile HOMO interaction prevails. The big lobe of the Si hybrid orbital points to the rear in the superjacent MO (Scheme 11), and the inversion is therefore favored. However, this reaction is quite slow (10).

2. Br, Cl, and F as Leaving Groups

Stereochemical studies on nucleophilic displacements of halosilanes suggest the following generalizations:

1. Fluorosilanes lead either to retention or inversion according to the nature of the nucleophile (Table I, Nos. 12–14, and Table VII).

2. Bromo- and chlorosilanes react with inversion, whatever the nucleophile (Table I, Nos. 19–24). Retention is only observed for dibal in hexane (23, 72).

The Anh and Minot model explains these observations as follows:

1. When chlorine or bromine are replaced by fluorine as the leaving group, the electronegativity increases the s character at the silicon atom. The σ^*_{Si-X} MO shows a bigger lobe pointing between Si and X.

2. The valence orbitals of X become more contracted for the fluorine atom, and this decreases the unfavorable overlap between X and the nucleophile (Scheme 9).

3. At the same time, the Si—X bond shortens, increasing the unfavorable X—Nu overlap in an attack with retention (Scheme 9).

The two former factors favor a front-side attack leading to the retention. Moreover, numerical calculations indicate that in this example, the Si—X bond shortening does not compensate for the two former effects. Therefore, retention is favored.

3. OR and SR as Leaving Groups

As shown for halosilanes, replacing SR by an element of higher electronegativity in the same column of the periodic table (OR), leads to a stereochemical shift to retention (Table I, Nos. 7–11 and 15–18).

In the case of the Si—OR bond, we are faced with a leaving group (oxygen atom) of high electronegativity. Two simultaneous factors favor front-side attack of the nucleophile: (1) the σ^*_{Si-OR} MO shows a big lobe at silicon pointing between the Si and oxygen atoms and (2) valence orbitals around oxygen are contracted so that the unfavorable Nu—X overlap is decreased (Scheme 9).

As a consequence, these two effects favor retention of configuration. Experimental trends show that inversion is only observed for soft nucleophiles, whose valence orbitals are usually diffuse (71). In this latter case, the unfavorable out-of-phase Nu—OR overlap prevails and an attack at the rear becomes possible (Scheme 9). However, the reaction is slow. For instance, the reaction of CH_2=CH—CH_2MgBr on a methoxysilane is 10^5-fold slower than the same reaction on a fluorosilane (8).

In contrast, the Si—SR bond is cleaved mainly with inversion. Although the Si—SR bond is longer, the unfavorable out-of-phase Nu—SR overlap prevails because of the diffuse valence orbitals around the sulfur atom. Furthermore, the σ^*_{Si-SR} shows a smaller lobe between the Si and S atoms than does the σ^*_{Si-OR} MO. As a consequence, the probability of a rear-side attack with inversion is enhanced. For instance, alkyl Grignard reagents cleave the Si—SR bond with inversion instead of retention, as is the case with alkoxysilanes (5).

Si◯ F⊃

SCHEME 12.   Shape of the σ^*_{Si-F} MO.

4. *A Comparison of F and SR as Leaving Groups*

The comparison of SR and F is more complex since fluorine and sulfur are not in the same column of the periodic table and have very different properties. On the basis of the older model involving pK_a values, F should be considered a good leaving group (pK_a of HF = 3.17) and MeS a poor leaving group (pK_a of MeSH $\simeq$ 10) (*1, 2*). However, experimentally X = SR and F give analogous results (Table I, Nos. 12–14 and 15–18). The Anh and Minot treatment provides a rationalization for this.

The valence orbitals around F are contracted. The σ^*_{Si-F} MO is centered on silicon (Scheme 12). These two factors favors retention. But because the Si—F bond is short [1.54 Å (*41*)], the unfavorable X—Nu overlap increases. Both inversion and retention may be possible. In contrast, valence orbitals on sulfur are diffuse: the σ^*_{Si-SR} MO has more p character (Scheme 13). These two factors favors inversion. But the Si—SR bond is long [2.14 Å (*41*)]. The unfavorable SR—Nu overlap in a RN attack decreases and retention may be also possible.

5. *Summary*

In conclusion, the influence of the leaving group on stereochemistry can be predicted from a knowledge of the shape of the σ^*_{Si-X} MO. The stereochemistry is the result of a fine balance between the electronegativity of the leaving group X, the size of the valence orbitals around X, and the length of the Si—X bond. The general trends are summarized in Table XIII.

This agrees well with the stereochemical data, and in particular, for the general trend observed experimentally.

$$Cl, Br > F, SR > OR > H$$
$$IN \longrightarrow RN$$

This frontier-orbital approximation does not give a complete picture of the influence of the leaving group on the stereochemistry. For instance, it does not account for the aptitude of the Si—X bond to be stretched under

SCHEME 13.   Shape of the σ^*_{Si-SR} MO.

TABLE XIII

INFLUENCE OF THE LEAVING GROUP ON THE STEREOCHEMISTRY AS A CONSEQUENCE
OF THE SHAPE OF THE σ^*_{Si-X} MO

X	Shape of the σ^*_{Si-X} MO	Bond length (Å)	Stereo-chemistry
H		1.50	RN
OR		1.63	
F		1.54	
SR		2.14	
Cl		2.01	IN

the influence of a nucleophile. The best illustration of this observation is given by the close parallel between the ability of the Si—X bond to be displaced with inversion and its tendency to be stretched (Section I,A,3) in pentacoordinated structures:

$$\text{OAc, Cl, Br} > \text{F} > \text{SR} > \text{H, OR}$$
$$\text{IN} \longrightarrow \text{RN}$$

However, this approach gives a qualitatively rational explanation for most of the experimental data.

B. Stereochemical Changes with the Nucleophile

In the foregoing discussion (Section II,B), we stressed the dominant influence of the electronic character of the nucleophile on the stereochemistry at silicon. We wish now to propose the following concerning this effect.

1. Carbon Nucleophiles

a. Alkyllithiums. These compounds react mainly with retention. Inversion is only observed with the best Cl or Br leaving groups (Table I, Nos. 1, 7, 12, 15, and 19). The cryptand effect results in an increase of retention ratio (Table XIV) (8). A parallel increase of the rate is observed.

TABLE XIV

CRYPTAND EFFECT ON THE STEREOCHEMISTRY AND RATE OF
NUCLEOPHILIC DISPLACEMENTS AT SILICON

Substrate	Nucleophile	Predominant stereochemistry (%)	$t_{1/2}$ ratios[a]
Ph | 1-Np—Si—OMe | Me	n-BuLi/KLi$^+$/Et$_2$O	RN (100)	2
	n-BuLi/TMDA/Et$_2$O	RN (90)	
	n-BuLi/Et$_2$O	RN (86)	
	n-BuLi/heptane	RN (68)	10^2–10^3
	n-BuLi/KLi$^+$/heptane	—	
Ph | 1-Np—Si—Me | F	n-BuLi/KLi$^+$/Et$_2$O	RN (98)	5
	n-BuLi/TMDA/Et$_2$O	RN (82)	
	n-BuLi/Et$_2$O	RN (80)	
	n-BuLi/heptane	RN (70)	10^3–10^4
	n-BuLi/KLi$^+$/heptane	—	
Ph | 1-Np—Si—SPh | Me	n-BuLi/KLi$^+$/Et$_2$O	RN (86)	
	n-BuLi/Et$_2$O	RN (65)	

[a] KLi$^+$ = a cryptand specific of Li$^+$. $t_{1/2}$ ratios = ratio $t_{1/2}$ of a reaction carried out without complexing reagent to that of a reaction carried out in the presence of added TMEDA or KLi$^+$.

For instance, in the case of R_3Si—OMe, we get a 10^2–10^3-fold rate acceleration.

Alkyllithiums have a negative charge on carbon which is in part transferred to the Li$^+$ cation. When the Li$^+$ cation is trapped, the ion pair is much less tight and the negative charge is contracted on the reactive carbon center. In the latter case, we can suppose smaller valence orbitals on carbon (Scheme 14). The out-of-phase overlap with the leaving group is diminished, and therefore the retention is favored. This is in agreement with the experimental data.

This cryptand effect which results in increases of retention ratio (8) and of the rate, also can be explained by a higher level of anion HOMO (Scheme 10). As a consequence, the frontier-orbital interaction is increased. Therefore, retention is kinetically favored.

b. Alkyl Grignards. These reagents are also quite interesting. Compared to organolithiums, the stereochemistry shifts generally toward in-

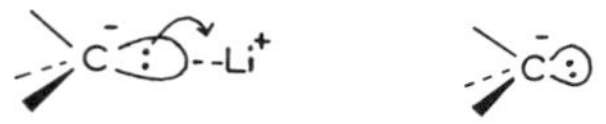

SCHEME 14. Supposed shapes of the HOMO of alkyllithium and of naked alkylanions.

SCHEME 15. Supposed shapes of the HOMO of alkyllithium and alkyl Grignard reagents.

version of configuration (Table VII). The carbon–magnesium bond is moderately covalent, and as a consequence the electrons of the nucleophile are those of the C—Mg bond, which are localized in an MO between the two atoms. It is probable that this MO is more voluminous than the small valence orbitals around the nucleophilic carbon of an alkyl organolithium (Scheme 15). Thus the greater aptitude of Grignard reagents to give inversion as compared to organolithiums can be understood.

Similar arguments allow an explanation of the change of stereochemistry observed with Grignard reagents when increasing the solvent basicity (Table VIII). The addition of a basic solvent (THF, DME) implies a modification of the carbon–magnesium MO, in which the latter become more contracted on the carbon atom (Scheme 16). As a consequence the nucleophile is smaller in size and the stereochemistry is shifted to retention (Tables VI and VIII).

c. Allyllithium. This is a softer nucleophile than alkyllithium, so it can lead to inversion of configuration (Table XV) (*56*). When the Li^+ cation is complexed, the proportion of inversion is always increased. These observations can be explained as follows. In the case of allyllithium, Schleyer *et al.* (*73*) have shown that the valence orbitals are centered toward Li^+ and therefore have some sp^3 character. In contrast to the naked allyl anion, the valence orbitals have a pure p character, and thus they are more diffuse. In the latter case, the unfavorable out-of-phase overlap with the leaving group is increased: the rear-side attack of the nucleophile is promoted, and therefore more inversion is observed when Li^+ is trapped (Scheme 17).

2. Nucleophilic Displacements by Hydrides

The alanes $AlH_{(3-n)}Y_n$ (Y = OR, SR, or I) are interesting nucleophiles. The hydrogen atom has no unshared electron pairs, and therefore the

SCHEME 16. Supposed shape of the RMgX HOMO (influence of a basic solvent).

TABLE XV

STEREOCHEMICAL BEHAVIOR OF ALLYLLITHIUM

	Predominant stereochemistry with			
Nucleophile[a]	Ph | 1-Np—Si—OMe | Me	Ph | 1-Np—Si—Me | F	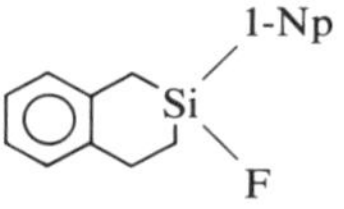 1-Np / Si \\ F	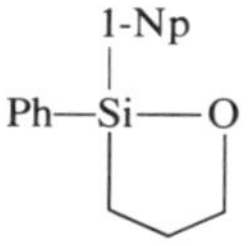 1-Np | Ph—Si—O
Allyllithium/Et$_2$O	85% RN	71% IN	87% RN	$[\alpha]_D = -5°$ IN
Allyllithium/KLi$^+$/Et$_2$O	75% RN	90% IN	75% RN	$[\alpha]_D = -10°$ IN

[a] KLi$^+$ = a cryptand specific of Li$^+$.

electrons of the nucleophile are those of the Al—H bond. Moreover, hydrogen has no lone pair and the unfavorable Nu–leaving group repulsion is therefore reduced.

In a noncoordinating solvent such as hexane (Table XII), the valence orbitals of the nucleophile are contracted. The unfavorable overlap with the leaving group is feeble and whatever its nature, retention is favored (57). In solvents such as THF or TMEDA, i.e., those able to coordinate at Al, the structure of the nucleophile HOMO is modified by diffusion of the valence orbitals around H and an increase in the electron density (Scheme 18).

Since the valence orbitals around hydrogen are more diffuse, the out-of-phase repulsion with the leaving group becomes dominant. Rear-side attack at silicon is favored and inversion is observed with the best leaving group (Br, Cl, F, SR) (23, 57). In contrast, we get retention with OR groups.

The above discussion shows clearly that, in the case of the alanes, the more the negative charge is displaced toward hydrogen, the more voluminous the valence orbitals are and the more inversion is favored.

It seems interesting to introduce at this point the case of the naked AlH$_4^-$ anion. In the latter, the negative charge is completely localized on the hydrogen atoms. Therefore, we can reasonably suppose diffuse valence orbitals around this species (Scheme 19). This explains the general

SCHEME 17.   Supposed shapes of the allyllithium HOMO and of the allyl anion HOMO.

SCHEME 18. Shape of the σ_{Al-H}.

SCHEME 19. Supposed AlH_4^- valence orbitals.

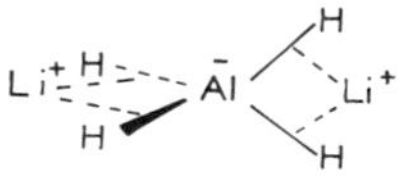

SCHEME 20

shift of the stereochemistry to inversion observed when AlH_4^- anions are used (Table XVI).

In contrast, when using $LiAlH_4$ or $LiAlH_4/LiBr$ (Table XVI) an increase of the RN ratio is observed (51). Li^+ and AlH_4^- are tightly associated in aggregates of unknown structure. Our results and those of others (74) suggest aggregates in which the Li^+ cations prevent the delocalization of the negative charge. One possibility could be that shown in Scheme 20. Such assumptions are supported by the crystal structures of $LiAl(Et)_4$ (75) and $KAl(Me)_4$ (76). The structure of the latter compound consists of isolated K^+ and $Al(Me)_4^-$ ions: the bonding between the central atom and ligands shows a highly polar character which causes a considerable

TABLE XVI

STEREOCHEMICAL BEHAVIOR OF LiAlH₄: INFLUENCE OF ION-PAIR DISSOCIATION
AS CONTROLLING FACTOR

	Predominant stereochemistry with	
Nucleophile[a]	Ph | 1-Np—Si—OMe | Me	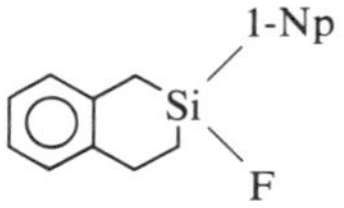 1-Np / ⬡Si \ F
$LiAlH_4/LiBr/Et_2O$	98% RN	95% RN
$LiAlH_4/Et_2O$	95% RN	Racemic
$LiAlH_4/KLi^+/Et_2O$	75% IN	65% IN

[a] KLi^+ = cryptand specific of the Li^+ cation.

weakening of the Al—C bond. We can find here a good model for the structure of naked AlH_4^-: the negative charge is highly delocalized around the H atoms. On the other hand, the structure of $LiAl(Et)_4$ consists of linear chains of alternating lithium and aluminium: there is some evidence of weak covalent interaction involving lithium. Such an interaction would explain the behavior of $LiAlH_4/LiBr$ aggregates in preventing the delocalization of the negative charge to hydrogen atoms.

3. Coupling Reactions with Alkoxides

In Section II,B we reported significant data concerning the influence of the solvation in the case of nucleophilic substitutions by alkoxides at Si—OMe, Si—F, and Si—SMe bonds (55). The most important fact that emerges from these data is that the stereochemistry changes from retention to inversion when the alcohol content of the medium is increased. We propose the following analysis:

1. The RO^-M^+ species have a localized negative charge on the oxygen atom. They act as hard anions with contracted valence orbitals. Overlap with the leaving group is negligible, and front-side attack leading to retention is promoted (Scheme 9). In contrast, in protic alcoholic solutions, RO^- anions are highly solvated and thus possess a large effective bulk. Moreover, the negative charge is dispersed, producing a softer species. Considerable orbital overlap can take place with the leaving group (Scheme 9), and the stereochemistry is shifted toward inversion of configuration (Table IX).

2. It has been shown that in protic solutions, the HOMO of hard anions such as RO^- lies at a lower energy (71). Thus increasing the alcohol content of the medium (Table IX) implies a stabilization of the RO^- species, and the relative importance of the nucleophile HOMO–substrate superjacent MO interaction increases (Scheme 10). Because of the structure of the latter (Scheme 11), inversion is favored (Table IX).

4. Phenoxide Reagents

In the case of the p-methoxyphenoxide anion, Taft et al. (77) have shown that the oxygen atom has a high degree of sp^3 character. This nucleophile is quite similar to hard alkyl anions from an electronic point of view, i.e., it is a hard nucleophile with contracted valence orbitals around oxygen, unfavorable out-of-phase overlap with the leaving group is minimized (Scheme 9), and a front-side attack leading to retention is therefore possible. The stereochemical data are summarized in Table XI.

In contrast, the oxygen atom of the p-nitrophenoxide anion has a high degree of sp^2 character. The p character of the nucleophilic center, as we

SCHEME 21. Supposed shapes of the HOMO the *p*-methoxyphenoxide anion and of the sodium *p*-methoxyphenoxide.

assumed in the case of the allyl anion, implies a delocalization of the negative charge over the aromatic system (77). The valence orbitals around oxygen are diffuse and sizeable overlap with the leaving group therefore prevails. A rear-side attack leading to inversion is favored (Scheme 9).

The effect of the counter cation (Table XI) may be in a manner similar to those proposed for alkyllithium and allyllithium reagents.

1. Sodium *p*-methoxyphenoxide shows behavior close to that of alkyllithium reagents. In this species, the negative charge on oxygen is in part transferred to the Na^+ cation. When the Na^+ cation is trapped (naked anion), the negative charge is concentrated on the oxygen reactive center (Scheme 21). In the latter case, because the valence orbitals are smaller, out-of-phase overlap with the leaving group is diminished and retention is therefore favored (Scheme 9). This is in agreement with the experimental data, which show that retention increases when the ion pair is separated (Table XI).

2. In contrast, sodium *p*-nitrophenoxide is similar to the allyllithium compounds. The counter cation reduces the p character of the oxygen and prevents the delocalization of the negative charge over the aromatic system. The valence orbitals are centered toward Na^+, whereas for the naked anion they are more disposable or, in other words, more diffuse (Scheme 22). Therefore, in the latter case, unfavorable out-of-phase overlap with the leaving group is increased, and rear-side attack of the nucleophile is promoted, leading to inversion (Table XI).

C. Stereochemical Changes with Angle Strain at Silicon

Increased angle strain at silicon always leads to a significant change of the stereochemistry toward retention for exocyclic leaving groups (Table XVII). This general trend is particularly increased with the most angle-

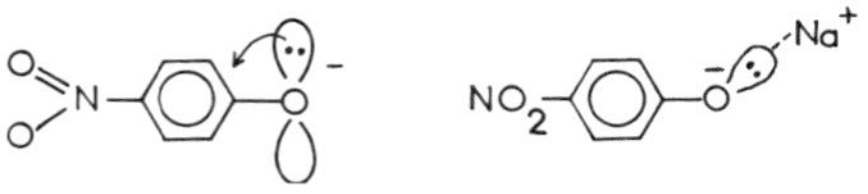

SCHEME 22. Supposed shapes of the HOMO of the *p*-nitrophenoxide anion and of the sodium *p*-nitrophenoxide.

TABLE XVII

Influence of an Angle Strain at Silicon on the Stereochemistry

Substrate	Angle (deg) (c_1–Si–c_2)	LiAlH$_4$	RLi (R = alkyl)	RLi (R = aryl)	RLi (R = benzyl or allyl)	RMgX (R = alkyl)	RMgX (R = aryl)	RMgX (R = allyl or benzyl)	Reference
22 (Si—Cl)	<90[a]	RN	—	—	—	—	RN	—	78, 79
(Si—F)	<90	RN	RN	—	—	RN	—	—	78, 79
(Si—OR)	<90	RN	RN	—	—	RN	—	—	78, 79
23 (Si—F)	<90	—	RN	—	—	RN	—	—	78
(Si—OR)	<90	RN	RN	—	—	RN	—	—	78
(Si—NMe$_2$)	<90	—	—	—	—	RN	—	—	78
24 (Si—Cl)	93.4	RN	—	—	—	—	—	—	80
25 (Si—Cl)	90[b]	IN	—	—	—	—	—	—	81
(Si—F)	90	RN	—	—	—	—	—	—	81

	(deg)								Ref
26 (Si—Cl)	92–96[c]	IN	—	IN	—	IN	—	IN	82
(Si—F)	92–96	Racemic	—	—	—	—	—	—	82
(Si—H)	92–96	—	—	RN	—	—	—	—	82
27 (Si—Cl)	105[d]	IN	RN	RN	IN	IN	RN	IN	35, 36
(Si—F)	105	Racemic	RN	RN	RN	RN	RN	IN	35, 36
(Si—OMe)	105	Racemic	RN	RN	RN	RN	RN	RN	35, 36
(Si—H)	105	—	RN	RN	RN	—	—	—	35, 36
R₃Si—Cl[e]	109[f]	IN	IN	IN	IN	IN	IN	IN	1, 2
R₃Si—F	109	IN	RN	—	IN	IN	IN	IN	1, 2
R₃Si—OMe	109	RN	RN	RN	RN	—	—	—	1, 2
R₃Si—H	109	—	RN	RN	—	—	—	—	1, 2

[a] Vilkov et al. (83).

[b] No experimental value is reported in literature; we give an approximate value calculated with the following bond lengths (85), Si—Ph = 1.87 Å, Si—1-Np = 1.85 Å, $\diagdown$C=C$\diagup$ (Aryl) = 1.38 Å.

[c] Dzhaparidze; Durig and Willes; and Durig et al. (84).

[d] Vidal et al. (85).

[e] R₃Si—X = 1-NpPhMeSiX.

[f] Okaya and Asida (86).

301

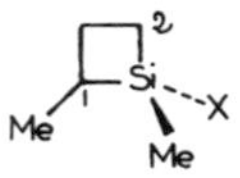

22 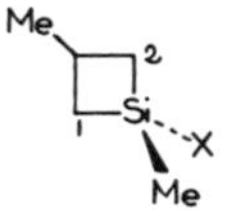   **23** 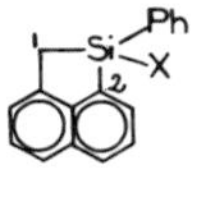   **24**

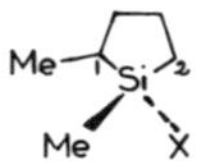

25 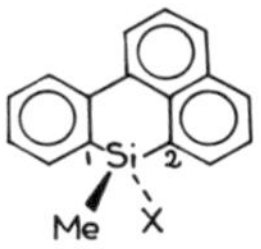   **26** 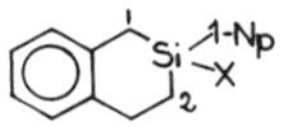   **27**

strained silacyclobutanes. These latter react mainly with retention, whatever the nature of the nucleophile. It is also noteworthy that a small angular strain is enough to cause such a stereochemical change. 1-Naphthyl-2-sila-2 tetrahydro-1,2,3,4-naphthalene shows significant deviations compared to 1-naphthylphenylmethylsilane (Table XVIII). Furthermore, experimental data show a gradual change of the stereochemistry from inversion to retention when angle strain at silicon is increased (Table XVII), as indicated in the following scheme:

IN ─────────────────────────────────→ RN
109°   105°   93–96°   90°

acyclic   six-membered   five-membered   silacyclobutanes
silanes   rings   rings

In contrast, a cyclic strained geometry influences the stereochemistry with a general shift to inversion for endocyclic leaving groups (*37, 38*). Relevant examples are summarized in Table XIX.

A possible interpretation generally invoked in phosphorus chemistry assumes that the four- or five-membered rings are unable to occupy the diequatorial position of a trigonal bipyramidal intermediate and must occupy the apical–equatorial position (*48*). Such a geometry implies an attack of the nucleophile at 90° to the leaving group and so leads to retention (Scheme 23).

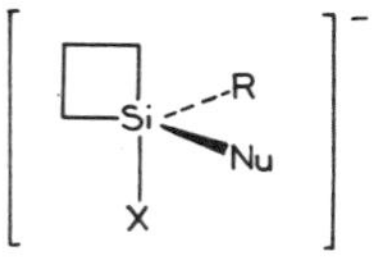

SCHEME 23

TABLE XVIII

Stereochemical Behavior of 27: Comparison with 1-NpPhMeSi—X (X = Cl, F, OMe, H)[a]

RM	Si—Cl		SiF		SiOMe		Si—H	
EtLi	63% RN	(100% IN)	96% RN	(90% RN)	94% RN	(90% RN)	100% RN	
n-BuLi	82% RN	(59% IN)	98% RN	(80% RN)	96% RN	(86% RN)	100% RN	(100% RN)
CH₂=CHCH₂Li	86% IN	(100% IN)	95% RN	(71% IN)	97% RN	(85% RN)	85% RN	(89% RN)
PhCH₂Li	99% IN	(100% IN)	70% RN	(85% IN)	75% RN	(79% IN)	96% RN	(90% RN)

Predominant stereochemistry[a]

[a] The stereochemical data reported with 1-NpPhMeSiX are given in parenthesis.

TABLE XIX

STEREOCHEMICAL BEHAVIOR OF THE ENDOCYCLIC OR LEAVING GROUP

| | | Predominant stereochemistry | |
| | | | Ph
\|
1-Np—Si—⌐
\|
O‿ |
Nucleophile	Product	1-NpPhMeSiOMe	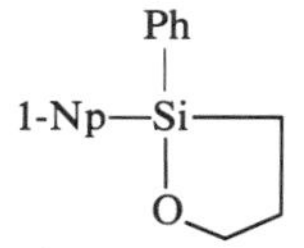
LiAlH$_4$	R$_3$Si—H	90% RN	$[\alpha]_D = +13°$ RN
LiAlH$_4$/4CuI/THF	R$_3$Si—H	55% RN	$[\alpha]_D = -18°$ IN
CH$_2$=CHCH$_2$Li	R$_3$Si—CH$_2$—CH=CH$_2$	85% RN	$[\alpha]_D = -5°$ IN
p-CH$_3$OC$_6$H$_4$CH$_2$Li	R$_3$Si—C$_6$H$_4$—(p-OCH$_3$)	90% RN	$[\alpha]_D = -4°$ IN

Such a proposal agrees with the general shift toward retention of exocyclic leaving groups. However, it fails to explain the following facts:

1. Silacyclopentanes can give inversion in some cases (Table XVII).

2. For endocyclic leaving groups, it would agree with a complete change of the stereochemistry to inversion. For instance, a five-membered ring with one electronegative group attached at silicon, would occupy the preferred equatorial–apical position of a trigonal bipyramidal intermediate with the electronegative group in the apical position (Scheme 24). As a consequence, inversion of configuration would be the predominant stereochemistry. The experimental facts show that the dominant stereochemistry is retention (Table IV). A change to inversion is only observed in borderline cases (Table XIX).

3. In the case of the six-membered ring (Tables XVII and XVIII), the angular value $[\angle C_1$–Si–$C_2 = 105°$ $(85)]$ cannot explain why inversion of configuration is so unfavorable compared to retention. This value suggests that the angular strain at silicon is quite similar in intermediates like 28 and 29 (Scheme 25). Thus both inversion and retention would be expected.

The above data can be easily explained in terms of change of the hybridization of the Si—R bond around the tetracoordinated silicon atom as

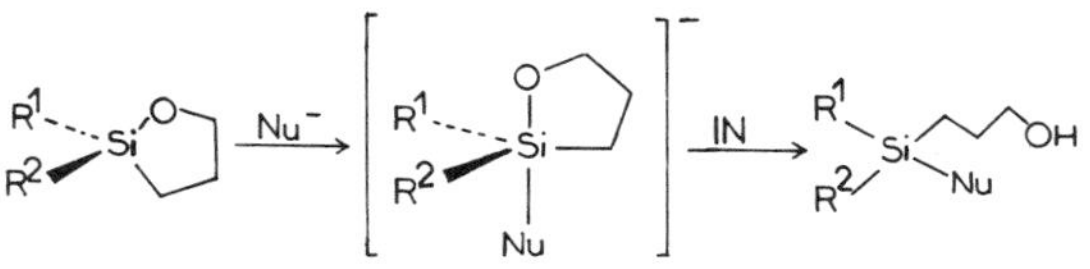

SCHEME 24

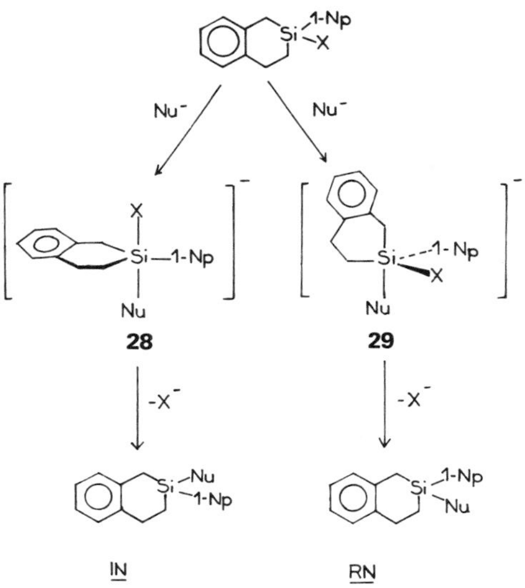

Scheme 25

proposed by Anh and Minot (6) (Scheme 26). If the angle becomes smaller than the tetrahedral value, the four hybrid atomic orbitals of Si are no longer equivalent. The two used for making Si—C-1 and Si—C-2 bonds have less s character than an sp^3 hybrid orbital, whereas the two remaining atomic orbitals (Si—R^1 and Si—X) acquire more s character. The above authors have shown that an increase of s character implies an easier front-side attack at silicon and, therefore, a greater ratio of retention. It follows that if the Si atom is included in a strained ring whereas X remains extracyclic, the percentage of retention will increase. In contrast, if Si and X are both in the ring, inversion of configuration will be favored.

The above conclusions explain the general effect of the angular strain on the stereochemistry and agree well with all the experimental facts:

1. Silacyclobutanes are highly strained rings. This means that in the Si—X bond (X = leaving group), the hybrid atomic orbital has a large s character and is quite dissymetric. It follows that the favorable overlap between Si and the incoming nucleophile is increased for a retention attack (Scheme 9). Even coupling reactions between a chlorosilacyclo-

Scheme 26

butane and $LiAlH_4$ or Grignard reagents occur with complete retention of configuration (*78, 79*).

2. When Si and OR are both in a five-membered ring, the σ^*_{Si-OR} MO has more p character, and front-side attack is less favorable (Scheme 9). However the experimental data indicate a shift of the stereochemistry to inversion only in borderline cases since the angular tension is not very high (Table XIX).

3. In the case of the six-ring 1-naphthyl-2-sila-2-tetrahydro-1,2,3,4-naphthalene (**27**), the small angular distorsion at silicon implies little increase of the s character in the Si—X bond. As a consequence, this change is only enough to shift the stereochemistry toward retention in the case of hard nucleophiles with contracted valence orbitals, such as alkyllithiums. For instance, the latter cleave the Si—X bond with predominant retention (Table XVII) instead of inversion, as with acyclic chlorosilanes (Table I). In contrast, the softer Grignard reagents are borderline nucleophiles. They cleave the Si—Cl bond of **27** with inversion, i.e., the normal stereochemical outcome. When used in a basic solvent such as DME (Table VI), the valence orbitals become more contracted on the carbon atom (Scheme 16). The nucleophile is now of smaller size, and the small shift of s character into the Si—X bond is sufficient to change the stereochemistry. MeMgBr in DME cleaves the Si—Cl bond (**27**) with predominant retention. Similar hybridization arguments may rationalize the stereochemical behavior of allyllithium and of the corresponding Grignard reagent. The former cleaves the Si—F bond of **27** with retention, the latter with inversion (Table XVII). In a basic solvent (Table VI), the opposite stereochemistry, i.e., retention, is observed with allylMgBr.

Finally, it is interesting to note that similar hybridization arguments can explain the order of reactivity of bicyclic systems. The following order was reported (*87*):

| **30** | **31** | **32** | **33** |

Because of the cage structure of the above bicyclic systems, only frontside attack at silicon is possible. Thus an increase of the reaction rate would be observed for the Si—X bonds that have a high degree of s character. Our analysis is as follows:

1. In the case of the silabicyclo[2.2.1]heptane ($\angle C_1-Si-C_2$ = 94.4°; Scheme 27), a highly strained molecule, the large angular distorsion at the silicon bridgehead leads to a significant increase of s character on silicon. Thus the σ^*_{Si-X} LUMO shows a big lobe on silicon directed toward the X leaving group. The rate of nucleophilic cleavage of the Si—X bond is increased because the overlap between the nucleophile and the reaction center is quite favorable.

2. In the silabicyclo[2.2.2]octane, the angular distorsion at the silicon atom is not so large ($\angle C_1-Si-C_2$ = 103.6°) (88). As a consequence, the silicon atom has less s character than in the above example, and the rate increase is smaller.

3. Finally, the silaadamantane (32) shows a silicon bridgehead which is a normal sp³ hybridized atom (Scheme 27). Its reactivity is quite similar to that of acyclic compounds (87).

The hybridization arguments employed above may be compared with recent work of Barton et al. (89) relative to the structure of 1-methyl-1-silabicyclo[2.2.2]octatriene (34). This molecule was found to exhibit a

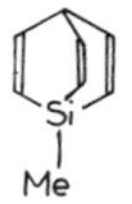

34 $<C_1-Si-C_2$ =98.8°

great deal of angular distortion at the silicon bridgehead ($<C_1-Si-C_2$ = 98.8°). Relative to a normal sp³ hybrid, there is a large shift of s character in the Si—Me bond; from the observed valence angles, the authors concluded that the localized orbital on silicon has 60% s character. This bonding model is entirely consistent with our rationalizations.

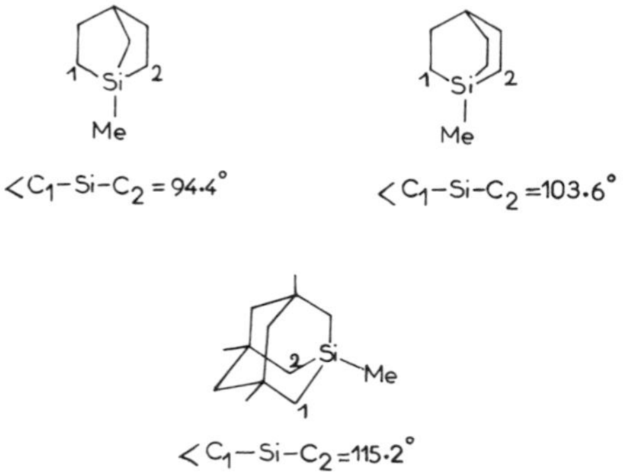

$<C_1-Si-C_2 = 94.4°$

$<C_1-Si-C_2 =103.6°$

$<C_1-Si-C_2 =115.2°$

SCHEME 27

IV
CONCLUDING COMMENTS

Studies in the stereochemistry of nucleophilic displacements at silicon over the past few years have been numerous. Significant progress has been achieved in establishing the controlling factors of the stereochemistry: (i) the lability of the leaving group, (ii) the electronic character of the nucleophile, and (iii) the bond angles at silicon.

Progress has also been achieved in developing a comprehensive description of these factors. Anh and Minot's concept, based on Salem's treatment of the Walden inversion, makes it possible to explain experimental trends using a perturbation argument that does not require the introduction of silicon d orbitals. According to this model the major interaction during the reaction occurs between the HOMO of the nucleophile and the LUMO of the substrate. This approach also accounts nicely for the influence of ring strain on the stereochemistry at silicon.

Finally, these mechanistic proposals can be easily generalized to other nucleophilic substitution reactions:

1. The stereochemical data reported in the case of germanium compounds by Eaborn *et al.* (*90*), Brook and Peddle (*91*), or Carré and Corriu (*92*) are quite parallel to those discussed here. The nature of the leaving group and the electronic character of the nucleophile, rationalized in terms of hard and soft reagents, are also the dominant factors that govern the stereochemistry.

2. Many data concerning nucleophilic displacements at phosphorus could also be rationalized using similar arguments. Phosphorus and silicon dynamic stereochemistries are very similar in many respects. The stereochemical outcome is strongly dependent on the leaving group lability (*47*). Bromine and chlorine are close, and both give more inversion than fluorine, although the latter is the most apicophilic group. In border-

$$
\begin{array}{ccc}
\text{Br} & \text{Cl} & \text{F}
\end{array}
$$
$$
\text{IN} \longrightarrow \text{RN}
$$

line cases, the stereochemistry is also governed by the nature of the nucleophile (*47*). Hard nucleophiles (the *p*-methoxyphenoxide anion) favor the retention of configuration. In contrast, softer nucleophiles (the *p*-nitrophenoxide anion) lead to inversion.

3. Nucleophilic activited racemizations have been observed in the case of chlorophosphonates (*93*) and halosilanes (*39*). They show similar rate laws:

$$v_{\text{rac}}^{\text{Si}} = k_{\text{rac}}^{\text{Si}}[R_3SiCl][Nu]^2$$

$$v_{\text{rac}}^{\text{P}} = k_{\text{rac}}^{\text{P}}[>\overset{\overset{\displaystyle O}{\|}}{P}\!-\!Cl][Nu]^2$$

Moreover, the process is governed by entropic factors. Strongly negative activation entropies are observed.

4. The very uncommon nucleophile-induced substitution process was observed both at silicon and phosphorus (*94*).

ACKNOWLEDGMENTS

Professor Corriu wishes to thank those co-workers whose names appear in the references cited. A special debt is owed to Drs. J. Massé, G. Royo, G. Lanneau, B. Henner, M. Leard-Henner, C. Guérin, C. Brelière, F. Larcher, J. M. Fernandez, and A. De Saxcé for their contributions. Collaboration with Professor Nguyên Trong Anh and Dr. C. Minot has been very fruitful. The stimulating interest of Drs. H. Felkin and J. Seyden-Penne is appreciated. Finally, the help of Dr. J. Jacques in determining the absolute configuration of optically active compounds is gratefully acknowledged.

The support of our work by C.N.R.S. is gratefully acknowledged.

REFERENCES

1. L. H. Sommer, "Stereochemistry, Mechanism and Silicon." McGraw-Hill, New York, 1965.
2. L. H. Sommer, *Intra-Sci. Chem. Rep.* **7,** 1 (1973).
3. R. H. Prince, *MTP Int. Rev. Sci.: Inorg. Chem., Ser. One* **9,** 353 (1972).
4. I. Fleming, *Compr. Org. Chem.* **3,** 542 (1979).
5. R. Corriu and C. Guérin, *J. Organomet. Chem.* **198,** 231–320 (1980).
6. (a) Nguyên Trong Anh and C. Minot, *J. Am. Chem. Soc.* **102,** 103 (1980); (b) Nguyên Trong Anh, *Top. Curr. Chem.* **88,** 145 (1980); (c) C. Minot, personal communication.
7. (a) R. Corriu and G. Royo, *J. Organomet. Chem.* **40,** 229 (1972); (b) R. Corriu, J. Massé, and G. Royo, *J. Chem. Soc. Chem. Commun.* p. 252 (1971).
8. R. Corriu, J. M. Fernandez, and C. Guérin, *J. Organomet. Chem.* **152,** 21 (1978).
9. R. Corriu and B. Henner, *J. Organomet. Chem.* **71,** 393 (1974).
10. L. H. Sommer and W. D. Korte, *J. Am. Chem. Soc.* **89,** 5802 (1967).
11. R. Corriu and G. Royo, *Bull. Soc. Chim. Fr.* p. 1497 (1972); R. Corriu and G. Royo, *Tetrahedron* **27,** 4289 (1972).
12. See Ref. 1, Chap. 6.
13. R. Corriu and G. Royo, *J. Organomet. Chem.* **14,** 291 (1968).
14. L. H. Sommer, C. L. Frye, M. C. Musolf, G. A. Parker, P. G. Rodewald, K. W. Michael, Y. Okaya, and R. Pepinsky, *J. Am. Chem. Soc.* **83,** 2210 (1961).
15. L. H. Sommer and J. D. Citron, *J. Am. Chem. Soc.* **89,** 5797 (1967).
16. G. J. D. Peddle, J. M. Shafir, and G. McGeachin, *J. Organomet. Chem.* **14,** 505 (1968).
17. L. H. Sommer, K. W. Michael, and W. D. Korte, *J. Am. Chem. Soc.* **89,** 868 (1967).
18. L. H. Sommer, C. L. Frye, and G. A. Parker, *J. Am. Chem. Soc.* **86,** 3276 (1964).

19. (a) See Ref. 1, Chap. 3; (b) L. H. Sommer, C. L. Frye, G. A. Parker, and K. W. Michael, *J. Am. Chem. Soc.* **86,** 3271 (1964).
20. L. H. Sommer, W. D. Korte, and P. G. Rodewald, *J. Am. Chem. Soc.* **89,** 862 (1967).
21. See Ref. 1, Chap. 4.
22. L. H. Sommer and J. McLick, *J. Am. Chem. Soc.* **88,** 5359 (1966); L. H. Sommer and J. McLick, *J. Am. Chem. Soc.* **89,** 5806 (1967).
23. R. Corriu, J. M. Fernandez, and C. Guérin, *J. Organomet. Chem.* **152,** 25 (1978).
24. L. H. Sommer, G. A. Parker, N. C. Lloyd, C. L. Frye, and K. W. Michael, *J. Am. Chem. Soc.* **89,** 857 (1967).
25. L. H. Sommer and J. E. Lyons, *J. Am. Chem. Soc.* **91,** 7061 (1969).
26. E. Colomer, R. J. P. Corriu, C. Marzin, and A. Vioux, *Inorg. Chem.,* in press (1982).
27. R. Corriu, G. Lanneau, and M. Leard, *J. Organomet. Chem.* **64,** 79 (1974).
28. R. Corriu and G. Lanneau, *Bull. Soc. Chim. Fr.* p. 3102 (1973).
29. R. Corriu, F. Larcher, and G. Royo, *J. Organomet. Chem.* **102,** C25 (1975); C. Brelière, R. Corriu, and G. Royo, *J. Organomet. Chem.* **148,** 107 (1978).
30. C. Brelière, R. Corriu, A. De Saxcé, F. Larcher, and G. Royo, *J. Organomet. Chem.* **164,** 19 (1979).
31. C. Brelière, R. Corriu, and G. Royo, *J. Chem. Soc. Chem. Commun.* p. 906 (1976).
32. C. Brelière, R. Corriu, A. De Saxcé, and G. Royo, *J. Organomet. Chem.* **166,** 153 (1979).
33. R. Corriu and B. Henner, *J. Chem. Soc. Chem. Commun.* p. 116 (1973); G. Chauvière, R. Corriu, and B. Henner, *J. Organomet. Chem.* **86,** C1 (1975); R. Corriu and B. Henner, *J. Organomet. Chem.* **102,** 407 (1975).
34. L. H. Sommer, G. A. Parker, and C. L. Frye, *J. Am. Chem. Soc.* **86,** 3280 (1964).
35. R. Corriu and J. Massé, *Tetrahedron Lett.* p. 5197 (1968); R. Corriu and J. Massé, *J. Organomet. Chem.* **34,** 221 (1972).
36. R. Corriu and J. Massé, *J. Chem. Soc. Chem. Commun.* p. 1373 (1968); R. Corriu and J. Massé, *J. Organomet. Chem.* **35,** 51 (1972).
37. R. Corriu, C. Guérin, and J. Massé, *J. Chem. Soc. Chem. Commun.* p. 75 (1975); R. Corriu, C. Guérin, and J. Massé, *J. Chem. Res. Synop.* p. 160 (1977); *J. Chem. Res., Miniprint* p. 1877 (1977).
38. R. Corriu, J. M. Fernandez, C. Guérin, and A. Kpoton, *Bull. Soc. Chim. Belg.* **89,** 783 (1980).
39. R. Corriu and M. Henner, *J. Organomet. Chem.* **64,** 1 (1974).
40. K. D. Onan, A. T. McPhail, C. H. Yoder, and R. W. Hillyard, *J. Chem. Soc. Chem. Commun.* p. 209 (1978).
41. E. A. V. Ebsworth, "Volatile Silicon Compounds," p. 54. Pergamon, Oxford, 1963.
42. S. Trippett, *Pure Appl. Chem.* **40,** 595 (1974).
43. R. Corriu, G. Royo, and A. De Saxcé, *J. Chem. Soc. Chem. Commun.* p. 893 (1980).
44. G. Van Koten and J. G. Noltes, *J. Am. Chem. Soc.* **98,** 5393 (1976); G. Van Koten, J. T. B. H. Jastrzebski, J. G. Noltes, W. M. G. F. Pontenagel, J. Kroon, and A. L. Spek, *J. Am. Chem. Soc.* **100,** 5021 (1978).
45. C. Brelière, F. Carré, R. Corriu, A. De Saxcé, M. Poirier, and G. Royo, *J. Organomet. Chem.* **205,** C1 (1981).
46. M. G. Voronkov, Y. L. Frolov, V. M. D'Yakov, N. N. Chipanina, C. I. Gubanova, G. A. Gavrilova, L. V. Klyba, and T. N. Aksamentova, *J. Organomet. Chem.* **201,** 165 (1980).
47. R. Corriu, J. P. Dutheil, G. Lanneau, and S. Ould-Kada, *Tetrahedron* **35,** 1889 (1979).
48. R. Luckenbach, "Dynamic Stereochemistry of Pentacoordinated Phosphorus and Related Elements." Thieme, Stuttgart, 1973; F. H. Westheimer, *Acc. Chem. Res.* **1,** 70 (1968); K. Mislow, *Acc. Chem. Res.* **3,** 321 (1970); P. Gillespie, P. Hoffman, H.

Keusacek, D. Marquarding, S. Pfohl, F. Ramirez, E. A. Tsolis, and I. Ugi, *Angew. Chem., Int. Ed. Engl.* **10**, 687 (1971); P. Gillespie, F. Ramirez, I. Ugi, and D. Marquarding, *Angew. Chem., Int. Ed. Engl.* **12**, 91 (1973).

49. C. Guérin, Ph.D. Thesis, Univ. Montpellier (1978).
50. F. Kanberg and E. L. Muetterties, *Inorg. Chem.* **7**, 155 (1968).
51. R. Corriu, J. M. Fernandez, and C. Guérin, *Tetrahedron Lett.* p. 3391 (1978).
52. R. Corriu and J. Massé, *Bull. Soc. Chim. Fr.* p. 3491 (1969).
53. O. Eisenstein, J. M. Lefour, C. Minot, Nguyên Trong Anh, and G. Soussan, *C. R. Hebd. Seances Acad. Sci., Ser. C.* **274**, 1310 (1972).
54. See Ref. 11a.
55. L. H. Sommer and H. Fujimoto, *J. Am. Chem. Soc.* **90**, 982 (1968).
56. R. Corriu and C. Guérin, *J. Organomet. Chem.* in press (1982).
57. (a) R. Corriu and C. Guérin, *J. Chem. Soc. Chem. Commun.* p. 74 (1977); (b) R. Corriu and C. Guérin, *J. Organomet. Chem.* **144**, 165 (1978).
58. T. Eicher, *in* "The Chemistry of the Carbonyl Groups" (S. Patai, ed.), p. 621. Wiley (Interscience), New York, 1966.
59. B. Deschamps, *Tetrahedron* **34**, 2009 (1978); R. Sauvètre and J. Seyden-Penne, *Tetrahedron Lett.* p. 3949 (1976); G. Kyriakakou, M. C. Roux-Schmitt, and J. Seyden-Penne, *Tetrahedron* **31**, 1883 (1975).
60. J. Botin, O. Eisenstein, C. Minot, and Nguyên Trong Anh, *Tetrahedron Lett.* p. 3015 (1972).
61. B. Deschamps, Nguyên Trong Anh, and J. Seyden-Penne, *Tetrahedron Lett.* p. 527 (1973).
62. J. Durand, Nguyên Trong Anh, and J. Seyden-Penne, *Tetrahedron Lett.* p. 2397 (1974).
63. J. M. Lefour and A. Loupy, *Tetrahedron* **34**, 1597 (1978).
64. G. H. Posner, *Org. React.* **19**, 1 (1972); J. F. Normant, *J. Organomet. Chem. Libr.* **1**, 219 (1976).
65. R. Sauvètre, M. C. Roux-Schmitt, and J. Seyden-Penne, *Tetrahedron* p. 2135 (1978).
66. E. F. Perozzi and J. C. Martin, *J. Am. Chem. Soc.* **101**, 1591 (1979).
67. J. A. Boudreau, C. Brown, and R. F. Hudson, *J. Chem. Soc. Chem. Commun.* p. 679 (1975).
68. L. Salem, *Chem. Br.* **5**, 449 (1969).
69. R. W. Gray, C. B. Chapleo, T. Vergnani, A. S. Dreiding, M. Liesner, and D. Seebach, *Helv. Chim. Acta* **58**, 1524 (1975).
70. R. G. Pearson, *J. Chem. Educ.* **45**, 581 (1975).
71. C. Minot and Nguyên Trong Anh, *Tetrahedron Lett.* p. 3905 (1975).
72. L. H. Sommer, J. McLick, and G. M. Golino, *J. Am. Chem. Soc.* **94**, 669 (1972).
73. F. Clark, E. D. Jemmis, and P. V. R. Schleyer, *J. Organomet. Chem.* **150**, 1 (1978).
74. A. Jean and M. Lequan, *Tetrahedron Lett.* p. 1517 (1970).
75. R. L. Gerteis, R. E. Dickerson, and T. L. Brown, *Inorg. Chem.* **3**, 872 (1964).
76. R. Wolfrum, G. Sauermann, and E. Weiss, *J. Organomet. Chem.* **18**, 27 (1969).
77. G. Kemister, A. Pross, L. Radom, and R. Taft, *J. Org. Chem.* **45**, 1056 (1980).
78. J. Dubac, P. Mazerolles and B. Serres, *Tetrahedron Lett.* p. 529 (1972); *Tetrahedron Lett.* p. 3495 (1972); J. Dubac, P. Mazerolles, and B. Serres, *Tetrahedron* **30**, 749 (1974); *Tetrahedron* **30**, 759 (1974).
79. B. G. McKinnie, N. S. Bhacca, F. K. Cartledge, and J. Fayssous, *J. Am. Chem. Soc.* **96**, 2637 (1974); B. G. McKinnie, N. S. Bhacca, F. K. Cartledge, and J. Fayssous, *J. Org. Chem.* **41**, 1534 (1976).
80. D. N. Roark and L. H. Sommer, *J. Am. Chem. Soc.* **95**, 969 (1973).
81. J. D. Citron, *J. Organomet. Chem.* **86**, 359 (1975).

82. F. K. Cartledge, J. M. Wolcott, J. Dubac, P. Mazerolles, and P. Fagoaga, *Tetrahedron Lett.* p. 3593 (1975); J. M. Wolcott and F. K. Cartledge, *J. Organomet. Chem.* **111**, C35 (1976); F. K. Cartledge, J. M. Wolcott, J. Dubac, P. Mazerolles, and M. Joly, *J. Organomet. Chem.* **154**, 187 (1978); F. K. Cartledge, J. M. Wolcott, J. Dubac, P. Mazerolles, and M. Joly, *J. Organomet. Chem.* **154**, 203 (1978).

83. L. S. Vilkov, B. S. Mastrioukov, J. V. Baourova, V. M. Vdovin, and L. Trinberg, *Dokl. Akad. Nauk SSSR* **177**, 1084 (1967).

84. K. G. Dzhaparidze, *Soobshch. Akad. Nauk Gruz. SSR* **29**, 401 (1962); J. R. Durig and J. N. Willes, Jr., *J. Mol. Spectrosc.* **32**, 320 (1969); J. R. Durig, W. J. Lafferty, and W. F. Kalinsky, *J. Phys. Chem.* **80**, 1199 (1976).

85. J. P. Vidal, J. Lapasset, and J. Falgueirettes, *Acta Crystallogr., Sect. B* **28**, 3137 (1972).

86. Y. Okaya and T. Ashida, *Acta Crystallogr* **20**, 461 (1966).

87. G. D. Homer and L. H. Sommer, *J. Am. Chem. Soc.* **95**, 7700 (1973).

88. R. L. Hilderbrandt, G. D. Homer, and P. Boudjouk, *J. Am. Chem. Soc.* **98**, 7476 (1976).

89. Quang Shen, R. L. Hilderbrandt, G. T. Burns, and T. J. Barton, *J. Organomet. Chem.* **195**, 39 (1980).

90. C. Eaborn, R. E. E. Hill, and P. Simpson, *J. Organomet. Chem.* **37**, 251 (1972).

91. A. G. Brook and G. J. D. Peddle, *J. Am. Chem. Soc.* **85**, 2338 (1963).

92. F. Carré and R. Corriu, *J. Organomet. Chem.* **65**, 343 (1974).

93. R. Corriu, G. Lanneau, and D. Leclercq, *Tetrahedron* **36**, 1617 (1980).

94. R. Corriu, G. Dabosi, and M. Martineau, *J. Chem. Soc. Chem. Commun.* p. 649 (1977); R. Corriu, G. Dabosi, and M. Martineau, *J. Organomet. Chem.* **154**, 33 (1978); R. Corriu, G. Dabosi, and M. Martineau, *J. Organomet. Chem.* **186**, 25 (1980).

ADVANCES IN ORGANOMETALLIC CHEMISTRY, VOL. 20

The Biological Methylation of Metals and Metalloids

JOHN S. THAYER

Department of Chemistry
University of Cincinnati
Cincinnati, Ohio

and

F. E. BRINCKMAN

Chemical and Biodegradation Processes Group
Center for Materials Science
National Bureau of Standards
Washington, D.C.

ISBN 0-12-031120-8

I

INTRODUCTION

The term "biological methylation" originated with Challenger (*1–3*), although the concept had previously been mentioned by Hofmeister (*4*). Challenger summarized the concept as follows (*3*):

> In the strictest sense, the term 'biological methylation' implies either (1) the transfer, under biological conditions, of an intact methyl group from a compound *A* to a second compound *B*, or (2) the fission, under biological conditions, of some compound *C*, not necessarily containing a methyl group, so as to eliminate a molecule such as formaldehyde or formic acid, a 'one-carbon fragment'; this is thereupon captured by a compound *D*, and afterwards the resulting group is reduced to —CH$_3$. In (1) compound *A* is known as a *methyl donor*, and the process is a true *transmethylation*; in (2), compound *C* is called a *methyl source* [Emphasis added].

For the purposes of this chapter, we shall use the term "biological methylation" (in recent years this has frequently been contracted to "biomethylation") in Challenger's first sense to describe methyl-transfer (transmethylation) reactions involving metals and metalloids in biological systems. We shall not differentiate between endo- and exocellular transmethylation (*5, 6*), and we shall include some purely chemical studies of "model" reactions. The relationship of these reactions to the larger area of organometallic chemistry has been reviewed (*7, 8*).

At this point, the reader may ask "Why biomethylation exclusively? Why no mention of bioethylation, biophenylation, etc.?" There is presently no firm evidence available for the biosynthesis of organometals having organic groups larger than methyl sigma bonded to a metal or metalloid, although some sporadic comments have appeared in the literature. Marked differences occur between transmethylation rates and corresponding rates for other organic groups under simulated biological conditions, as will be discussed later. Nevertheless, even for the 10^8-fold range suggested by such experiments (*9, 10*), formation of such organometals, even in competition with the observed methyl derivatives, should be discernable with the recent advances in analytical techniques. Up to now, no such species have been observed, except for a few isolated cases that will be discussed later.

Historically, the investigations into biological methylation have followed two separate and distinct lines of research whose relationships have only recently become apparent and which remain to be fully developed. The first and older line had its origin in the "arsenic rooms" of the nineteenth century—rooms in which people succumbed to arsenic poisoning by breathing a volatile compound exuded by mold growing on wallpaper. The early reports in this area have been reviewed (2). Gosio showed that this compound contained arsenic, was not AsH_3, and proposed the formula $(C_2H_5)_2AsH$ (11). In a series of papers spanning a quarter century (1–3, 12), Challenger and co-workers showed that: (a) this volatile compound (termed Gosio-gas) was actually trimethylarsine, $(CH_3)_3As$; (b) this toxic gas formed from arsenic trioxide by direct action of the mold; (c) similar compounds could be formed from the oxides of selenium and tellurium in the same manner; and that (d) the compound S-adenosylmethionine served as the biogenic methyl donor. Challenger's research turned out to be related to independent work on organosulfur biochemistry, which has resulted in a substantial literature on the chemistry of S-adenosylmethionine and organic biomethylation (13–16).

The second line of research, like the first, grew out of reported human poisonings at Minamata Bay and Niigata in Japan during the late 1950s. These poisonings arose from ingestion of fish and shellfish containing methylmercuric compounds (17–22), which were probably derived, at least in part, through biomethylation of mercuric salts by aquatic organisms (22). Subsequent investigation indicated that methylcob(III)alamin was the likely biogenic methyl donor (23, 24) and that methyl transfer could be either enzymatic or nonenzymatic, occurring in contaminated sediments, the ambient water column, or the intestines of the fish themselves (22, 25, 26). Discovery of the accidental discharge of traces of methylmercuric compounds formed as side products from mercury catalysts used in Japanese industrial processes obscured the importance of the biological pathway. Subsequently, Swedish, American, and Canadian researchers would provide compelling evidence for the role of natural biomethylation in making uncontrolled mercury discharges so threatening to human and ecological well-being (20, 21, 27).

Out of these two lines of research has grown an extensive literature. Reactions involving the vitamin B_{12} derivative methylcob(III)alamin (hereafter referred to as methylcobalamin or CH_3B_{12}) (28) have been more extensively studied, probably because of the ready commercial availability of this biogenic reactant and the large number of model cobalt compounds showing strikingly similar chemistries (9). In this chapter we shall discuss chemical kinetic studies reported for methylcobalamin, S-adenosylmethionine, and other aqueous methyl donors that may serve as models for natural transmethylation processes. Most current research has concentrated on the extent and scope of these reactions and their relationship to the larger, rapidly evolving area of environmental organometallic chemistry (29).

II

METHYLCOBALAMIN: CHEMICAL RATE STUDIES

A. *Mercuric Salts*

1. *Hg(II)–Acetate System*

Although three early papers briefly discussed reactions between methylcobalamin and mercury compounds (*30–32*), the most systematic investigation has come from Wood and co-workers (*33*). They proposed the mechanism shown in Fig. 1, with values for the various rate constants presented in Table I. Species **2** and **3**, in which the benzimidazole nitrogen no longer bonds to the cobalt atom, are termed "base-off" compounds, whereas **1** is "base-on" methylcobalamin and **4** is aquocob(III)alamin, the usual product of aqueous transmethylation by **1**. Each one of these species has a unique ultraviolet–visible spectrum, which allows quantitative studies by spectrophotometric techniques to be made (*28, 32, 33*). The mercuric acetate–**1** exchange is so rapid that it must be studied using stopped-flow kinetic techniques (*33*).

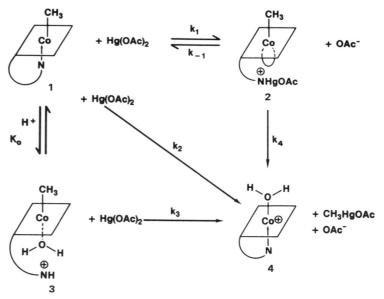

FIG. 1.   Proposed mechanism for the mercuric acetate—methylcobalamin reaction (*33*).

TABLE I

RATE CONSTANTS[a] FOR MERCURIC ACETATE–METHYLCOBALAMIN REACTION

Constant	Reference 33	Reference 34	Other
K_0	—	0.0021	—
k_1	1.2 (3000)[b]	4000	—
k_{-1}	(43)[b]	54	—
k_2	370	380	300 35; 349 36; 370 37
$k_3{}^c$	0.12	0.113	0.064 35
k_4	1.1	—	—
$K \ (= k_1/k_{-1})$	70	74	—

[a] Units of M^{-1} sec^{-1}. See Fig. 1 for the reactions involved.

[b] The value of 1.2 was given in DeSimone et al. (33). Recalculation in Robinson et al. (34), using the values from DeSimone et al. (33), gave the two values in parentheses.

[c] This term depends on the concentration of $Hg(OAc)_2$. The values listed are for 1.00 M $Hg(OAc)_2$.

The reaction between ethylcobalamin and $Hg(OAc)_2$ has also been reported (33). The mechanism is the same as for the methyl analog; however, k_2 is much smaller (0.18 M^{-1} sec^{-1}), and the transethylation may be studied by standard techniques. No reaction was reported for n-propylcobalamin and mercuric acetate (33).

Other groups have also studied the transmethylation of mercuric acetate; their values agree reasonably well with those of DeSimone et al., and are listed in Table I (34–37). One group found no formation of 2 in the $Hg(OAc)_2$–1 reaction in buffered acetic acid (36).

2. Hg(II) and Other Anions

The presence of chloride ion greatly slows transmethylation of Hg(II). Values of 0.430 M^{-1} sec^{-1} (in 0.1 M KCl) (36) and 4.0 M^{-1} sec^{-1} (for $HgCl_2$) (34) have been reported. Comparison of various anions (38) showed that the rate of transmethylation varied OAc$^-$ > Cl$^-$ ⩾ SCN$^-$ ⩾ Br$^-$ > CN$^-$. Exchange between methylcobalamin and $Hg(NO_3)_2$ in nitric acid gave rapid formation of 2 with subsequent transmethylation (36). Reported values for $\log(k_1/k_{-1})$ and $\log K_0$ were 4.82 and 2.69, respectively.

3. Solvent Effects

Sodium dodecylsulfonate (5) markedly slowed the transmethylation of $Hg(OAc)_2$ (34). This was attributed to enhancement in the formation of 3 because it interacted more strongly than **1** with **5**. The measured trans-methylation rate constant in 0.01 M **5** was 2.33 M^{-1} sec^{-1} (*34*). Compounds **1** and **5** will also interact to form the base-off complex when **5** is present as micelles (*39*). A corresponding micellar system of hexadecyl-trimethylammonium bromide completely inhibits this transmethylation (*34*), whereas the reaction proceeds approximately 18,000 times slower in dodecylammonium propionate-solubilized water pools in benzene than in bulk water (*34*).

B. *Platinum Salts*

An early paper reported that reaction of PtCl$_6^{2-}$ and **1** obeyed the rate law (*30*) expressed in Eq. (1) through the formation of a methylplatinum intermediate that decomposed in the presence of excess chloride to form chloromethane. Fanchiang *et al.* made a more detailed study of this reaction and proposed the reaction mechanism shown in Fig. 2, from which

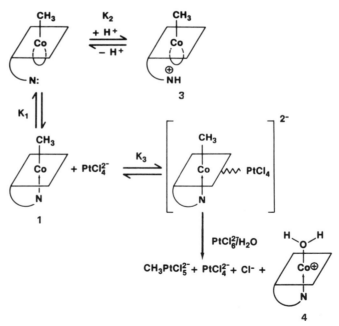

FIG. 2. Proposed mechanism for the hexachloroplatinate(IV)–methylcobalamin reaction (*40*).

they derived the rate law shown in Eq. (2) (*40*):

$$\text{rate} = k[\text{PtCl}_6^{2-}][\text{PtCl}_4^{2-}][\text{CH}_3\text{B}_{12}] \tag{1}$$

$$\text{rate} = \frac{d[\text{H}_2\text{O} \cdot \text{B}_{12}]}{dt} = \frac{kK_2K_3[\text{PtCl}_6^{2-}][\text{PtCl}_6^{2-}][\text{CH}_3\text{B}_{12}]_{\text{TOT}}}{K_2 + K_1[\text{H}^+] + K_2K_3[\text{PtCl}_4^{2-}]} \tag{2}$$

An independent investigation by Taylor *et al.* supports this same basic mechanism (*41–43*). The major difference between the two reports is that Fanchiang *et al.* propose $\text{CH}_3\text{PtCl}_5^{2-}$ as the methyl–platinum product (*40*), whereas the Taylor group propose $\text{CH}_3\text{PtCl}_3^{2-}$ (*42, 43*). Additional support for the formation of $\text{CH}_3\text{PtCl}_3^{2-}$ comes from its direct observation by proton NMR spectroscopy in the PtCl_4^{2-}–$(\text{CH}_3)_3\text{Pb}^+$ exchange reaction (*44*). Decomposition products were CH_3Cl, C_2H_6, and Pt.

The corresponding reaction of PtCl_4 with **1** gives two methylplatinum products, proposed to be $\text{CH}_3\text{PtCl}_3^{2-}$ and $\text{CH}_3\text{Pt}(\text{OH}_2)\text{Cl}_2^-$ (*45*). Taylor and Hanna also reported that solid PtO_2 reacts with **1** (*41*), and this has been independently confirmed (*46*). Some groups have claimed that K_2PtCl_4 does not react with **1** (*30, 40*), but there is one report that indicates this is a slow, autocatalytic reaction (*47*), and that it accelerates in the presence of Pt powder. K_2PtBr_6 reacts with **1** about six times as fast as the chloro analog under the same conditions (*46a*).

C. Palladium Salts

Potassium tetrachloropalladate(II) reacts with **1** to give CH_3Cl and metallic Pd as products (*48*). This reaction apparently follows the same mechanism as the Hg(II) system, with reported rate constants K (= k_1/k_{-1}) = 150 and $k_2 = 7.7 \times 10^{-3}\ M^{-1}\ \text{sec}^{-1}$. Additional support for the proposed methylpalladium(II) intermediate comes from the report of such a species, detected by proton NMR spectroscopy, in the reaction of PdCl_4^{2-} with $(\text{CH}_3)_3\text{Sn}^+$ (*44*). There has been a separate study on the base-off complex formed between PdCl_4^{2-} and **1** (*49*). Attempts to react K_2PdCl_6 with **1** proved futile, due to the extremely rapid decomposition of the Pd salt under reaction conditions (*46*).

D. Lead Compounds

Solid PbO_2 and solid Pb_3O_4 react with **1** (*41*). When ^{14}C-labeled methyl-cobalamin is used, the labeled carbon becomes an unidentified, volatile product. Reaction is more rapid in dilute acid than in water, and the rate increases as the PbO_2 particle size decreases (*46*). The concentration of dissolved Pb, as measured by atomic absorption spectroscopy, increases rapidly during this reaction (*46*). Plumbous salts react, albeit quite slowly and erratically, with **1** to give $(CH_3)_4Pb$ and $(CH_3)_3Pb^+$ (*30, 41, 46, 47*).

E. *Tin Compounds*

An extensive study of reactions between Sn(II) compounds and **1** showed that the reaction followed second-order kinetics, with a rate constant of $1.4 \pm 0.1 M^{-1} sec^{-1}$ (*50*); CH_3SnX_3 was the product. According to this group, Sn(IV) compounds did not react; however, a separate investigation indicates that there is a slow reaction between finely divided SnO_2 and **1** (*46*).

F. *Thallium Compounds*

One early paper reported that Tl(III) reacted with **1** to give an uncharacterized methylthallium(III) product (*31*). A second-order reaction occurs between $Tl(OAc)_4^-$ and **1** in buffered acetic acid, with a rate constant of $72.5 M^{-1} sec^{-1}$ (*46a*); an earlier reported value of $1.60 M^{-1} sec^{-1}$ is too low (*47*). As in the analogous Hg(II) systems (*46a, 51*), halide ions retard transmethylation of Tl(III). Thallous ion reacts only very slowly with **1** to give $(CH_3)_2Tl^+$ (*46*).

G. *Miscellaneous Inorganic Compounds*

Sodium bismuthate reacts with **1** (*47, 52*), as does $BiONO_3$ (*46*). Methyl exchange between Cr(II) and **1** follows second-order kinetics (*53*); the rate constant was $360 \pm 30 M^{-1} sec^{-1}$, and values for the activation parameters ΔH^* and ΔS^* were 15.9 ± 0.9 kJ mol^{-1} and -144 J $mol^{-1} K^{-1}$, respectively. Ferric compounds are reported to demethylate **1** (*30, 31, 37*). Cupric nitrate gives no noticeable reaction by itself with **1** (*46, 54*), but demethylation proceeds readily in the presence of high chloride or bromide concentrations (>2 *M*). Methyl chloride and methyl bromide formed as products. When ethanol was used as solvent, methyl ethyl ether also formed as product (*54*). An unstable intermediate CH_3CuCl may form as part of this reaction (54, 55). The salts $Ni(NO_3)_2$, $AgNO_3$, and $InCl_3$ react very slowly with **1**, whereas $Cd(NO_3)_2$ does not appear to react at all (30, 38, 46). The products of these reactions have not been characterized, but the increasing availability of new analytical methods should lend impetus to the detailed investigation of these unusual systems.

H. Organometallic Compounds

Reports that $(CH_3)_2Hg$ forms under natural conditions suggested that CH_3HgX, an organometallic compound, might undergo further methylation (22, 23); methylation of organoarsenicals was reported by Challenger (1, 3). Bertilsson and Neujahr reported that CH_3HgX, C_6H_5HgX, and $CH_3OCH_2CH_2HgX$ reacted with **1**, but at rates markedly slower than Hg(II) salts (32). Another group stated that CH_3Hg^+ reacted with CH_3B_{12} "several orders of magnitude slower" than mercuric acetate did (33). Kinetic studies on various water-soluble organometals indicate that all reactions follow second-order kinetics; rate constants are presented in Table II (56). Methylarsonate ion and dimethylarsinate (cacodylate) ion did not react with **1**. Many organometals have little or no solubility in water. Heterogeneous mixtures of these compounds in the presence of methylcobalamin undergo transmethylation, with the rate of reaction apparently being directly proportional to the quantity of solid present (47, 52).

TABLE II

RATE CONSTANTS FOR
ORGANOMETAL—METHYLCOBALAMIN REACTIONS

Organometal	$K(M^{-1} \text{ sec}^{-1})$	Reference
CH_3HgOAc	6.88×10^{-2}	56
$(CH_3)_2TlOAc$	3.94×10^{-4}	56
$CH_3Tl(OAc)_2$	1.40×10^{-2}	47
$(CH_3)_3PbOAc$	3.46×10^{-4}	56
$C_6H_5Pb(OAc)_3$	5.93×10^{-2}	46
$(CH_3)_3SnOAc$	4.50×10^{-5}	56
$(C_2H_5)_3SnOAc$	2.42×10^{-5}	47
$(CH_3)_3TeI$	1.62×10^{-4}	56
$(CH_3)_4PI$	5.85×10^{-4}	56
$(CH_3)_4AsI$	1.53×10^{-3}	56
$(CH_3)_4SbI$	1.38×10^{-4}	56

III

OTHER AQUEOUS TRANSMETHYLATIONS: CHEMICAL RATE STUDIES

A. *Methylsulfonium Salts*

Despite the importance of S-adenosylmethionine (**6**) (Fig. 3) in biological methylation, very little research has been done in the laboratory on its reaction with arsenic or other metalloids, nor has there been much effort to explore possible methylsulfur model compounds. When $(CH_3)_3S^+PF_6^-$ and As_2O_3 were mixed in water, they did not react at 25°C and pH 6.0–6.5, but did undergo reaction at 80°C and pH 12 (*57, 58*). Methylarsonate formed first and was subsequently converted to cacodylate and trimethylarsine. The half-life of the reaction appears to be inversely proportional to temperature (*57*). Sodium methylsulfate ("magic methyl") transfers a methyl group to arsenite ion and to phenylarsonate, but investigations were complicated by solvolysis of the reagent (*57*). S-Methylmethionine (**7**) (Fig. 3) reacts with methylarsonate at pH 5.8 and 25°C (*57, 58*), but an analogous 2-mercaptoethanesulfonate (**8**) (Fig. 3), derived from Coenzyme M (*8*), does not. Preliminary studies on the $(CH_3)_3S^+I^-$---As_2O_3 system indicate a slow and irregular transmethylation (*46*).

B. *Methylmetal Compounds*

Various groups have reported on methyl exchange among metals other than methylcobalamin in aqueous solution. This area has been discussed in some detail by Jewett *et al.* (*51*) and by Pratt and Craig (*59*). Table III lists representative examples of such systems, along with the reported

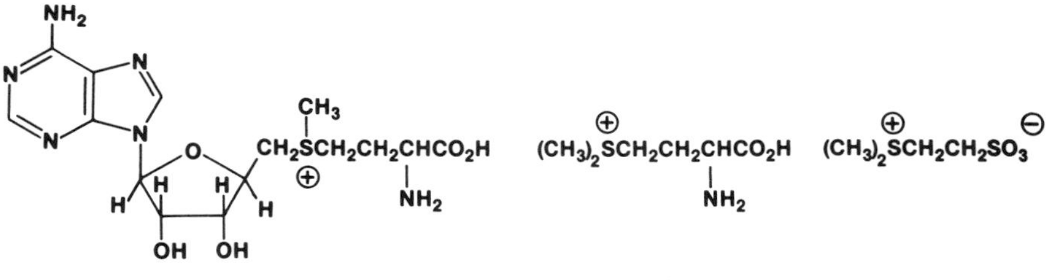

FIG. 3. Structures of S-adenosylmethionine(**6**), S-methylmethionine(**7**), and coenzyme M cofactor(**8**).

TABLE III

METHYL EXCHANGE BETWEEN METALS IN AQUEOUS MEDIA[a]

Substrate	Methylating agent	$K(M^{-1}/sec^{-1})$	Reference
$Hg(ClO_4)_2$	Methylcobaloxime	54	60
	Methylcobaloxime	65	61
$Tl(ClO_4)_3$	Methylcobaloxime	2.2	60
$HgBr_2$	$(CH_3)_3AuP(C_6H_5)_3$	1450	62
CH_3HgBr	$(CH_3)_3AuP(C_6H_5)_3$	0.41	62
$Hg(ClO_4)_2$	$[CH_3Cr[15]aneN_4(H_2O)]^{2+}$	3.1×10^6	63
CH_3HgClO_4	$[CH_3Cr[15]aneN_4(H_2O)]^{2+}$	1.63×10^3	63
Hg^{2+}	$(CH_3)_3SiCH_2CH_2CO_2^-$	0.32	66
$HgCl_2$	$(CH_3)_3SiCH_2CH_2CO_2^-$	1.8×10^{-3}	66
$HgCl_2$	$(CH_3)_3Sn^+$	1.74×10^{-2}	51
$HgCl_3^-$	$(CH_3)_3Sn^+$	7.3×10^{-3}	51
$HgCl_2$	$(CH_3)_3SnCl$	0.116	51
$HgCl_3^-$	$(CH_3)_3SnCl$	0.130	51
$PdCl_4^{2-}$	$(CH_3)_3Sn^+$	0.072	10, 44
$PtCl_4^{2-}$	$(CH_3)_3Sn^+$	$<7.4 \times 10^{-4}$	44
$IrCl_6^{2-}$	$(CH_3)_3Sn^+$	$\sim 9 \times 10^{-4}$	44

[a] Rates not adjusted to common ionic strength.

rate constants (9, 10, 44, 60–63). The species $(CH_3)_3SiCH_2CH_2CO_2^-$ and $(CH_3)_3SiCH_2CH_2CH_2SO_3^-$, used as proton NMR reference standards in aqueous solutions, will react with mercuric salts to form methylmercuric ion (64). Hexamethyldisiloxane will react with $Hg(NO_3)_2$ to give products arising from methyl transfer (65). Nies and Bellama have studied this reaction for various combinations of methylsilicon compounds and mercuric salts (66, 67).

All systems of this type for which kinetic studies have been reported show a second-order rate dependence, first order in each reactant. The methyl transfer for such systems may follow this pattern (68):

$$CH_3-D + M^{n+} \rightarrow [D--CH_3 -M] \rightarrow D^+ + CH_3M^{(n-1)+} \qquad (3)$$

Changes in the "electrophilicity" of the acceptor enormously affects the rate of exchange (69). This trend is illustrated by Fig. 4, which shows the rates of transmethylation of some methylmetals toward chloromercury species; in certain cases there is a 10^7-fold diminution of rate constant as salinity increases. This strongly suggests that the frequent salinity gradients found in natural waters will markedly influence the rates of naturally occurring transmethylations (*vide infra*).

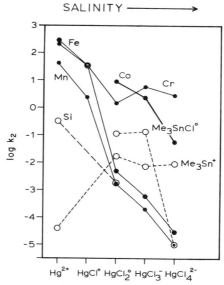

FIG. 4. Relationship between bimolecular rate constants (k_2) for aqueous trans-methylation of $HgCl_n$ species and solution salinity. Abbreviations: Si = $(CH_3)_3SiCH_2CH_2$-$CH_2CO_2^-$; Cr = 3-PyH^+—CH_2—$Cr(OH_2)_5^{3+}$; Fe = 3-PyH^+—CH_2Fe=$(CO)_2(-C_5H_5)$; Co = 4-PyH^+—$CH_2Co(CN)_5^{3-}$; Mn = 3-PyH^+—$CH_2Mn(CO)_5$, where PyH^+ is the pyridinium group.

IV

BIOLOGICAL TRANSMETHYLATION: RATE STUDIES

A. Background and Scope

Systems where *in vivo* methylation is the principal step are not as readily susceptible to quantitative measurements as the chemical systems just discussed. Various obstacles remain to be overcome in order to develop a complete picture of such *in vivo* processes. These are as follows:

(a)   relating the metabolic contribution of individual microbial or microbial community methylators to the overall process,

(b)   assessing the relative contributors of intermediate exo- and endocellular metabolites to the overall methylation,

(c)   developing and refining quantitative characterization techniques to determine the nature and concentrations of those biologically available methyl–acceptor substrates and/or methylated products required for kinetic interpretation.

Survey experiments during the past decade have partially achieved these goals. Either pure isolated cultures of microorganisms, or uncharacterized mixed cultures grown with nutrient stimulants in tissues, sediments or soils provide biologically active "methylators." When run against sterile controls, these cultures provide some measure of overall methylation rates as a function of extent (or rate) of biological growth and also concentration of substrate. However, some mixed microfloral communities can involve both methylating *and* demethylating bacteria at different elapsed periods, as Spangler *et al.* showed for methylmercuric species (*70, 71*). Consequently, common variables of temperature, light, nutrients, salinity, etc. will favor one microbial species' growth over another, thereby leading to marked variations in rates of CH_3Hg^+ production or degradation under both laboratory and field conditions.

B. *Mercury*

For various reasons, and especially the Minamata disaster, most kinetic studies on *in vivo* transmethylation have concentrated on mercury. The great majority of these have involved individual species of microorganisms or microbial communities as the actual methylators. There have been several reviews on the role of microorganisms in transmethylation (*5, 6, 72*) and other cyclic processes involving metals or metalloids (*25, 73–75*).

Rowland *et al.* found that intestinal flora of rats would methylate $HgCl_2$ at a rate of 0.90 ng h^{-1}/g of cecal contents (*76*). They also found that this rate increased initially, passed through a maximum, and decreased sharply after 60 h. In addition, metabolism of mercury in the small intestine seemed to involve formation of a methylmercury sulfur compound (*77*). The same authors showed that CH_3HgCl will be volatilized in the presence of H_2S (*78*). In this connection, it may be recalled that the major methylmercuric compound isolated from toxic Minamata Bay shellfish was the *thiomethyl* derivative CH_3HgSCH_3 (*21*). Subsequently, Craig and Bartlett argued that $(CH_3Hg)_2S$ is the active intermediate in this process, later decomposing to $(CH_3)_2Hg$ and β-HgS (*79*). Although *Clostridium cochlearium* T-2 seemed to form methylmercuric compounds from HgS, they were actually methylating traces of $HgCl_2$ impurity (*80*). Brook trout do not methylate Hg(II) compounds (*81*), nor do fish tissues and organs (*82*). However, the indigenous microflora of animal intestines (as has al-

ready been noted) does cause biomethylation of mercury and arsenic (83).
This biotransformation [Eq. (4)] has been reported for such microflora of

$$^{203}Hg^{2+} \rightarrow CH_3^{203}Hg^+ \tag{4}$$

six freshwater fish species (26). Kozak and Forsberg, in their investigation of Hg(II) interaction with cow rumen microflora, found that methylation did not occur, and that most of the mercuric compound passed through unchanged, with a small quantity reduced to Hg^0 (84). An enzyme system isolated from a strain of *Pseudomonas* catalyzes reductive cleavage of the mercury–carbon linkage in methylmercuric compounds (85, 86). Sulfhydryl groups are involved, and the reaction proceeds as follows:

$$CH_3Hg \xrightarrow{RS^-} CH_3HgSR \xrightarrow{E_1, RS^-} CH_4 + Hg(SR)_2 \xrightarrow[Hg^0]{E_2} 2RS^- \tag{5}$$

where E_1 and E_2 represent the enzymes involved. Ethyl- and phenylmercuric compounds are decomposed in the same way (86).

In this connection should also be considered the findings of Fang (87). He studied the metabolism of $C_6H_5{}^{203}HgOAc$ in various higher aquatic organisms. In all species, dearylation and the production of Hg(II) was the major metabolic route, but *ethyl*mercuric chloride was consistently found as a minor metabolite, along with varying quantities of $CH_3Hg(II)$. Wood *et al.* have pointed out the biosynthesis of ethylcobalamin in nature (88), and this compound will undergo transethylation to Hg(II) (9, 33). Taken in combination, these processes form an interesting microbiological cycle which may presage important insights into biomethylation of other elements.

C. Arsenic

After a long fallow period, other workers have recently begun to extend Challenger's arsenic biomethylation work (1–3). Cox and Alexander have studied this reaction by the mold *Candida humicola* (89–91). The overall reaction may be summed up by the reaction shown in Eq. (6). Both cell extracts and whole cells of *Methanobacterium* strain M.o.H. converted arsenate to dimethylarsine (92). Investigations on *C. humicola* and other molds determined that the rate of $(CH_3)_3As$ production follows the growth rate pattern of the mold, and decreases when the mold reaches the resting phase (93). The methylated arsenic intermediates were identified by use of

[74]As-labeled starting materials (94). When C. humicola was preconditioned by treatment with cacodylate ion, the rate of subsequent trimethylarsine production from arsenate or cacodylate increased, possibly due to a permanent As build-up in cell walls (95).

$$O{=}As(OH)_3 \rightarrow (CH_3)_3As \qquad (6)$$

Human volunteers who drank wine containing inorganic arsenic excreted methylarsonic and cacodylic acids in their urine (96). These two compounds also occur in the urine of copper smelters, the elevated arsenic content presumably coming from copper ores (97). Cacodylate ion forms when $H_3^{74}AsO_4$ is administered orally to dogs or hamsters (98–100). A Japanese group reported that various small animals will form methylarsonate and cacodylate from H_3AsO_4 (101). On the basis of presently available data, it remains uncertain whether arsenic biomethylation occurs from the bodily processes of the mammals themselves or from microflora of the intestinal tract.

D. Selenium and Tellurium

Challenger's reports that oxygen derivatives of Se and Te could undergo biomethylation have been confirmed and extended by more recent work (102–106). In a report on selenium biomethylation by sludge microorganisms, the products were dimethylselenide [$(CH_3)_2Se$], dimethyldiselenite [$(CH_3)_2Se_2$], and dimethylselenone [$(CH_3)_2SeO_2$] (106). In a competitive experiment, C. humicola generated $(CH_3)_2Se$ exclusively from selenate in the presence of phosphate, tellurate or arsenate; however, generation of $(CH_3)_3As$ from arsenate was inhibited by selenate (90, 103). C. humicola also generated an unidentified gas, presumably $(CH_3)_2Te$, from tellurate (90, 103).

Mouse liver extracts formed dimethylselenide from sodium selenite and had a specific requirement for glutathione (107). Trimethylselenonium ion has been reported as a major urinary metabolite of selenium in rats (108). By analogy with its sulfur congener, $(CH_3)_3Se^+$ may serve as a potential ex vivo methylator; no reports, however, are yet available. Similarly, Naganuma and Imura detected $(CH_3Hg)_2Se$ as a reaction product from methylmercuric(II) species and selenite in rabbit blood (109), possibly parallel to the formation of $(CH_3Hg)_2S$ from the bacterial metabolites H_2S and CH_3Hg^+ in aqueous media (79) (vide supra). The unusual monotonic relationship of tissue-bound methylmercuric ion to total Se content in marine animals (110) and human organs (111) has received widespread discussion, and specific attraction of Se toward an electrophilic CH_3Hg^+ is usually assumed (*112*).

E. *Tin*

Several years ago Brinckman, Iverson, and co-workers reported that a species of *Pseudomonas* found in the sediments of Chesapeake Bay would form methyltin compounds; in the presence of both mercury and tin, methylmercuric and methyltin compounds formed (*5, 113, 114*). Based on their laboratory and field observations, these authors proposed a "cross-over" mechanism (Fig. 5) wherein an appropriate methyltin species generated by *Pseudomonas* reacted with environmentally formed Hg(II) to form methylmercuric ion (*114*).

In recent years organotin compounds have been increasingly used in various technological applications that would facilitate their entry into the environment (*115, 116*). Methyltin compounds have been found in natural waters at very low concentrations, as have butyltin compounds (*117–119*). The latter probably come from widespread commercial use of butyltin compounds in antifouling paints or agricultural biocades (*113, 114, 118*).

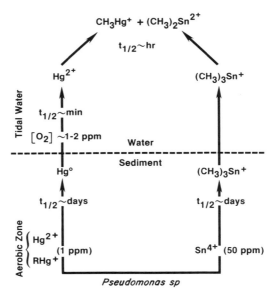

FIG. 5. "Crossover" model for combined biotic/abiotic methylations in the environment.

Considerable effort should be forthcoming to ascertain the relative contributions of biogenic and anthropogenic organotin sources to global aquatic burdens (*10, 120*).

Very recently there have been various independent reports on tin biomethylation. Coleman *et al.* reported that $(CH_3)_3SnOH$ will be converted to $(CH_3)_4Sn$ by estuarine sediments from San Francisco Bay, possibly proceeding by both biotic and abiotic pathways, with the latter being catalyzed by reducing or sulfur-bearing ligands (*121*). Various inorganic Sn(II) and Sn(IV) compounds, along with some organotin compounds, will undergo biomethylation by microorganisms from the sediments of Plastic Lake (Ontario, Canada) (*122*). Butyl- and phenyltin compounds gave only $(CH_3)_3Sn^+$; there was no evidence for mixed organomethyltin compounds. Hallas and Cooney reported that microorganisms from Chesapeake Bay sediments displayed varying tin resistance and volatilization, probably through biomethylation (*123, 124*). Three reviews touching on certain aspects of environmental organotin compounds and their speciation have recently appeared (*10, 125, 126*). This subarea is currently in a state of very rapid development and change.

F. *Lead*

An initial report (*127*) that both Pb(II) and $(CH_3)_3PbOAc$ could undergo biomethylation was questioned (*128*), and an alternative abiotic rearrangement proposed. Subsequent work apparently confirms the original report of biomethylation (*129–131*). $(CH_3)_4Pb$, the final product, has a deleterious effect on *Scenedesmus quadricauda* and other freshwater algae (*132, 133*). Craig, from a study on the effects of Lake Minnetonka sediments on $(CH_3)_3PbOAc$, proposed that $(CH_3)_4Pb$ forms primarily, and perhaps exclusively, by abiotic chemical disproportionation (*134*). Another group reached a different conclusion; having determined the nature and source of atmospheric decomposition of tetramethyllead (*135*), they interpreted the extensive airborne transport of this volatile species near Lancaster (England) as emanating from biogenic sources in tidal flats of the Morecambe Bay estuary (*136*). A conclusive answer to the relative importance of biotic versus abiotic pathways for $(CH_3)_4$ formation still remains to be determined (*8*). The recent remarkable observation of abiotic oxidative methylation of Pb(II) in aqueous solution further complicates the intertwined observations in this area. Ahmad *et al.* (*137*) found the reaction shown in Eq. (7) and pointed out that CH_3I occurs in appreciable quantities in marine waters (*138*). The authors suggested another possible cross-over type mechanism involving both biogenic iodine and abiotic lead and leading to toxic methyllead compounds.

$$Pb^{2+} (aq) + CH_3I (1) \rightarrow (CH_3)_4Pb (g) \qquad (7)$$

G. Other Metals and Metalloids

Inorganic thallium compounds undergo biomethylation under laboratory conditions (131, 139). Unlike previously discussed metals, biomethylation of Tl stops at $(CH_3)_2Tl^+$, probably due to the extreme instability of $(CH_3)_3Tl$ to water (140). There have been no reports of methylthallium compounds in natural waters.

Attempts to detect methylgermanium or methylantimony compounds in various natural water samples proved negative (141),[1] although the possibility of biogeochemical cycling involving the latter species has been discussed (142).

Methylphosphorus compounds have been detected in some waters, but their origin is uncertain (143). Some isolated studies (144, 145) demonstrate the biogenesis of $(CH_3)_2S$, $(CH_3)SH$, and $(CH_3)_2S_2$, but their relationship to each other and the Se or Te analogs remains unknown.

H. Soil and Sediment Transmethylation Studies

Numerous papers have appeared on this topic, largely, but not exclusively, dealing with mercury. In his studies on mercury transmethylation in soils (146–151), Rogers found that these contained a chemically isolable abiotic methylating factor that was stable at 122°C, soluble in 0.5 N OH⁻, and removable by dialysis. He also found that the rate of methylation depended on mercury concentration, soil texture, soil type, and soil pH. The rate of methylation increased initially, passed through a maximum, and subsequently declined, thereby following the same pattern as reported for methylation by intestinal microflora. A similar, more limited study was reported by Beckert et al. (152). Humic acid isolated from pond sediments reduced Hg^{2+} to gaseous Hg^0 by a pH-dependent first order mechanism, with $k = 0.009$ h⁻¹ (153).

Mercury-containing river sediments released methylmercuric ion under aerobic conditions at a rate of 5.06 g m⁻² d⁻¹ at 20°C; the rate was lower under anaerobic conditions (154). Methylmercuric ion production from

---

[1] Methylstibonic and dimethylstibinic acids have recently been discovered in certain natural waters (141a).

$HgCl_2$ in the presence of estuarine sediments showed distinct seasonal variation, whose rhythm apparently relates to the seasonal ability of the microorganisms to carry out biomethylation (155, 156). The ability of different river sediments to perform biomethylation depended markedly on the nature of the sediments (157, 158); thus, organic sediments reacted appreciably faster than coarse sand. Activated sludge will methylate inorganic mercury compounds under both sterile and nonsterile conditions, suggesting the presence of both biotic and abiotic pathways (159).

Marine sediments, in contrast, apparently perform transmethylation solely through a biological mechanism, which may proceed under both anaerobic and aerobic conditions (160). This observation provides a challenge to microbial ecologists and environmental chemists in their attempts to develop a reliable model for predicting the type of mercury [or other metal(loid)] species biologically formed in marine sediments. Estuarine sediments provide a special case because of their frequent and rapid fluctuations in oxygen, metal, and saline content. Thus Blair et al. (161) isolated several mercury-tolerant bacteria from Chesapeake Bay. Although most of them produced only Hg^0, one strain, an obligate anaerobe, generated both Hg^0 and CH_3Hg^+. Of two facultative isolates, one generated Hg^0 and CH_3Hg^+ under anaerobic conditions and Hg^0 exclusively under aerobic conditions, whereas the other produced only Hg^0 anaerobically and neither mercury species under aerobic conditions. The presence of sulfide in sediments affects the pathway of methylation: at sulfide levels below 160 ppm, CH_3Hg^+ was the favored product, whereas above that level $(CH_3)_2Hg$ and HgS formed (162).

Five soils from southeastern Montana were treated with ^{203}Hg-labeled CH_3HgCl to determine its effect on Hg volatilization (163). This volatilization increased at higher soil temperatures and decreased by addition of nitrate, glucose, and insufficient or excessive water. When various arsenic compounds were added to soils, AsH_3 was the major product; CH_3AsH_2 and $(CH_3)_2AsH$ also formed from methylarsonate and cacodylate (164). Application of $[^{14}CH_3]$methylarsonate to soils revealed that up to 10% of the label emerged as CO_2 (compared to 0.7% in sterilized soils), and arsenate was the only inorganic As product (165). Mixed bacterial–fungal communities obtained from freshwater pond sediments converted cacodylic acid to gaseous $(CH_3)_3As$ under both aerobic and anaerobic conditions (166). The same type of inocula in liquid medium selectively demethylated cacodylate ion (but not methylarsonate ion) to form arsenite (167). Up to 11% of cacodylate ion was converted to arsenite ion in 51 days.

Methylation of tin or lead compounds in sediments has previously been noted. Reports of corresponding transformations for other heavy metals or metalloids remain unknown, though many elements of technological and environmental significance can be incorporated into bacterial or fungal cellular metabolisms through absorption or redox reactions (5, 168, 169). Among such elements, the following merit special attention because of their scarcity, essentiality, or toxicity: Ag (170), Au (171, 172), Cu (169, 171), Zn (169, 173, 174), Cd (169, 175), Hg (171, 172, 176), Tl (177), Sn (178), Pb (179–181), Sb (182), Se (183), Mn (169, 171), Fe (171, 184), Co (169, 175), Ni (169, 171), Pd (172), Pt (172), and even the radioactive actinides Am (185), Pu (185, 186), and U (186, 187). Present limitations on characterization or speciation of very low concentrations of metal-containing metabolites have restricted progress in this field, especially with regard to the determination of possible organometal intermediates (8, 10). Nevertheless, the enormous range of cellular selectivity and bioamplification of metals (in some cases $>10^5$ times higher than the surrounding waters) causes these processes to generate intense activity among researchers seeking to exploit them in potential bioengineering applications, such as microbiological extraction and recovery of ores and metals (188) or removal of toxic metals from aqueous process streams (186).

V
MECHANISMS OF TRANSMETHYLATION

A. General Considerations

At some point during any transmethylation, biotic or abiotic, the linkage between the methyl group and the atom holding it must break. As has been pointed out elsewhere (47, 68, 189–191), this linkage might cleave in three different ways (cf. Fig. 6): as a positive or negative ion (*heterolytic cleavage*) or as a free radical (*homolytic cleavage*). In view of their high reactivity, it is highly unlikely that CH_3^+, CH_3^-, or $CH_3 \cdot$ actually form as distinct moieties in such a polar and reactive solvent as water. Available evidence indicates that the most probable form is a bridging methyl group in a bimolecular transition state (69, 192):

S_E2 (closed) S_E2 (open)

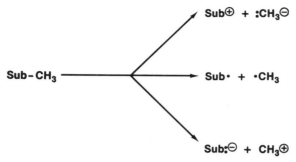

FIG. 6.   Modes of cleavage for methyl–substrate bonds.

Stereochemical retention of configuration around the saturated α-carbon can only result from the closed transition state, whereas the open transition state can lead to inversion (*10, 193*). Configurational inversions occur in many biological substrates, and much future work on biomethylation will likely involve the use of substituted methyl groups and the study of any optical changes that occur. Determination of the relative rates of cleavage for a series of normal and branched alkyl derivatives enables a distinction between retention or inversion in the S_E2 (open) pathway (*194*).

In the following discussion, we shall classify, for the sake of convenience, the various mechanisms according to the nature of the methyl group being transferred.

B. *Electrophilic Transmethylation*

1. *Methylcobalamin*

The mechanisms of transmethylation involving methylcobalamin have been extensively discussed by Wood and co-workers (*189–191*). They proposed a classification based on the standard reduction potentials for the elements (Table IV) and having three types of reaction:

Type 1. The methyl group is transferred as an anion to the electrophilic metal substrate.

Type 2. The methyl group is transferred as a radical to the metal(loid) substrate.

Redox switch. Two different oxidation states of the metal must be present in order for methylation to occur.

TABLE IV

Standard Reduction Potentials and
Proposed Transmethylation Mechanisms

Couple	$E^0(V)$	Mechanism 189
Bi_2O_4/BiO^+	+1.59	—
PbO_2/Pb^{2+}	+1.46	Type 1
Tl^{3+}/Tl^+	+1.26	Type 1
MnO_2/Mn^{2+}	+1.21	—
SeO_4^{2-}/SeO_3^{2-}	+1.15	—
$V(OH)_4^+/VO^{2+}$	+1.00	—
$Pd^{++}/Pd(0)$	+0.987	Type 1
$AuCl_4^-/AuCl_2^-$	+0.93	Redox Switch
$Hg^{2+}/Hg(0)$	+0.854	Type 1
$AuBr_4^-/AuBr_2^-$	+0.81	—
Fe^{3+}/Fe^{2+}	+0.771	—
$PtCl_6^{2-}/PtCl_4^{2-}$	+0.760	Redox Switch
$PtCl_4^{2-}/Pt(0)$	+0.75	—
$PtBr_6^{2-}/PtBr_4^{2-}$	+0.64	—
Sb_2O_5/SbO^+	+0.64	—
$PdCl_4^{2-}/Pd(0)$	+0.59	Type 1
$H_2AsO_4^-/H_3AsO_3$	+0.559	—
$Co(OH)_3/Co(OH)_2$	+0.17	—
Sn^{4+}/Sn^{2+}	+0.154	Type 2
Tl_2O_3/Tl^+	+0.02	—
$(CysS)_2/CysH$	−0.22	Type 2
$Ni^{2+}/Ni(0)$	−0.23	—
Cr^{3+}/Cr^{2+}	−0.41	Type 2
AsO_4^{3-}/AsO_3^{3-}	−0.67	—

The first type has been most extensively studied, especially for mercury (33). If electron-donating ligands, such as halides or organic groups, are attached to the mercury atom, the rate of reaction with methylcobalamin slows down markedly, presumably due to the lowered electrophilic nature of the metal (69). Similarly, when the solvent is made less polar, the reaction rate also decreases (34, 54). Anything favoring the formation of base-off methylcobalamin will slow the transmethylation rate, hence the retardation of this reaction in the presence of micelles (34, 38).

Available evidence suggests that the same mechanism applies to various other metals. Representative metals such as Tl, Pb, Bi, Sn, and Sb in their highest oxidation states react with methylcobalamin by this mechanism, and certain transition metals in high oxidation states (e.g., MnO_2, PtO_2, Co_2O_3, and Ni_2O_3), which are quite reactive toward methylcobalamin, may likewise follow this pathway (46, 47, 54). Unlike mercury, most of these metals do not form stable monomethylmetal compounds, decomposing via reductive elimination [Eq. (8)]. Aquocobalamin forms as the

$$CH_3B_{12} + MX^{n+} \rightarrow [CH_3MX^{(n-1)+}] \rightarrow CH_3X + M^{(n-2)+} \tag{8}$$

corrinoid product. The decomposition of $CH_3Tl(OAc)_2$ in polar solvents was studied by Pohl and Huber (195), who found that the reaction followed first-order kinetics and proposed the following mechanism:

$$CH_3Tl(OAc)_2 \rightleftharpoons CH_3TlOAc^+ + OAc^-$$
$$CH_3TlOAc^- + X^- \rightarrow CH_3X + Tl^+ + OAc^- \tag{9}$$

The species X^- is either acetate or another anion. Comparative studies found that halides facilitate demethylation (131). Other metals in higher oxidation states probably behave in similar fashion, although some (e.g., Pb) may also undergo disproportionation.

In contrast to representative metals, some transition metals show an enhancement of the transmethylation rate in the presence of halide. Scovell reported that excess Cl^- speeds up the $CH_3B_{12}-PdCl_4^{2-}$ reaction rate (48). Although $Cu(OH_2)_4^{2+}$ shows no reaction toward CH_3B_{12}, the rate accelerates enormously at high chloride ion concentrations (46a, 54) and even more at high bromide ion concentrations (46a). In the transmethylation of hexachloroplatinate(IV), proposed by Wood as following the "Redox Switch" mechanism, the reported evidence, although less complete than may be desired, supports the suggestion that the methyl group transfers as an anion rather than as a radical (40–43, 190). Addition of Br^- to this system accelerates the reaction, and hexabromoplatinate ion reacts faster still (46a). As noted in Section II,B, there remains the question of whether the final product is a methylplatinum(II) or a methylplatinum(IV) compound, or possibly a mixture of species. The mechanism proposed for this system is shown in Fig. 2. It may be noted that K_2PtCl_4 is rather unreactive toward CH_3B_{12}, but in the presence of Pt powder, the reaction goes rapidly (46, 47). The standard reduction potential for the $PtCl_4^{2-}/Pt$ couple is virtually identical to that for the $PtCl_6^{2-}/PtCl_4^{2-}$ couple.

2. Methylmetals

In every case where a mechanism has been proposed for aqueous transmethylations involving methylmetals, the mechanism has been S_E2 (44, 51, 59–66, 196). Available evidence strongly indicates that, from a chemical point of view, methylcobalamin is not *sui generis* but is rather one of numerous water-soluble, labile methylmetal compounds whose exocellular properties may be compared in purely chemical terms. Reported effects of solvents and ligands on both methyl donor and substrate are consistent with, and supportive of, the proposed S_E2 mechanism.

$$M^{n+} \xrightarrow[\text{stabilizing ligand(s)}]{\text{biotic}} CH_3M^{(n-1)+} \underset{m \text{ steps}}{\overset{X^-}{\rightleftharpoons}} \begin{array}{l} CH_3X + M^{(n-2)+} \\ \\ (CH_3)_m M^{(n-m)+} \end{array} \tag{10}$$

The fate of intermediates in transmethylation, especially ones that have not been isolated, may be considered at this point. One possible scheme is proposed in Eq. (10) where n represents the maximum valence of the metal. At enzymatic sites where biomethylation occurs, appropriate coordination of the metal with ligands presumably stabilizes the intermediate by suppressing reductive elimination. In exocellular reactions, a sufficient solution concentration of biogenic ligands such as thiols (197) or comparably effective donors [e.g., CN^- (198)] may provide the necessary stabilization to enable successive methylations. It is not clear, however, how such complexation, which also lowers the electrophilic character of the intermediate methylmetal (10, 69), could readily provide requisite stability without severe diminution of the overall rate of methylation. This question remains actively debated (29) and will offer a research challenge for some time to come.

C. *Free Radical Transmethylation*

A free radical mechanism has been proposed for the reactions of CH_3B_{12} with Cr(II) (53), thiols (199), and Sn(II) (50, 190). Wood has proposed the following mechanism for the transmethylation of Sn(II):

$$:SnCl_3^- + Fe(III) \rightarrow [\cdot SnCl_3] + Fe(II) \tag{11}$$
$$[\cdot SnCl_3] + CH_3B_{12} \rightarrow CH_3SnCl_3 + B_{12r}$$

A similar mechanism has been proposed for the catalyzed methylation of $AuCl_4^-$ (190, 191):

$$Fe(CN)_6^{4-} + AuCl_4^- \rightarrow Fe(CN)_6^{3-} + [AuCl_4^{2-}]$$
$$[AuCl_4^{2-}] + CH_3B_{12} \rightarrow [CH_3AuCl_3^-] + B_{12r} + Cl^- \tag{12}$$
$$[CH_3AuCl_3^-] \rightarrow AuCl_2^- + CH_3Cl$$

The presence of Au(I) may likewise catalyze the reaction of Au(III) with methylcob(III)alamin; Wood has included this couple under his Redox Switch category. Like $PtCl_6^{2-}$, the reaction of $AuCl_4^-$ with CH_3B_{12} is enhanced by the presence of Br^- (46). It remains to be proven that the Au reaction does involve methyl radical transfer, presumably with subsequent oxidation of B_{12r} [a cob(II)alamin compound] to the observed aquocobalamin. However, the available evidence for the formation of Au(II) intermediates is more extensive and convincing (200) than for Pt(III) intermediates (201).

Photolysis of acetate ion in the presence of Hg(II) gives methylmercuric compounds as products (44, 202–205):

$$HgX_2 + CH_3CO_2^- \xrightarrow{h\nu} CH_3HgX + CO_2 + X^- \tag{13}$$

The rate of reaction is enhanced by the presence of strongly colored solids (162, 163). Jewett reported (44) that photolysis of thallous acetate in D_2O gave the following reaction:

$$Tl^+ + CH_3CO_2^- + D_2O \xrightarrow{h\nu} CH_3D + CO_2 + TlOD \tag{14}$$

Irradiation in the absence of thallous ion did not give this reaction. A methylthallium(I) intermediate may form, which upon deuterolysis gives the observed products.

Since acetate ion (and similar potential methyl donors) can occur in natural waters, photolysis by sunlight may well be a potential, ubiquitous route to methylmetal compounds (204–206). One such example might be the reported photolysis of some aliphatic α-amino acids to form CH_3Hg^+ (207):

$$RCH(NH_2)CO_2H + HgCl_2 \xrightarrow{h\nu} CH_3HgCl \tag{15}$$

where R = CH_3, i-C_3H_7, i-C_4H_9, or sec-C_4H_9. Methylmercuric chloride formed in all cases, regardless of the amino acid used. The presence of the colored Cu^{2+} ion accelerated photomethylation. Photomethylation of $Hg(OAc)_2$ was also sensitized by sulfur atoms produced during concurrent solubilization of HgS (202), thereby providing a possible abiotic methylation route comparable to the reported biomethylation (208). If not biomethylation in the narrowest sense, these reactions certainly represent potential environmental methylation pathways!

D. *Nucleophilic Transmethylation*

This pathway involves transfer of a methyl group as a cation, requiring that the methyl acceptor be nucleophilic [Eq. (16)]. This mechanism has been most studied for arsenic compounds. The reactions summarized in Fig. 7 were originally proposed by Challenger (*2, 3, 12*) and subsequently developed by Cullen and co-workers (*58, 93–95*). One noteworthy feature of this mechanism is that each methylation is followed by reduction involving removal of oxygen. If organoarsenic compounds (e.g., $RAsO_3H_2$), are used as starting materials, the mixed arsines $RAs(CH_3)_2$ form as products. S-Adenosylmethionine (**6**) serves as methyl donor. Methylarsonic and cacodylic acids do not react with methylcobalamin, although $(CH_3)_4As^+$ reacts slowly (*56*). Arsenic(III) oxide reacts with trimethylsulfonium ion (*46, 57, 58, 92*). Although $(CH_3)_3As$ is the highest methylated arsenic species so far observed by direct biomethylation of arsenic, there is no reason in principle why $(CH_3)_4As^+$ should not also form by this pathway. Recently, this species has been identified by low-resolution mass spectrometry and proton NMR in the "marine arsenic" metabolite

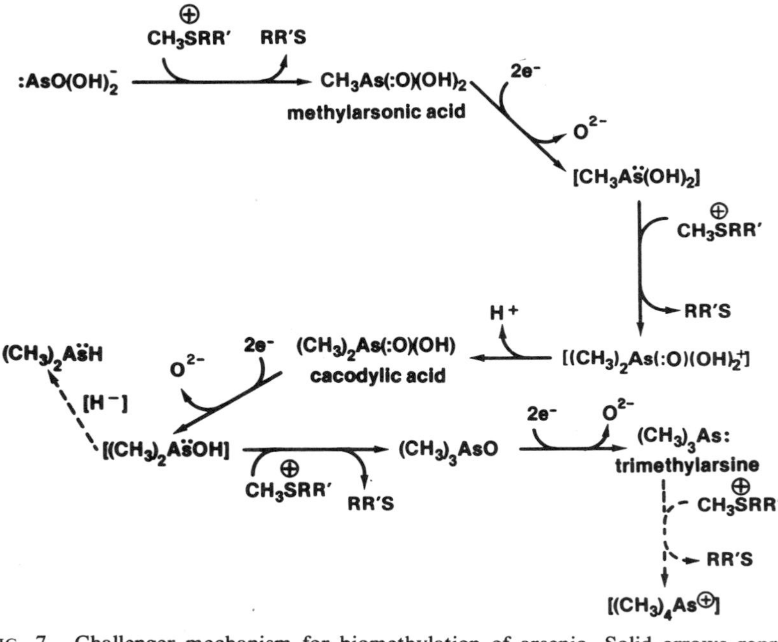

FIG. 7. Challenger mechanism for biomethylation of arsenic. Solid arrows represent pathways proposed originally; dashed arrows represent additional pathways proposed by the present authors.

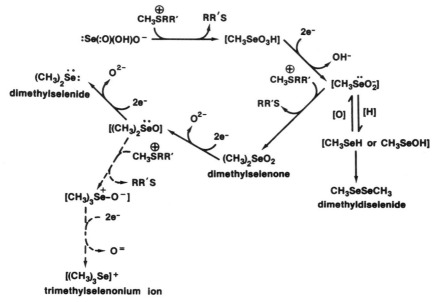

FIG. 8.   Proposed mechanism for the biomethylation of selenium. Adapted from Reamer and Zoller (*106*). Copyright 1980 by the American Association for the Advancement of Science. Dashed arrows represent pathways proposed by the present authors.

isolated from sea halibut and shrimp (*209*). Arsenobetaine [(CH$_3$)$_3^+$-AsCH$_2$CO$_2^-$] has been isolated from commercially imported rock lobsters (*210*).

$$CH_3X^+ + :Y \rightarrow X: + CH_3Y^+ \tag{16}$$

Detailed studies on the analogous selenium reaction have recently appeared (*102, 106*). The proposed mechanism is illustrated in Fig. 8. The primary reactions are basically the same as for arsenic; differences appear to be attributable to the differences in the chemistry of the two elements. The intermediate CH$_3$SeO$_2^-$ may gain a second methyl group, leading to (CH$_3$)$_2$SeO$_2$ and (CH$_3$)$_2$ Se, or it may be directly reduced to (CH$_3$)$_2$Se$_2$. This difference from arsenic may arise from the greater electronegativity of Se and the greater stability of Se—Se bonds. (CH$_3$)$_3$Se$^+$ has also been reported as a selenium metabolite (*108*).

In comparative studies with *C. humicola* (*90*), Cox and Alexander reported that phosphate will inhibit (CH$_3$)$_3$As formation from arsenite, arsenate, or methylarsonate but not from cacodylate. Selenite, selenate, and tellurate inhibited $(CH_3)_3As$ formation from arsenate, but the reverse was not true. There were no reports of any methylphosphorus compounds from phosphite or phosphate, or any methylantimony compounds from antimonate (*90*). Challenger found traces of a volatile Sb compound in some methylation experiments (*12*), but there was insufficient material to establish the nature of the species formed. $(CH_3)_2Te$ forms from tellurite and tellurate (*vide infra*).

The generality of this mechanism remains to be determined. Compound **6** might be compared to the better known reagent methyl iodide. In a comparative study on $(CH_3)_3As$ and $(CH_3)_3Sb$ (*142, 211*), Parris and Brinckman found that the As compound reacts several times faster with CH_3I under comparable conditions. There is still no evidence for the biomethylation of antimony, though the possibility is very much present (see refs *5, 12*, and *142* for discussion on this point). Methylstibonoate has never been reported (*212*). Various inorganic Sn(II) (*122*) and Pb(II) (*137*) compounds are known to react with CH_3I. The possibility that these compounds, and possibly Tl(I) as well, might undergo transmethylation with **6** cannot be ruled out on the basis of present knowledge.

VI

OTHER ASPECTS OF BIOLOGICAL TRANSMETHYLATION

A. *The Challenge of Trace Chemical Speciation*

Water-soluble methylmetals frequently show marked lability and undergo facile transformations. To a lesser degree this also applies to higher alkylmetals that may have anthropogenic pathways into natural waters. Jewett *et al.* (*51*) pointed out that many water-soluble organometals are polar, highly hydrated species of the type $(CH_3)_nM(OH_2)_m^{(z-n)+}$, where z is the oxidation number of the metal M. The partition of such species between aqueous solution and the atmosphere, or between aqueous and lipid phases, depends on the value of z, the nature of the organic moieties on the metal, the overall charge on the molecule, and the presence of anions, most especially chloride since most organometal chlorides tend to be covalent, frequently volatile materials with high lipid solubility. This partitioning ability in turn depends on two factors:

1. The potential for uptake of such organometals into viable organisms (or the release therefrom).

2. The underlying chemistry of the organometals, which determines their tendency to *speciate* and which determines their molecular identification and analysis (qualitative and quantitative) in environmental or biological matrices.

Alkylmetal toxicity, for example, depends very much both on n and on the nature of the alkyl groups on the metal (7). To some extent this behavior also reflects the lipophilicity of the organometal species, by analogy with water–lipid partitioning of organic molecules (213). A number of empirical relationships based upon volatility (214) or solubility (215), as well as particular "quantitative structure–activity relationships" related to linear free-energy correlations between organic substituents on metals and their hydrophobic properties, have been proposed (216, 217). Recent advances in semiempirical theories for solvophobic behavior of organic moieties on well-behaved reverse bonded-phase and ion-exchange liquid chromatographic substrates (218–220) suggest that, in the near future, independent laboratory measurements should enable predictions of bioaccumulation potential or the ease of organismal uptake of aqueous organometals.

The second factor involves development of measurement techniques for studying organometal species at the extremely low concentrations found in biological and environmental samples. Since most bioactive organometals do not partition favorably out of biological fluids, investigators must either establish reliable methods to isolate and characterize such compounds, without alteration, in those fluids; or to develop specific chemical procedures for the formation of volatile (and usually hydrophobic) derivatives that can be studied by gas or head-space analyses. For these reasons, analytical development in the field of bioactive organometals has emphasized liquid or gas chromatography, with a battery of specialized detection schemes, frequently involving some form of element-specific analysis (10, 221, 222). Tandem combinations of gas chromatographic–flame photometric emission (10, 117–119), gas chromatographic–atomic absorption (161, 166, 223, 224), or gas chromatographic–mass spectrometry (225) have proven successful for organotin, -mercury, -lead, -arsenic, -selenium, and -antimony compounds. Combinations of high performance liquid chromatography with flameless AA (226–228) or induction-coupled plasmas (229, 230) have also proven successful, especially when they have employed single- or simultaneous multielement detection (222, 230).

Analysts, however, must exercise caution in interpreting results, since the work-up techniques may cause misleading alterations in the compound(s) under study. A case in point involves the gas chromatographic speciation of As(III) and As(V) metabolites in soils or fluids [cf. Eq. (6)], where reduction with borohydride can give a single product from either oxidation state of arsenic (*231*):

$$(CH_3)_2AsO_2^- \atop (CH_3)_2As^+ \searrow \nearrow \xrightarrow{BH_4^-} (CH_3)_2AsH \qquad (17)$$

In such a case, direct As-specific liquid chromatography of sufficient sensitivity could distinguish between these possibilities (*167, 232, 233*).

In addition to the above, analytical processes must be developed that are capable of discerning key, possibly short-lived, metabolites or transport species in complex media at extremely low concentrations. Present kinetic measurements of microbial methylation (or other transformations) are performed at the parts per million (micrograms per milliliter) to parts per billion (nanograms per milliliter) levels with analysis times (or "windows") of 10 min to 1 h. Future investigations of biomethylation rates and mechanisms, including other closely related or competing processes, will require smaller analytical windows provided by increasingly sensitive and element-selective detectors. Speciation of transient metal(loid)-containing metabolites *in situ,* using analysis times of seconds, appear achievable by new coupled techniques (*234*), such as laser excitation spectrophotometry employing new optical detectors (*235*).

The extensive research into biomethylation has shown that there is widespread biological production of methylmetal(loid)s not predicted by conventional organometal experience. Also, previous views regarding the stability of some organometal species in water have proven inadequate for the interpretation of new experimental data. Metabolic or enzymic pathways for metal(loid) redox reactions utilizing one- or two-electron transfer from biogenic methyl or hydride ligands represent one exciting new development:

$$[Metal—H] \overset{\text{Metal (oxidized form)}}{\underset{\text{Metal (reduced form)}}{\rlap{\nearrow}\rlap{\searrow}}} [Metal—CH_3] \qquad (18)$$

Intermediate metal hydrides or methylmetals may decompose by the pathways shown in Fig. 6. Heterolytic cleavage giving CH_3^+, H^+, or CH_3^-, H^-, or homolytic cleavage giving $CH_3\cdot$ or $H\cdot$ will reduce the metal atom by 2, 0, or 1 electron(s), respectively. Present chemical evidence for reductive demethylation (*40–44, 48*) or oxidative methylation (*122, 131, 137*) of various metals has previously been described. Apparent enzymatic formation of both methyl and hydridic derivatives of arsenic (*92, 164*) and tin (*10, 119*) suggest that such transformations can occur by a presently unknown mechanism [Eq. (19)]. Such mixed species would not be ex-

$$\text{As(V)} \xrightarrow{\textit{Pseudomonas} \text{ spp.}} (CH_3)_n AsH_{3-n}$$
$$\text{Sn(IV)} \xrightarrow{\textit{Pseudomonas} \text{ spp.}} (CH_3)_n SnH_{4-n}$$

(19)

pected to survive long in an aerobic aqueous environment; nevertheless, present analytical techniques have permitted their direct detection by *in vitro* experiments and in field samples. Quite probably analogous methylmetal hydrides of other elements should likewise be detectable. Biotic processes thus far identified as involving CH_3— or H— ligands result either in reduction of the element or no redox at all. Possible reverse processes for both ligands may occur, but present evidence for biological oxidative methylation is limited to Tl(I), Pb(II), and possibly Sn(II), which form $(CH_3)_2Tl^+$ (*131, 139*), $(CH_3)_3Pb^+$ (*127, 129*), and CH_3Sn^{3+} (*50*), respectively. In contrast, Holm and Cox showed that six different bacterial cultures oxidized elemental Hg to Hg^{2+} with no formation of CH_3Hg^+ (*236*).

B. *Biogeochemical Cycling*

Mechanisms for the cycling of carbon, nitrogen, oxygen, and sulfur through the environment have been known for many years. These always include crucial contributions from organismal metabolism. In recent years, people have gradually come to realize that other elements, particularly the metals, also have ecological cycles; for example, cycles are recognized for such strategic industrial metals as iron and manganese (*237*). Bacteria apparently play a major role in these cycles, and their contributions figure heavily in ore and minerals recovery (*5, 6, 72, 238–240*).

Investigations into biomethylation have added further dimensions to environmental cycling of elements. Wood has proposed possible cycles for arsenic, mercury, and tin (*88, 189, 241, 242*) in which biomethylation plays an important role. More recently, Craig has extended, in detail and to other elements, a visualization of the relationships among biomethylation, biogeochemical cycles, and anthropogenic inputs, with emphasis on global dispersions (*8*). Mackenzie *et al.* have summarized predictive global cycles for trace levels of mercury, arsenic, and selenium that include their atmophilic (volatilization) properties in biological fluxes (*243*).

Biological methylation will continue to receive attention in connection with the environmental flux and transport of elements. Even a single methyl group introduced onto a metal(loid) will cause sharp changes in the metal's volatility and water–lipid solubility. Permethylmetals (in which every metal valence is occupied by a methyl group) are gases or volatile liquids at ambient temperatures and are virtually insoluble in water. They escape into the atmosphere, to be distributed across the globe. The presence of sunlight can further affect this process by causing photochemical dissociation:

$$(CH_3)_nM \rightarrow M + CH_4 + C_2H_6 \tag{20}$$

Figure 9 shows a generalized cycling pattern that metals might follow (244). The net effect of biomethylation is to open up new pathways for metal transfer through water, air, and/or food chains. Of special concern to environmentalists will be the translocation of toxic elements from natural or man-made sources through aquatic media to susceptible biota.

In addition to the above, there may also be less obvious indirect effects on geochemical cycles stemming from biomethylation. It has been more or less tacitly assumed that only those methylmetal compounds relatively stable to air, heat, light, or water need be considered in conjunction with biomethylation. However, methylcobalamin can act as a reducing agent. Many initially formed methylmetal compounds undergo reductive elimination, particularly in the presence of halides. Moreover, various refractory metal oxides (e.g., PbO_2, Pb_3O_4, PtO_2, MnO_2) react with CH_3B_{12} *as solids* (41, 46). Their solubility in water increases sharply as they are

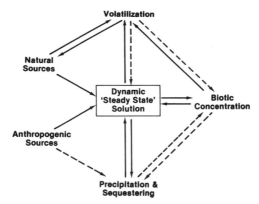

FIG. 9.   Ecological circulation of metals and metalloids.

reduced to lower oxidation states that dissolve more easily; this provides a potential mechanism for enhancing the circulation of such metals. The same reaction may be extendible to the corresponding sulfides (245).

Sequestration of certain elements in biota that subsequently become mineralized has proven significant for determining origins and locations of valuable mineral ore bodies with fossil fuel deposits. Coal and kerogen (oil shale) both contain substantial concentrations of metal(loid)s in uncharacterized forms (246, 247). The observation that marine algae, the modern counterparts of ancient precursors to fossil kerogen, will methylate and accumulate arsenic (209, 248, 249) raises some intriguing questions. Do primordial methylarsenic compounds reside in kerogen or coal? Do such fossil fuels contain other methylmetal compounds? Preliminary evidence from study of product water generated during the commercial pyrolysis of oil shale reveals that methyl- and phenylarsenic compounds are present along with arsenate (250). Although these organoarsenicals may form during the thermal process, it seems just as probable that they were there to begin with; subsequent studies relating to increased production of synthetic fuels from organic fossil materials will have to include investigation of involvement of organometals in processing and use.

VII
CONCLUSIONS

Humans became aware of biological methylation and its consequences primarily as a result of two series of widespread poisonings: first, the many cases of "Gosio-gas" poisonings, spread over many decades and countries; and second, the hundreds of tragic cases of "Minamata Disease" in Japan. In both instances, naturally occuring microorganisms (fungi and bacteria, respectively) converted anthropogenic materials by biochemical processes into, ultimately, methyl derivatives. In both instances, these methyl derivatives were absorbed by humans: trimethylarsine by breathing and methylmercuric compounds by ingestion of fish and mollusks whose tissues had previously concentrated them. Other cases of direct methylmercury poisoning have occured from the use of these compounds as fungicides on seeds that were subsequently ground into flour or fed to swine and thereafter eaten.

At present, arsenic, mercury, selenium, and tin have been unequivocally established as undergoing biomethylation. There is substantial, though not completely unequivocal, evidence for the biomethylation of lead, tellurium, and thallium, at least under laboratory conditions. Various other metals will react with methylcobalamin under laboratory conditions and thus may be considered potential candidates for biomethylation.

This process has been proposed as a mechanism for detoxification (245)—a biological mechanism for the removal of toxic metals by converting them to methyl derivatives that are subsequently removed by volatilization or extrusion in solvents. Although plausible, this suggestion remains to be directly confirmed.

Research efforts into biological methylation have been intense but irregular, leaving gaping lacunae, abundant speculation, and a relative sparseness of facts. These efforts have uncovered or suggested linkages to such intriguing topics as organic group exchange among aqueous organometals, the speciation of metal compounds, and the role of microorganisms in the ecological cycling of metals and metalloids. There remains ample scope for further research.

ADDENDUM

Introduction

The Tenth International Conference on Organometallic Chemistry, held at Toronto, Canada, in August 1981, included a symposium on the Environmental Aspects of Organometallic Chemistry. Thirteen papers were presented, including seven papers on biomethylation and related topics.

Recent work indicates that various methylarsenic compounds can be accumulated by marine organisms. These include arsenobetaine (210, 251), trimethylarsoniumlactate [$(CH_3)_3\overset{+}{As}CH_2CH(OH)CO_2^-$] and its phosphatidyl derivatives (252–254), and arsenosugars (255). These compounds form from arsenate salts, and provide strong evidence for the biological formation of As–C bonds other than As–CH_3.

Methylcobalamin: Chemical Rate Studies

The reaction between $SnCl_2$ and methylcobalamin in 1.0 M HCl is first order in each reactant, with a rate constant of 1.04 ± 0.10 M^{-1} sec^{-1} (256). The reaction is hypothesized as passing through a methyltin(III) radical intermediate. Ethylcobalamin and chloromethylcobalamin also react with $SnCl_2$, with rate constants of (1.66 ± 0.20) × $10^{-3} M^{-1}$ sec^{-1} and 7 × 10^{-3} M^{-1} sec^{-1}, respectively (256). Reaction between $CdCl_2$ and methylcobalamin at pH 9.6 produced traces of an unidentified volatile Cd compound (257).

Biological Transmethylation: Rate Studies

Arsenic

The marine facultative anaerobe bacterium *Serratia marinorubra* and the yeast *Rhodotorula rubra* both methylate arsenate ion to methylarsonate, but only the latter produces cacodylic acid (*258*). Human volunteers who ingested 500 μg doses of As as sodium arsenite, sodium methylarsonate, and sodium cacodylate excreted these compounds in their urine (*259*). Of these three, approximately 75% of the sodium arsenite is methylated, while 13% of methylarsonate is methylated. Rat liver subcellular fractions methylated sodium arsenate *in vitro*, providing the first direct evidence for possible mammalian methylation independent of symbiotic bacteria (*260*). Shariatpanahi *et al.* have reported kinetics studies on arsenic biotransformation by five species of bacteria (*261*). They found that the As(V)-As(III) reduction followed a pattern of two parallel first-order reactions, while the methylation reactions all followed first-order kinetics. Of the five species tested, only the *Pseudomonas* produced all four metabolites (arsenite, methylarsonate, cacodylate, trimethylarsine) (*261*).

Lead

The role of potential biomethylation of lead remains controversial and clouded. In a detailed study using ^{14}C and ^{210}Pb labeling, Reisinger *et al.* reported no evidence for biomethylation of inorganic or organic lead compounds (*262*), but reported that tetraalkylleads could form through disproportionation reactions involving sulfur compounds. This conclusion is likewise supported by the work of Jarvie (*128, 263*), who has suggested that $(R_3Pb)_2S$ forms as an intermediate, followed by disproportionation (*263*). A similar reaction has been proposed for tin (*264*)

$$3[(CH_3)_3Sn]_2S \rightarrow 3(CH_3)_4Sn + [(CH_3)_2SnS]_3 \qquad (21)$$

and reported for mercury (*79*)

$$(CH_3Hg)_2S \rightarrow (CH_3)_2Hg + HgS \qquad (22)$$

In this connection it might be noted that $(CH_3Hg)_2Se$ forms in biological systems from CH_3HgCl and selenite; it is a very labile compound and undergoes rapid decomposition (*265*). Both Sn(II) and Pb(II) will react with various alkyl halides (*266*).

Soil and Sediment Transmethylation

The rate of methylation of mercury in anaerobically incubated estuarine sediments proved to be inversely related to salinity (*267*); this is consistent with results reported in Section II,A. Methylmercuric ion forms in sediments upon addition of $HgCl_2$, with a lag phase of 1 month (*268*). Biomethylation by lake water columns and by sediments coincided, apparently being related to overall microbial activity, and showed periodic fluctuations (*269*). Topping and Davies have demonstrated that mercury can be methylated in the water column of a sea loch (*270*). As has previously been noted, tin compounds can be methylated by sediments (*121–124*), and this is also true for lead (*134–136, 271*). The relative proportions of biotic and abiotic methylation processes for such systems still remain to be determined.

Other Aspects of Biological Transmethylation

The role of methylation and demethylation in the natural cycle of arsenic has been discussed (*272, 273*), as well as environmental transformations of various alkylmetals (*274*).

ACKNOWLEDGMENTS

The authors would like to thank Drs. Joseph J. Cooney and Laurence E. Hallas of the Chesapeake Biological Laboratory and Dr. Jon M. Bellama of the Department of Chemistry (all of the University of Maryland) for providing copies of unpublished manuscripts. The authors are indebted to Drs. Thomas D. Coyle, Warren P. Iverson, and Greg J. Olson, all of the National Bureau of Standards, for valuable comments and discussions. The authors also thank Dr. William H. Zoller (Department of Chemistry, University of Maryland) and *Science* magazine for granting copyright release.

REFERENCES

1. F. Challenger, *J. Soc. Chem. Ind.* **54**, 657 (1935).
2. F. Challenger, *Chem. Rev.* **36**, 326 (1945).
3. F. Challenger, *Q. Rev. Chem. Soc.* **9**, 225 (1955).
4. F. Hofmeister, *Arch. Exp. Pathol. Pharmakol.* **33**, 198 (1894).
5. W. P. Iverson and F. E. Brinckman, *in* "Water Pollution Microbiology" (R. Mitchell, ed.), Vol. 2, pp. 201–232. Wiley, New York, 1978.
6. A. O. Summers and S. Silver, *Annu. Rev. Microbiol.* **32**, 637 (1978).
7. J. S. Thayer, *J. Organomet. Chem.* **76**, 265 (1974); J. S. Thayer, *Adv. Organomet. Chem.* **13**, (1975).
8. P. J. Craig, *in* "The Handbook of Environmental Chemistry" (O. Hutzinger, ed.), Vol. 1, Part A, pp. 169–227. Springer-Verlag, Berlin and New York, 1980.

9. D. Dodd and M. D. Johnson, *Organomet. Rev.* **52**, 1 (1973).
10. F. E. Brinckman, *J. Organomet. Chem. Lib.*, **12**, 343 (1981).
11. B. Gosio, *Ber. Dtsch. Chem. Ges.* **30**, 1024 (1897).
12. F. Challenger, *in* "Organometals and Organometalloids: Occurrence and Fate in the Environment" (F. E. Brinckman and J. M. Bellama, eds.), ACS Symposium Series, No. 82, pp. 1–22. Am. Chem. Soc., Washington, D.C., 1978.
13. V. DuVigneaud, "A Trail of Research in Sulfur Chemistry and Metabolism." Cornell Univ. Press, Ithaca, New York, 1952.
14. S. K. Shapiro and F. Schlenck, eds., "Transmethylation and Methionine Biosynthesis." Univ. of Chicago Press, Chicago, Illinois, 1965.
15. F. Salvatore, E. Borek, V. Zappia, H. G. Williams-Ashman, and F. Schlenck, eds., "The Biochemistry of Adenosylmethionine." Columbia Univ. Press, New York, 1977.
16. E. Usdin, R. T. Borchardt, and C. R. Creveling, eds., "Transmethylation." Elsevier/North-Holland, New York, 1979.
17. D. McAlpine and S. Araki, *Lancet* **ii**, 629 (1958).
18. K. Irukayama, M. Fujiki, F. Kai, and T. Kondo, *Kumamoto Med. J.* **15**, 1 (1962).
19. K. Kojima and M. Fujita, *Toxicology* **1**, 43 (1973).
20. N. Nelson *et al.*, *Environ. Res.* **4**, 1 (1971).
21. P. A. D'Itri and F. M. D'Itri, "Mercury Contamination: A Human Tragedy," p. 15. Wiley (Interscience), New York, 1977.
22. S. Jensen and A. Jernelöv, *Nature (London)* **223**, 753 (1969).
23. J. M. Wood, F. S. Kennedy, and C. G. Rosen, *Nature (London)* **220**, 173 (1968).
24. L. Landers, *Nature (London)* **230**, 452 (1971).
25. K. Beijer and A. Jernelöv, *in* "The Biogeochemistry of Mercury in the Environment" (J. O. Nriagu, ed.), p. 203. Elsevier/North-Holland, Amsterdam, 1979.
26. J. W. M. Rudd, A. Furutani, and M. A. Turner, *Appl. Environ. Microbiol.* **40**, 777 (1980).
27. R. Hartung and B. D. Dinman, eds., "Environmental Mercury Contamination." Ann Arbor Sci. Publ., Ann Arbor, Michigan, 1972.
28. J. M. Pratt, "Inorganic Chemistry of Vitamin B_{12}," pp. 150–153. Academic Press, New York, 1972.
29. F. E. Brinckman and J. M. Bellama, eds., "Organometals and Organometalloids: Occurrence and Fate in the Environment," ACS Symposium Series, No. 82. Am. Chem. Soc., Washington, D.C., 1978.
30. G. Agnes, S. Bendle, H. A. O. Hill, F. R. Williams, and R. J. P. Williams, *Chem. Commun.* p. 850 (1971).
31. G. Agnes, H. A. O. Hill, J. M. Pratt, S. C. Ridsdale, F. S. Kennedy, and R. J. P. Williams, *Biochim. Biophys. Acta* **252**, 207 (1971).
32. L. Bertilsson and H. Y. Neujahr, *Biochemistry* **10**, 2805 (1971).
33. R. E. DeSimone, M. W. Penley, L. Charbonneau, S. G. Smith, J. M. Wood, H. A. O. Hill, J. M. Pratt, S. Ridsdale, and R. J. P. Williams, *Biochim. Biophys. Acta* **304**, 851 (1973).
34. G. C. Robinson, F. Nome, and J. M. Fendler, *J. Am. Chem. Soc.* **99**, 4969 (1977).
35. G. N. Schrauzer, J. M. Weber, T. M. Beckham, and R. K. Y. Ho, *Tetrahedron Lett.* p. 275 (1971).
36. V. C. W. Chu and D. W. Gruenwedel, *Bioinorg. Chem.* **7**, 169 (1977).
37. H. A. O. Hill, J. M. Pratt, S. Ridsdale, F. R. Williams, and R. J. P. Williams, *Chem. Commun.* p. 341 (1970).
38. H. Yamamoto, T. Yokoyama, C. L. Chen, and T. Kwan, *Bull. Chem. Soc. Jpn.* **48**, 844 (1975).

39. L. S. Backman and D. G. Brown, *Biochim. Biophys. Acta* **428**, 720 (1976).
40. Y. T. Fanchiang, W. P. Ridley, and J. M. Wood, *J. Am. Chem. Soc.* **101**, 1442 (1979).
41. R. T. Taylor and M. L. Hanna, *J. Environ. Sci. Health Part A* **A11**, 201 (1976).
42. R. T. Taylor and M. L. Hanna, *Bioinorg. Chem.* **6**, 281 (1976).
43. R. T. Taylor, J. A. Happe, and R. Wu, *J. Environ. Sci. Health, Part A,* **A13**, 707 (1978).
44. K. L. Jewett, Ph.D. Thesis, Univ. of Maryland (1978).
45. R. T. Taylor, J. A. Happe, M. L. Hanna, and R. Wu, *J. Environ. Sci. Health, Part A* **14**, 87 (1979).
46. J. S. Thayer, unpublished observations.
46a. J. S. Thayer, *Inorg. Chem.* **20**, 3573 (1981).
47. J. S. Thayer, *in* "Organometals and Organometalloids: Occurrence and Fate in the Environment" (F. E. Brinkman and J. M. Bellama, eds.), ACS Symposium Series, No. 82, pp. 188–204. Am. Chem. Soc., Washington, D.C., 1978.
48. W. M. Scovell, *J. Am. Chem. Soc.* **96**, 3451 (1974).
49. A. M. Yurkevich, E. G. Chauser, and I. P. Rudakova, *Bioinorg. Chem.* **7**, 315 (1977).
50. L. J. Dizikes, W. P. Ridley, and J. M. Wood, *J. Am. Chem. Soc.* **100**, 1010 (1978).
51. K. L. Jewett, F. E. Brinckman, and J. M. Bellama, *in* "Organometals and Organometalloids: Occurrence and Fate in the Environment" (F. E. Brinckman and J. M. Bellama, eds.), ACS Symposium Series, No. 82, pp. 158–187. Am. Chem. Soc., Washington, D.C., 1978.
52. J. S. Thayer, *Proc. Int. Symp. Controlled Release Bioact. Mater., 5th* pp. 2.25–2.30. Univ. of Akron Press, Akron, Ohio, 1978.
53. J. H. Espenson and T. D. Sellers, *J. Am. Chem. Soc.* **96**, 94 (1974).
54. N. Yamamoto, T. Yokoyama, and T. Kwan, *Chem. Pharm. Bull.* **23**, 2185 (1975).
55. N. A. Clinton and J. K. Kochi, *J. Organomet. Chem.* **56**, 243 (1973).
56. J. S. Thayer, *Inorg. Chem.* **18**, 1171 (1979).
57. T. Antonio, A. K. Chopoa, W. R. Cullen, and D. Dolphin, *J. Inorg. Nucl. Chem.* **41**, 1220 (1979).
58. B. C. McBride, H. Merilees, W. R. Cullen, and W. Pickett, *in* "Organometals and Organometalloids : Occurrence and Fate in the Environment" (F. E. Brinckman and J. M. Bellama, eds.), ACS Symposium Series, No. 82, pp. 99–114. Am. Chem. Soc., Washington, D.C., 1978.
59. J. M. Pratt and P. J. Craig, *Adv. Organomet. Chem.* **11**, 331 (1973).
60. P. Abley, E. R. Dockal, and J. Halpern, *J. Am. Chem. Soc.* **95**, 3166 (1973).
61. A. Adin and J. H. Espenson, *Chem. Commun.* p. 653 (1971).
62. B. J. Gregory and C. K. Ingold, *J. Chem. Soc. B* p. 276 (1969).
63. G. J. Samuels and J. H. Espenson, *Inorg. Chem.* **19**, 233 (1980).
64. R. E. DeSimone, *J. Chem. Soc. Chem. Commun.* p. 780 (1972).
65. J. S. Thayer, *Synth. React. Inorg. Met.-Org. Chem.* **8**, 371 (1978).
66. J. D. Nies, Ph.D. Thesis, Univ. of Maryland (1978).
67. J. D. Nies and J. M. Bellama, unpublished observations.
68. J. K. Kochi, *in* "Organometals and Organometalloids: Occurrence and Fate in the Environment" (F. E. Brinckman and J. M. Bellama, eds.), ACS Symposium Series, No. 82, pp. 205–234. Am. Chem. Soc., Washington, D.C., 1978.
69. M. D. Johnson, *Act. Chem. Res.* **11**, 57 (1978).
70. W. J. Spangler, J. L. Spigarelli, J. M. Rose, and H. M. Miller, *Science* **180**, 192 (1973).
71. W. J. Spangler, J. L. Spigarelli, J. M. Rose, R. S. Flippen, and H. M. Miller, *Appl. Microbiol.* **25**, 488 (1973).
72. A. Jernelöv and A. L. Martin, *Annu. Rev. Microbiol.* **29**, 61 (1975).
73. D. Perlman, *Adv. Appl. Microbiol.* **7**, 103 (1965).

74. T. M. Lexmond, F. A. M. Dellaan, and M. J. Frissel, *Neth. J. Agric. Sci.* **24**, 79 (1976); *C.A.* **86**, 46947 (1977).
75. J. J. Bisogni, *Top. Environ. Health* **3**, 211 (1976).
76. I. R. Rowland, M. J. Davies, and P. Grasso, *Arch. Environ. Health* **31**, 24 (1977).
77. I. R. Rowland, M. J. Davies, and P. Grasso, *Xenobiotica* **8**, 37 (1978).
78. I. R. Rowland, M. J. Davies, and P. Grasso, *Nature (London)* **265**, 718 (1977).
79. P. J. Craig and P. D. Bartlett, *Nature (London)* **275**, 635 (1978).
80. M. Yamada and K. Tonomura, *J. Ferment. Technol.* **50**, 901 (1972).
81. J. W. Huckabee, S. A. Janzen, B. G. Blaylock, Y. Talmi, and J. J. Beauchamp, *Trans. Am. Fish. Soc.* **107**, 848 (1978).
82. A. Pennacchioni, R. Marchetti, and G. F. Gaggino, *J. Environ. Qual.* **5**, 451 (1976).
83. Y. K. Chau and P. T. S. Wong, *in* "Organometals and Organometalloids: Occurrence and Fate in the Environment" (F. E. Brinckman and J. M. Bellama, eds.), ACS Symposium Series, No. 82, pp. 39–53. Am. Chem. Soc., Washington, D.C., 1978.
84. S. Kozak and C. W. Forsberg, *Appl. Environ. Microbiol.* **38**, 626 (1979).
85. K. Tonomura and F. Kanzaki, *Biochim. Biophys. Acta* **184**, 227 (1969).
86. T. Tezuka and K. Tonomura, *J. Biochem. (Tokyo)* **80**, 79 (1976).
87. S. C. Fang, *Arch. Environ. Contam. Toxicol.* **1**, 18 (1973).
88. J. M. Wood, *Naturwissenschaften* **62**, 357 (1975).
89. D. P. Cox and M. Alexander, *Bull. Environ. Contam. Toxicol.* **9**, 84 (1973).
90. D. P. Cox and M. Alexander, *Appl. Microbiol.* **25**, 408 (1973); D. P. Cox and M. Alexander, *Microb. Ecol.* **1**, 136 (1974).
91. D. P. Cox, *in* "Arsenical Pesticides" (E. A. Woolson, ed.), ACS Symposium Series, No. 7, pp. 81–96. Am. Chem. Soc., Washington, D.C., 1975.
92. B. C. McBride and R. S. Wolfe, *Biochemistry* **10**, 4312 (1971).
93. W. R. Cullen, C. L. Froese, A. Lui, B. C. McBride, D. J. Patmore, and M. Reimer, *J. Organomet. Chem.* **139**, 61 (1977).
94. W. R. Cullen, B. C. McBride, and A. W. Pickett, *Can. J. Microbiol.* **25**, 1201 (1979).
95. W. R. Cullen, B. C. McBride, and M. Reimer, *Bull. Environ. Contam. Toxicol.* **21**, 157 (1979).
96. E. A. Crecelius, *EHP, Environ. Health Perspect.* **19**, 147 (1977).
97. T. J. Smith, E. A. Crecelius, and J. C. Reading, *EHP, Environ. Health Perspect.* **19**, 89 (1977).
98. S. M. Charbonneau, G. K. H. Tam, F. Bryce, Z. Zawidzka, and E. Sandi, *Toxicol. Lett.* **3**, 107 (1979).
99. G. K. H. Tam, S. N. Charbonneau, G. Lacroix, and F. Bryce, *Bull. Environ. Contam. Toxicol.* **22**, 69 (1979).
100. S. M. Charbonneau, J. G. Hollins, G. K. H. Tam, F. Bryce, J. M. Ridgeway, and R. F. Willes, *Toxicol. Lett.* **5**, 175 (1980).
101. Y. Odanka, O. Matano, and S. Goto, *Bull. Environ. Contam. Toxicol.* **24**, 452 (1980).
102. Y. K. Chau, P. T. S. Wong, B. A. Silverberg, P. L. Luxon, and G. A. Bengert, *Science* **192**, 1130 (1976).
103. R. W. Fleming and M. Alexander, *Appl. Microbiol.* **24**, 424 (1973).
104. J. W. Doran and M. Alexander, *Soil Sci. Soc. Am. J.* **40**, 687 (1976).
105. L. Barkes and R. W. Fleming, *Bull. Environ. Contam. Toxicol.* **12**, 308 (1974).
106. D. C. Reamer and W. H. Zoller, *Science* **208**, 500 (1980).
107. H. E. Ganther, *Biochemistry* **5**, 1089 (1966); H. S. Hsieh and H. E. Ganther, *Biochim. Biophys. Acta* **497**, 205 (1977).
108. H. E. Ganther, *in* "Selenium" (R. A. Zingaro and W. C. Cooper, eds.), p. 570. Van Nostrand-Reinhold, New York, 1974.

109. A. Naganuma and N. Imura, *Res. Commun. Chem. Pathol. Pharmacol.* **27**, 163 (1980).
110. H. C. Freeman, G. Shum, and J. F. Uthe, *J. Environ. Sci. Health, Part A* **A13**, 235 (1978); J. H. Koeman, W. H. M. Peters, C. H. Koudstaal-Hol, and J. J. M. de Goey, *Nature (London)* **245**, 385 (1973).
111. L. Kosta, A. R. Byrne, and V. Zelenko, *Nature (London)* **254**, 238 (1975); S. Nishigaki, *Nature (London)* **258**, 324 (1975).
112. Y. Sugiura, Y. Hojo, Y. Tamai, and H. Tanaka, *J. Am. Chem. Soc.* **98**, 2339 (1978).
113. C. Huey, F. E. Brinckman, S. Grim, and W. P. Iverson, *Proc. Int. Conf. Transp. Persistent Chem. Aquat. Ecosyst.* pp. II-74–II-78. Nat. Res. Counc., Ottawa, 1974.
114. F. E. Brinckman and W. P. Iverson, *in* "Marine Chemistry in the Coastal Environment" (T. M. Church, ed.), ACS Symposium Series, No. 18, pp. 319–342. Am. Chem. Soc., Washington, D.C., 1975.
115. J. J. Zuckerman, R. P. Reisdorf, H. V. Ellis, and R. R. Wilkinson, *in* "Organometals and Organometalloids: Occurrence and Fate in the Environment" (F. E. Brinckman and J. M. Bellama, eds.), ACS Symposium Series, No. 82, pp. 388–424. Am. Chem. Soc., Washington, D.C., 1978.
116. J. J. Zuckerman, ed., "Organotin Compounds: New Chemistry and Applications," Advances in Chemistry Series, No. 157. Am. Chem. Soc., Washington, D.C., 1976.
117. R. S. Braman and M. A. Tompkins, *Anal. Chem.* **51**, 12 (1979).
118. V. F. Hodge, S. L. Seidel, and E. D. Goldberg, *Anal. Chem.* **51**, 1256 (1979).
119. J. A. Jackson, W. R. Blair, F. E. Brinckman, and W. P. Iverson, *Environ. Sci. Technol.*, **16**, in press (1982).
120. R. J. Lantzy and F. T. Mackenzie, *Geochim. Cosmochim. Acta* **43**, 511 (1979).
121. W. M. Coleman, A. B. Cobet, and H. E. Guard, *Abstr. Int. Conf. Organomet. Coord. Chem. Germanium, Tin Lead, 3rd, Univ. Dortmund (BRD)* p. 26 (1980).
122. I. Ahmad, Y. K. Chau, P. T. S. Wong, A. J. Carty, and L. Taylor, *Nature (London)* **287**, 710 (1980).
123. L. E. Hallas and J. J. Cooney, *Appl. Environ. Microbiol.* **41**, 466 (1981).
124. L. E. Hallas and J. J. Cooney, *Abstr. Annu. Meet. Am. Soc. Microbiol., 80th* p. 181 (1980).
125. P. J. Craig, *Environ. Technol. Lett.* **1**, 225 (1980).
126. A. G. Davies and P. J. Smith, *Adv. Inorg. Chem. Radiochem.* **23**, 1 (1980).
127. P. T. S. Wong, Y. K. Chau, and P. Luxon, *Nature (London)* **253**, 263 (1975).
128. A. W. P. Jarvie, R. N. Markall, and H. R. Potter, *Nature (London)* **255**, 217 (1975).
129. U. Schmidt and F. Huber, *Nature (London)* **259**, 157 (1976).
130. J. P. Dumas, L. Pazdernik, S. Bellonick, D. Bouchard, and G. Vaillancourt, *Water Pollut. Res. Can.* **12**, 91 (1977).
131. F. Huber, U. Schmidt, and H. Kirchmann, *in* "Organometals and Organometalloids: Occurrence and Fate in the Environment" (F. E. Brinckman and J. M. Bellama, eds.), ACS Symposium Series, No. 82, pp. 65–81. Am. Chem. Soc., Washington, D.C., 1978.
132. B. A. Silverberg, P. T. S. Wong, and Y. K. Chau, *Arch. Environ. Contam. Toxicol.* **5**, 305 (1977).
133. P. T. S. Wong and Y. K. Chau, *Proc. Int. Conf. Manage. Control Heavy Met. Environ.* pp. 131–134 (1979).
134. P. J. Craig, *Environ. Technol. Lett.* **1**, 17 (1980).
135. R. M. Harrison and D. P. H. Laxen, *Environ. Sci. Technol.* **12**, 1384 (1978).
136. R. M. Harrison and D. P. H. Laxen, *Nature (London)* **275**, 738 (1978).
137. I. Ahmad, Y. K. Chau, P. T. S. Wong, A. J. Carty, and L. Taylor, *Nature (London)* **287**, 716 (1980).
138. J. E. Lovelock, R. J. Maggs, and R. J. Wade, *Nature (London)* **241**, 194 (1973).

139. F. Huber and H. Kirchmann, *Inorg. Chim. Acta* **29**, L249 (1978).
140. C. R. Hart and C. K. Ingold, *J. Chem. Soc.* p. 4372 (1964).
141. R. S. Braman and M. A. Tompkins, *Anal. Chem.* **50**, 1088 (1978).
141a. M. O. Andreae, J. F. Asmondé, P. Foster, and L. van'tdack, *Anal. Chem.* **53**, 1766 (1981).
142. G. E. Parris and F. E. Brinckman, *Environ. Sci. Technol.* **10**, 1128 (1976).
143. A. Verweij, H. L. Boter, and C. E. A. M. Degenhardt, *Science* **204**, 616 (1979).
144. R. A. Rasmussen, *Tellus* **26**, 254 (1974); F. B. Hill, V. P. Aneja, and R. M. Felder, *J. Environ. Sci. Health, Part A* **A13**, 199 (1978).
145. J. E. Lovelock, R. J. Maggs, and R. A. Rasmussen, *Nature (London)* **237**, 452 (1972); B. C. Nguyen, A. Gaudry, B. Bonsang, and G. Lambert, *Nature (London)* **275**, 637 (1978).
146. R. D. Rogers, *J. Environ. Qual.* **5**, 454 (1976).
147. R. D. Rogers, *C.A.* **86**, 47020 (1977).
148. R. D. Rogers, *C.A.* **87**, 37915 (1977).
149. R. D. Rogers, *J. Environ. Qual.* **6**, 463 (1977).
150. R. D. Rogers and J. C. McFarlane, *J. Environ. Qual.* **8**, 255 (1979).
151. R. D. Rogers, *Soil. Sci. Soc. Am. J.* **43**, 289 (1979).
152. W. F. Beckert, A. A. Moghissi, F. H. F. Au, E. W. Bretthauer, and J. C. McFarlane, *Nature (London)* **249**, 674 (1974).
153. J. J. Alberts, J. E. Schindler, R. W. Miller, and D. E. Nutter, *Science* **184**, 895 (1974).
154. S. H. Wang, T. C. Ho, and K. C. Liu, *C.A.* **90**, 11952 (1979).
155. B. H. Olson, *Microb. Ecol.* p. 416 (1978).
156. C. L. So, *Mar. Pollut. Bull.* **10**, 267 (1979).
157. A. Kudo, H. Akagi, D. C. Mortimer, and D. R. Miller, *Nature (London)* **270**, 419 (1977).
158. H. Akagi, D. C. Mortimer, and D. R. Miller, *Bull. Environ. Contam. Toxicol.* **23**, 372 (1979).
159. K. Tanaka, F. Fukaya, S. Fukui, and S. Kanno, *C.A.* **85**, 130109 (1976).
160. I. Berdicevsky, H. Shoyerman, and S. Yannai, *Environ. Res.* **20**, 325 (1979).
161. W. Blair, W. P. Iverson, and F. E. Brinckman, *Chemosphere* **3**, 167 (1974).
162. P. D. Bartlett and P. J. Craig, *C.A.* **92**, 134908 (1980).
163. E. R. Landa, *Soil Sci.* **128**, 9 (1979).
164. C. N. Cheng and D. D. Focht, *Appl. Environ. Microbiol.* **38**, 494 (1979).
165. D. W. von Endt, P. C. Kearney, and D. D. Kaufman, *J. Agric. Food Chem.* **16**, 17 (1968).
166. F. E. Brinckman, G. E. Parris, W. R. Blair, K. L. Jewett, W. P. Iverson, and J. M. Bellama, *EHP, Environ. Health Perspect.* **19**, 11 (1977).
167. F. E. Brinckman, W. P. Iverson, and K. L. Jewett, unpublished observations.
168. F. E. Brinckman, W. P. Iverson, and W. Blair, *Proc. Int. Biodegradation Symp., 3rd* pp. 919–936. Appl. Sci. Publ., London, 1976.
169. P. R. Norris and D. P. Kelly, *Dev. Ind. Microbiol.* **20**, 299 (1979).
170. R. C. Charley and A. T. Bull, *Arch. Microbiol.* **123**, 239 (1979).
171. T. J. Beveridge and R. G. E. Murray, *J. Bacteriol.* **127**, 1502 (1976).
172. A. M. Chakrabarty, *Annu. Rev. Genet.* **10**, 7 (1976).
173. W. H. N. Paton and K. Budd, *J. Gen. Microbiol.* **72**, 173 (1972).
174. M. L. Failla, C. D. Benedict, and E. D. Weinberg, *J. Gen. Microbiol.* **94**, 23 (1976).
175. P. R. Norris and D. P. Kelly, *J. Gen. Microbiol.* **99**, 317 (1977).
176. A. D. Murray and D. K. Kidby, *J. Gen. Microbiol.* **86**, 66 (1975).

177. P. Norris, W. K. Man, M. N. Hughes, and D. P. Kelly, *Arch. Microbiol.* **110**, 279 (1976).
178. W. R. Blair, G. J. Olson, F. E. Brinckman, and W. P. Iverson, *Proc. Int. Conf. Heavy Metals Environ.*, in press (1982).
179. T. G. Tornabene and H. W. Edwards, *Science* **176**, 1334 (1972).
180. R. M. Aickin and A. C. R. Dean, *Microbios Lett.* **9**, 55 (1979).
181. C. L. Haber, T. G. Tornabene, and R. K. Skogerboe, *Chemosphere* **9**, 21 (1980).
182. N. N. Lyalikova, *Mikrobiologiya* **43**, 941 (1974); *C. A.* **82**, 251 (1974).
183. T. L. Gerrard, J. N. Telford, and H. H. Williams, *J. Bacteriol.* **119**, 1057 (1974).
184. D. E. Caldwell and S. J. Caldwell, *Geomicrobiol. J.* **2**, 39 (1980).
185. J. P. Giesy and D. Paine, *Prog. Water Technol.* **9**, 845 (1977).
186. S. E. Shumate, G. W. Strandberg, and J. R. Parrott, *Biotechnol. Bioeng. Symp.* **8**, 13 (1979).
187. T. Horikoshi, A. Nakajima, and T. Sakaguchi, *J. Ferment. Technol.* **57**, 191 (1979).
188. D. P. Kelly, P. R. Norris, and C. L. Brierley, *in* "Microbial Technology: Current State and Future Prospects" (A. T. Bull, D. C. Ellwood, and C. Ratledge, eds.), pp. 263–308. Cambridge Univ. Press, London and New York, 1979.
189. W. P. Ridley, L. J. Dizikes, and J. M. Wood, *Science* **197**, 329 (1977).
190. Y. T. Fanchiang, W. P. Ridley, and J. M. Wood, *in* "Organometals and Organometalloids: Occurrence and Fate in the Environment" (F. E. Brinckman and J. M. Bellama, eds.), ACS Symposium Series, No. 82, pp. 54–64. Am. Chem. Soc., Washington, D.C., 1978.
191. Y. T. Fanchiang, W. P. Ridley, and J. M. Wood, *Adv. Inorg. Biochem.* **1**, 147 (1979).
192. C. K. Ingold, "Structure and Mechanism in Organic Chemistry," 2nd ed. Cornell Univ. Press, Ithaca, New York, 1970; O. A. Reutov and I. P. Beletskaya, "Reaction Mechanisms of Organometallic Compounds." Elsevier North-Holland, Amsterdam, 1968.
193. F. R. Jensen and D. D. Davis, *J. Am. Chem. Soc.* **93**, 4048 (1971); F. R. Jensen, V. Madan, and D. H. Buchanan, *J. Am. Chem. Soc.* **93**, 5284 (1974).
194. M. H. Abraham and P. L. Grellier, *J. Chem. Soc. Perkin Trans. 2* p. 1132 (1973).
195. U. Pohl and F. Huber, *J. Organomet. Chem.* **116**, 141 (1976).
196. J. H. Weber and M. W. Witman, *in* "Organometals and Organometalloids: Occurrence and Fate in the Environment" (F. E. Brinckman and J. M. Bellama, eds.), ACS Symposium Series, No. 82, pp. 247–262. Am. Chem. Soc., Washington, D.C., 1978.
197. W. B. Jensen, "The Lewis Acid-Base Concepts: An Overview," Chaps. 7 and 8. Wiley (Interscience), New York, 1980.
198. L. R. Freeman, P. Angelini, G. J. Silverman, and C. Merritt, *Appl. Microbiol.* **29**, 560 (1975).
199. T. Frick, M. D. Francia, and J. M. Wood, *Biochim. Biophys. Acta* **428**, 808 (1976).
200. R. J. Puddephat, "The Chemistry of Gold," pp. 69–75. Elsevier, Amsterdam, 1978.
201. U. Belluco, "The Organometallic and Coordination Compounds of Platinum," p. 108. Academic Press, New York, 1974.
202. H. Akagi, Y. Fujita, and E. Takabatake, *Chem. Lett.* p. 171 (1975).
203. H. Akagi, Y. Fujita, and E. Takabatake, *Nippon Kagaku Kaishi* p. 1180 (1974).
204. H. Akagi and E. Takabatake, *Chemosphere* **2**, 131 (1973).
205. K. L. Jewett, F. E. Brinckman, and J. M. Bellama, *in* "Marine Chemistry in the Coastal Environment" (T. M. Church, ed.), ACS Symposium Series, No. 18, pp. 319–342. Am. Chem. Soc., Washington, D.C., 1975.
206. E. G. Janzen and B. J. Blackburn, *J. Am. Chem. Soc.* **91**, 4481 (1969).
207. K. Hayashi, S. Kawai, T. Ohno, and Y. Maki, *J.C.S. Chem. Commun.* p. 158 (1977).

208. T. Fagerström and A. Jernelöv, *Water Res.* **10**, 333 (1976).
209. W. R. Penrose, H. B. S. Conacher, R. Black, J. C. Meranger, W. Miles, H. M. Cunningham, and W. R. Squires, *EHP, Environ. Health Perspect.* **19**, 53 (1977).
210. J. S. Edmonds, K. A. Francesconi, J. R. Cannon, C. L. Roston, B. W. Skelton, and A. H. White, *Tetrahedron Lett.* p. 1543 (1977).
211. G. E. Parris and F. E. Brinckman, *J. Org. Chem.* **40**, 3801 (1975).
212. G. O. Doaks and L. D. Freedman, "Organometallic Compounds of Arsenic, Antimony and Bismuth." Wiley, New York, 1970.
213. W. R. Nealy, D. R. Branson, and G. E. Blau, *Environ. Sci. Technol.* **8**, 1113 (1974).
214. J. C. McGowan, *J. Appl. Chem.* **16**, 103 (1966).
215. S. P. Wasik, *in* "Organometals and Organometalloids: Occurrence and Fate in the Environment" (F. E. Brinkman and J. M. Bellama, eds.), ACS Symposium Series, No. 82, pp. 314–326. Am. Chem. Soc., Washington, D.C., 1978; S. P. Wasik, R. L. Brown, and J. I. Minor, *J. Environ. Sci. Health, Part A* **11**, 99 (1976).
216. C. Hantsch and A. Leo, "Substituent Constants for Correlation Analysis in Chemistry and Biology." Wiley, New York, 1979.
217. P. R. Wells, "Linear Free Energy Relationships." Academic Press, New York, 1968.
218. C. Horvath, W. Melander, and A. Nahum, *J. Chromatogr.* **186**, 371 (1979).
219. J. K. Baker, *Anal. Chem.* **51**, 1693 (1979).
220. K. L. Jewett and F. E. Brinckman, *J. Chromatogr. Sci.* **19**, 583 (1981).
221. F. J. Fernandez, *At. Absorpt. Newl.* **16**, 33 (1977).
222. J. C. Van Loon, *Anal. Chem.* **51**, 1139A (1979).
223. Y. K. Chau and P. T. S. Wong, *in* "Environmental Analysis" (G. W. Ewing, ed.), pp. 215–225. Academic Press, New York, 1977.
224. G. E. Parris, W. R. Blair, and F. E. Brinckman, *Anal. Chem.* **49**, 378 (1977).
225. H. A. Meinema, T. Burger-Wiersma, G. Versluis-de Haan, and E. C. Gevers, *Environ. Sci. Technol.* **12**, 288 (1978).
226. F. E. Brinckman, K. L. Jewett, W. P. Iverson, and W. R. Blair, *J. Chromatogr. Sci.* **15**, 493 (1977).
227. R. A. Stockton and K. J. Irgolic, *Int. J. Environ. Anal. Chem.* **6**, 313 (1979).
228. T. M. Vickery, H. E. Howell, and M. T. Paradise, *Anal. Chem.* **51**, 1880 (1979).
229. P. C. Uden, B. D. Quimby, R. M. Barnes, and W. G. Elliot, *Anal. Chim. Acta* **101**, 99 (1978); D. M. Fraley, D. Yates, and S. E. Manahan, *Anal. Chem.* **51**, 2225 (1979).
230. M. Masatoshi, T. Uehiro, and K. Fuwa, *Anal. Chem.* **52**, 350 (1980).
231. T. A. Hinners, *Analyst* **105**, 751 (1980).
232. E. A. Woolson and N. Aharonson, *J. Assoc. Off. Anal. Chem.* **63**, 523 (1980).
233. F. E. Brinckman, K. L. Jewett, W. P. Iverson, K. J. Irgolic, K. C. Ehrhardt, and R. A. Stockton, *J. Chromatogr.* **191**, 31 (1980).
234. T. Hirschfeld, *Anal. Chem.* **52**, 297A (1980).
235. M. Ahmadjian and C. W. Brown, *Environ. Sci. Technol.* **7**, 452 (1973); W. F. Howard, W. H. Nelson, and J. F. Sperry, *Appl. Spectrosc.* **34**, 72 (1980).
236. H. W. Holm and M. F. Cox, *Appl. Microbiol.* **29**, 491 (1975).
237. R. G. Wetrel, "Limnology," p. 261. Saunders, Philadelphia, Pennsylvania, 1975.
238. E. D. Weinberg, ed., "Microorganisms and Minerals." Dekker, New York, 1977.
239. T. Fenchel and T. H. Blackburn, "Bacteria and Mineral Cycling." Academic Press, New York, 1979.
240. P. A. Trudinger and D. J. Swaine, "Biogeochemical Cycling of Mineral-Forming Elements." Elsevier/North-Holland, New York, 1979.
241. J. M. Wood, *Science* **183**, 1049 (1974).
242. J. M. Wood, W. P. Ridley, A. Cheh, W. Chudyk, and J. S. Thayer, *Proc. Int. Conf. Heavy Met. Environ.* pp. 49–68. Univ. of Toronto Press, Toronto, 1977.

243. F. T. Mackenzie, R. J. Lantzy, and V. Paterson, *Math. Geol.* **11**, 99 (1979).
244. W. R. Blair, F. E. Brinckman, T. D. Coyle, W. P. Iverson, J. A. Jackson, and R. B. Johannesen, *Abstr. Int. Conf. Organomet. Chem., 9th, Dijon* p. D-44 (1979).
245. G. E. Parris, *in* "Organometals and Organometalloids: Occurrence and Fate in the Environment" (F. E. Brinckman and J. M. Bellama, eds.), ACS Symposium Series, No. 82, pp. 23–38. Am. Chem. Soc., Washington, D.C., 1978.
246. S. P. Babu, ed., "Trace Elements in Fuel," Advances in Chemistry Series, No. 141. Am. Chem. Soc., Washington D.C., 1975; P. C. Uden, S. Siggia, and H. B. Jensen, eds., "Analytical Chemistry of Liquid Fuel Sources," Advances in Chemistry Series, No. 170. Am. Chem. Soc., Washington, D.C., 1978.
247. T. F. Yen and G. V. Chilingarian, eds., "Developments in Petroleum Science: Oil Shale," Vol. 5. Elsevier, Amsterdam, 1976.
248. M. O. Andrae and D. Klumpp, *Environ. Sci. Technol.* **13**, 738 (1979).
249. N. R. Bottino, F. R. Cox, K. J. Irgolic, S. Maeda, W. J. McShane, R. A. Stockton, and R. A. Zingaro, *in* "Organometals and Organometalloids: Occurrence and Fate in the Environment" (F. E. Brinkman and J. M. Bellama, eds.), ACS Symposium Series, No. 82, pp. 116–129. Am. Chem. Soc., Washington, D.C., 1978.
250. R. H. Fish, J. P. Fox, F. E. Brinckman, and K. L. Jewett, *Environ. Sci. Technol.* **16**, in press (1982).
251. R. V. Cooney, R. O. Mumma, and A. A. Benson, *Proc. Natl. Acad. Sci. U.S.A.* **75**, 4262 (1978).
252. M. O. Andreae and D. Klumpp, *Environ. Sci. Technol.* **13**, 738 (1979).
253. A. A. Benson and R. E. Summons, *Science* **211**, 482 (1981).
254. J. S. Edmonds and K. A. Francesconi, *Nature (London)* **289**, 602 (1981).
255. D. W. Klumpp and P. J. Peterson, *Mar. Biol.* **62**, 297 (1981).
256. Y. T. Fanchiang and J. M. Wood, *J. Am. Chem. Soc.* **103**, 5100 (1981).
257. J. W. Robinson and E. L. Kiesel, *J. Environ. Sci. Health Part A* **A16**, 341 (1981).
258. F. V. Vidal and V. M. V. Vidal, *Mar. Biol.* **60**, 1 (1980).
259. J. P. Buchet, R. Lauwerys, and H. Roels, *Int. Arch. Occup. Environ. Health* **48**, 7 (1981).
260. D. Y. Shirachi, J. U. Lakso, and L. J. Rose, *C.A.* **95**, 55773w (1981).
261. M. Shariatpanahi, A. C. Anderson, A. A. Abdelghani, A. J. Englande, J. Hughes, and R. F. Wilkinson, *J. Environ. Sci. Health Part B* **B16**, 35 (1981).
262. K. Reisinger, M. Stoeppler, and H. W. Nürnberg, *Nature (London)* **291**, 228 (1981).
263. A. W. P. Jarvie and A. Whitmore, *Abs. Tenth Int. Conf. Organomet. Chem.* 1D08 (1981).
264. P. J. Craig and S. Rapsomanikis, *Abs. Tenth Int. Conf. Organomet. Chem.,* 1D07 (1981).
265. A. Naganuma, Y. Kojima, and N. Imura, *Res. Commun. Chem. Pathol. Pharmacol.* **30**, 301 (1980).
266. Y. K. Chau, P. T. S. Wong, A. J. Carty, and L. Taylor, *Abs. Tenth Int. Conf. Organomet. Chem.* 1D09 (1981).
267. J. E. Blum and R. Bartha, *Bull. Environ. Contam. Toxicol.* **25**, 404 (1980).
268. J. Kuiper, *Ecotoxicol. Environ. Saf.* **5**, 106 (1981).
269. A. Furutani and J. M. Rudd, *Appl. Environ. Microbiol.* **40**, 770 (1981).
270. G. Topping and I. M. Davies, *Nature (London)* **290**, 243 (1981).
271. J. A. J. Thompson and J. A. Crerar, *Mar. Pollut. Bull.* **11**, 251 (1980).
272. J. G. Sanders, *Mar. Environ. Res.* **3**, 257 (1980).
273. T. R. Holm, M. A. Anderson, R. R. Stanforth, and D. G. Iverson, *Limnol. Oceanogr.* **25**, 23 (1980).
274. J. Saxena and P. H. Howard, *Adv. Appl. Microbiol.* **21**, 185 (1977).

Index

Cumulative List of Contributors

Cumulative List of Titles

Acetylene and Allene Complexes: Their Implication in Homogeneous Catalysis, **14**, 245

Activation of Alkanes by Transition Metal Compounds, **15**, 147

Alkali Metal Derivatives of Metal Carbonyls, **2**, 157

Alkali Metal–Transition Metal π-Complexes, **19**, 97

Alkyl and Aryl Derivatives of Transition Metals, **7**, 157

Alkylcobalt and Acylcobalt Tetracarbonyls, **4**, 243

Allyl Metal Complexes, **2**, 325

π-Allylnickel Intermediates in Organic Synthesis, **8**, 29

1,2-Anionic Rearrangement of Organosilicon and Germanium Compounds, **16**, 1

Application of ^{13}C-NMR Spectroscopy to Organo-Transition Metal Complexes, **19**, 257

Applications of ^{119m}Sn Mössbauer Spectroscopy to the Study of Organotin Compounds, **9**, 21

Arene Transition Metal Chemistry, **13**, 47

Arsonium Ylides, **20**, 115

Aryl Migrations in Organometallic Compounds of the Alkali Metals, **16**, 167

Biological Methylation of Metals and Metalloids, **20**, 313

Boranes in Organic Chemistry, **11**, 1

Boron Heterocycles as Ligands in Transition-Metal Chemistry, **18**, 301

Carbene and Carbyne Complexes, On the Way to, **14**, 1

Carboranes and Organoboranes, **3**, 263

Catalysis by Cobalt Carbonyls, **6**, 119

Catalytic Codimerization of Ethylene and Butadiene, **17**, 269

Catenated Organic Compounds of the Group IV Elements, **4**, 1.

Chemistry of Carbon-Functional Alkylidynetricobalt Nonacarbonyl Cluster Complexes, **14**, 97

Chemistry of Titanocene and Zirconocene, **19**, 1

Chiral Metal Atoms in Optically Active Organo-Transition-Metal Compounds, **18**, 151

^{13}C NMR Chemical Shifts and Coupling Constants of Organometallic Compounds, **12**, 135

Compounds Derived from Alkynes and Carbonyl Complexes of Cobalt, **12**, 323

Conjugate Addition of Grignard Reagents to Aromatic Systems, **1**, 221

Coordination of Unsaturated Molecules to Transition Metals, **14**, 33

Cyclobutadiene Metal Complexes, **4**, 95

Cyclopentadienyl Metal Compounds, **2**, 365

Diene-Iron Carbonyl Complexes, **1**, 1

Dyotropic Rearrangements and Related σ-σ Exchange Processes, **16**, 33

Electronic Effects in Metallocenes and Certain Related Systems, **10**, 79

Electronic Structure of Alkali Metal Adducts of Aromatic Hydrocarbons, **2**, 115

Fast Exchange Reactions of Group I, II, and III Organometallic Compounds, **8**, 167

Fischer–Tropsch Reaction, **17**, 61

Flurocarbon Derivatives of Metals, **1**, 143

Fluxional and Nonrigid Behavior of Transition Metal Organometallic π-Complexes, **16**, 211

Free Radicals in Organometallic Chemistry, **14**, 345

Heterocyclic Organoboranes, **2**, 257

α-Heterodiazoalkanes and the Reactions of Diazoalkanes with Derivatives of Metals and Metalloids, **9**, 397

High Nuclearity Metal Carbonyl Clusters, **14**, 285

Homogeneous Catalysis of Hydrosilation by Transition Metals, **17**, 407

Hydroformylation, **17**, 1

Hydrogenation Reactions Catalyzed by Transition Metal Complexes, **17**, 319

367